SPACE SCIENCE, EXPLORATION AND POLICIES

SOLAR SYSTEM

STRUCTURE, FORMATION AND EXPLORATION

SPACE SCIENCE, EXPLORATION AND POLICIES

Additional books in this series can be found on Nova's website under the Series tab.

Additional E-books in this series can be found on Nova's website under the E-book tab.

SPACE SCIENCE, EXPLORATION AND POLICIES

SOLAR SYSTEM

STRUCTURE, FORMATION AND EXPLORATION

MATTEO DE ROSSI
EDITOR

Nova Science Publishers, Inc.
New York

For permission to use material from this book please contact us:
Telephone 631-231-7269; Fax 631-231-8175
Web Site: http://www.novapublishers.com

LIBRARY OF CONGRESS CATALOGING-IN-PUBLICATION DATA

Solar system : structure, formation, and exploration / editor, Matteo de Rossi.
p. cm.
Includes index.
ISBN 978-1-62100-057-0 (hardcover)
1. Solar system. I. Rossi, Matteo De.
QB501.S646 2011
523.4--dc23

2011030091

Published by Nova Science Publishers, Inc. † New York

CONTENTS

PREFACE

In this book, the authors present topical research in the study of the structure, formation and exploration of the solar system. Topics discussed in this compilation include a quantum-like model to search the origin of the solar system structure; close binaries, eccentric exojupiters, and the solar system; harnessing energy from the sun by splitting water using Mn-oxo or Co-based catalytic systems to mimic photosynthesis; a relativistic positioning system exploiting pulsating sources for navigation across the solar system and the role of solar wind dynamics on interstellar dust in the solar system.

Chapter 1 - The origin of the solar system structure is explored by introducing a quantum-like model of gravitational field. In this model, the chaos behavior of a large number of atoms or molecules around the Sun is described in terms of the wave function satisfying formal Schrödinger equation. Numerical computing shows that the radial distribution density of the nebula particles in solar system takes on a character like wave curve with the decreasing amplitudes and elongating wavelengths. The locations of these wave peaks have a good relationship with the position celestial objects of the solar system.

Chapter 2 - Most of the stars are members of multiple systems, which can be traced to multi-stage fragmentation of turbulent protostellar clouds bearing a standard excess of angular momentum. Close binaries are products of the last fragmentation of a rapidly collapsed, differentially rotating protostar (PS). It breaks down as a result of rotational-exchange instability, when the angular momentum is carried out by turbulent viscosity as the outer Hayashi convection sets in in the PS. This culminates in a massive gas ring being detached from the PS, with ensuing immediate azimuthal fragmentation. Interaction of these ~3-10 fragments of mass ~1-10 M_J (M_J is a mass of Jupiter) with the PS and with one another leads to part of them becoming ejected out of the system, with some of them merging with the PS, and the others entering strongly eccentric orbits around the PS. These are exo-Jupiters. The convection-embraced matter outflows rapidly onto the most “lucky” fragment, which at the perigee of its low-eccentricity orbit comes close to the PS, so that the first Lagrangian point falls into its convective zone. In the close binary thus formed, the mass ratio of the components is determined by that of the Hayashi convective zone and the stable core of the PS. If convection extends over all of the PS (the $M \leq 1.5\ M_\odot$ case), overflow of all convective mass results in mass reversal of the components (a process well known for the Algol-type binaries) strong enough to produce a system with a mass ratio $q \sim 1{:}1000$. Here mass transfer from the fully convective PS stops after its mass has dropped to ~M_J because of the onset of water condensation in its gas that became adiabatically cooled in expansion. Eventually, a rapidly rotating PS converts into a Jupiter-like planet, while the "lucky" fragment that obtained almost all the initial PS mass becomes the "Sun". Cooling of matter in such a rapidly rotating and extremely dense *proto-Jupiter* (PJ) will initiate, starting from $q \sim 0.1\text{-}0.02$, condensation of nonvolatile compounds and formation of numerous (~104-105) Pluto-like and larger bodies moving in it along unwinding orbits and, hence, leaving it. Thus, PJ may be considered here as a very dense analog of a traditional protoplanetary disk. This is how a Solar-type planetary system forms. The latter four bodies survived in our system as Galilean satellites, the most massive ones retained heliocentric planetary orbits, the larger part of these 104-105 bodies were ejected out of the system, and a small part of them (~103) formed, primarily at a distance of 50÷300 AU, a trans-Neptunian cloud of interacting dwarf planets moving in disordered orbits. This prediction finds convincing support in the discovery made in the recent decade of planets of the type of Sedna, Xena etc. To sum up, the above approach is essentially a synthesis of the gas-dynamic scenario of the formation and evolution of binary stars with the classical pattern of the planets' formation in circumstellar disks. In contrast to the latter, however, it is capable of accounting for the totality of available data and offering predictions borne out by observations.

Chapter 3 - The sun, one member of the solar system, provides all of the energy to support the life on our planet over several billion years, and the solar energy is harvested and storied via photosynthesis by green plants, cyanobacteria, and algae on the large scale at room temperature and neutral pH. Use of solar energy is completely carbon-free and is able to eliminate current concerns on the energy crisis and global climate change. A deep understanding of photosynthesis is the key to provide a solid foundation to facilitate the transformation from carbon-based energy source to sustainably solar fuels. The energy from the sun can be stored via water splitting, which is a chemical reaction through chemical bond rearrangement to convert the energy-deficient water molecule to energy-rich oxygen and hydrogen molecules. Water splitting chemistry is driven by sunlight in the reaction center of photosystem II located in the thylakoid membranes of plant leaves. The three-dimensional structures of photosystem II with oxygen-evolving activity have been determined at an atomic and a molecular level in the past ten years. To mimic the water splitting of photosystem II oxygen evolving complex, appealing systems including earth abundant element catalytic materials were discovered. In this chapter, recent progress in solar fuel production emphasizing on the development of Mn-oxo complexes and Co-phosphate catalytic systems were summarized and discussed. These systems, including Mn-oxo tetramer/Nafion, Mn-oxo dimer/TiO_2, Mn-oxo oligomer/WO_3, Co-Pi/Fe_2O_3, and Co-Pi/ZnO were, show a compelling working principle by combing the active Mn-oxo and Co-based catalysts in water splitting with semiconductor hetero-nanostructures for effective solar energy harnessing. The protocols are suit for preparing earth-abundant metal/semiconductor catalysts and highly likely open a new area of fabricating next generation of highly efficient water splitting catalysts to store the energy from the sun. Grand challenges include the discovery of inexpensive, robust, and efficient water oxidation catalysts. It is particularly important forthe improvement in efficiency and durability of the water-splitting catalytic systemsfor their practical application as well as theutilization of visible and infrared light.

Chapter 4 - The authors introduce an operational approach to the use of pulsating sources, located at spatial infinity, for defining a relativistic positioning and navigation system, based on the use of null four-vectors in a flat Minkowskian spacetime. The authors describe their approach and discuss the validity of it and of the other approximations they have considered in actual physical situations. As a prototypical case, the authors show how pulsars can be used to define such a positioning system: the reception of the pulses for a set of different sources whose positions in the sky and periods are assumed to be known allows the determination of the user's coordinates and spacetime trajectory, in the reference frame where the sources are at rest. In order to confirm the viability of the method, the authors consider an application example reconstructing the world-line of an idealized Earth in the reference frame of distant pulsars: in particular the authors have simulated the arrival times of the signals from four pulsars at the location of the Parkes radiotelescope in Australia. After pointing out the simplifications they have made, the authors discuss the accuracy of the method. Eventually, the authors suggest that the method could actually be used for navigation across the Solar System and be based on artificial sources, rather than pulsars.

Chapter 5 - According to the conservation principle of angular momentum, the authors calculate in this chapter the revolution period and the distance between the Earth and the Moon in the equilibrium state of the tidal evolution in the Earth-Moon system. The difference of energy between the current state and the equilibrium state is used to compute the time needed to fulfil the equilibrium state. Then the long-term variations of the Earth-Moon distance and of the Earth rotation rate are further estimated.

Chapter 6 - Interstellar dust grains have been detected by the dust detectors onboard the *Ulysses* and *Galileo* spacecrafts. Motion of the interstellar dust particles in the Solar System is driven by gravitational and nongravitational forces. As for gravity, the action of the Sun is the dominant gravitational effect. Nongravitational forces are represented by solar electromagnetic radiation force, similar effect of the solar wind, and, Lorentz force for submicrometer-sized dust grains. Lorentz force originates from the action of interplanetary magnetic field on electrically charged grains and solar wind velocity plays a crucial role in this nongravitational force.

Chapter 7 - The discovery of several extrasolar systems, each one characterized by its own planetary distribution around a central star, made the scientific interest addressed to the analysis of models permitting to predict, or at least estimate, the orbital features of the extrasolar planets. The main purpose of this work is to describe a mathematical model, inspired by quantum mechanics, able to provide a probability distribution of planets placing in a star system, mainly driven by the central star mass. More in detail, for any given eigenvalue of the model discrete spectrum, a distinct probability distribution with respect to the central star distance can be built. As per the Solar System, it has been possible to prove that both inner and outer planets belongs to two different spectral sequences, each one originated by the minimum angular momentum owned by silicate/carbonate and icy planetesimals respectively. In both sequences, the peak of the probability distributions almost precisely coincided with the average planets distance from Sun; further more, the eigenvalue spectrum of the inner planets thickens in an accumulation point corresponding to the asteroids belt, thus showing a striking similarity to the real matter distribution in the Solar System. From this point of view, the Titius-Bode law for the Solar System planets distribution is nothing but an exponential interpolation of the eigenvalues of both inner and outer sequences.

Chapter 8 - The authors deal with regularities of the distances in the solar system. On starting with the Titius–Bode law, these prescriptions include, as "hidden parameters", also the numbering of planets or moons. The authors reproduce views of mathematicians and physicists of the controversy between the opinions that the distances obey a law and that they are of a random origin. Hence, the authors pass to theories of the origin of the solar system and demonstrations of the chaotic dynamics and planetary migration, which at present lead to new theories of the origin of the solar system and exoplanets. We provide a review of the quantization on a cosmic scale and its application to derivations of some Bode-like rules.

Chapter 9 - In this chapter, the authors consider a statistical theory of gravitating spheroidal bodies to explore and develop a model of forming and self-organizing the Solar system. It has been proposed the statistical theory for a cosmological body forming (so-called spheroidal body) by means of numerous gravitational interactions of its parts (particles). The proposed theory starts from the conception for forming a spheroidal body inside a gas-dust protoplanetary nebula; it permits us to derive the form of distribution functions, mass density, gravitational potentials and strengths both for immovable and rotating spheroidal bodies as well as to find the distribution function of specific angular momentum. As the specific angular momentums are averaged during conglomeration process, the specific angular momentum for a planet of the Solar system (as well as a planetary distance) can be found by means of such procedure. This work considers a new law for the Solar system planetary distances which generalizes the well-known Schmidt law. Moreover, unlike the well-known planetary distances laws the proposed law is established by a physical dependence of planetary distances from the value of the specific angular momentum for the Solar system. The proposed simple statistical approach to investigation of our Solar system forming describes only a natural self-evolution inner process of development of protoplanets from a dust-gas cloud. Naturally, this approach however does not include any dynamics like collisions and giant impacts of protoplanets with large cosmic bodies. Henceforth, the presented statistical theory will only be able to predict surely the protoplanet's positions according to the proposed ent[n/2] rule, i.e. the findings in this work are useful to predict if today's position or orbit of a considered planet coincides with its protoplanet's location or not. Though orbits of moving particles into the flattened rotating spheroidal body are circular ones initially, however, they could be distorted by collisions with planetesimals and gravitational interactions with neighboring originating protoplanets during evolutionary process of protoplanetary formation. Really, at first the process of evolution of gravitating and rotating spheroidal body leads to its flattening, after that the evolutionary process results in its decay into forming protoplanets. This work shows that the orbits of moving particles are formed by action of centrally-symmetrical gravitational field mainly on the later stages of evolution of gravitating and rotating spheroidal body, i.e. when the particle orbits become Keplerian ones. In this connection, this work investigates the orbits of moving planets and bodies in centrally-symmetrical gravitational field of gravitating and rotating spheroidal body during the planetary stage of its evolution (in particular, the angular shift of the Mercury's perihelion is estimated). This work shows that according to the proposed statistical theory of gravitating spheroidal bodies the turn of perihelion of Mercury' orbit is equal to 43.93" per century that well is consistent with conclusions of the general relativity theory of Einstein (this estimation is equal to 43.03") and astronomical observation data (43.11 ± 0.45").

In: Solar System: Structure, Formation and Exploration
Editor: Matteo de Rossi
ISBN: 978-1-62100-057-0

Chapter 1

A QUANTUM-LIKE MODEL TO EXPLORE THE ORIGIN OF THE SOLAR SYSTEM STRUCTURE

Qingxiang Nie[*]
College of Physics and Electronics, Shandong Normal University,
Jinan, P. R. China

ABSTRACT

The origin of the solar system structure is explored by introducing a quantum-like model of gravitational field. In this model, the chaos behavior of a large number of atoms or molecules around the Sun is described in terms of the wave function satisfying formal Schrödinger equation. Numerical computing shows that the radial distribution density of the nebula particles in solar system takes on a character like wave curve with the decreasing amplitudes and elongating wavelengths. The locations of these wave peaks have a good relationship with the position celestial objects of the solar system. By means of this model we can well explain many characteristics of the solar system structure as follows:

1) The planetary distance rule, including asteroid belt and Kuiper belt.
2) The mass distribution of planets.
3) Planetary energies and angular momentums.
4) The distributions of objects in the asteroid belt and the Kuiper belt.
5) The distance rule of the satellites .
6) The structures of planetary ring systems.
7) The rotations of planets,

INTRODUCTION

Since Immanuel Kant (1724—1804) first proposed the nebular hypothesis of the solar system origin in 1755, many kinds of theories on the formation of the solar system have been

[*] E-mail address: nqx2000cn@yahoo.com

proposed. But there are many issues which have not been satisfactorily explained yet, in particular, the formation of the solar system structure, including the distributions of planetary distance and mass, the distribution of satellites, asteroid belt and Kuiper belt, the structures of planetary rings, and so on. In this chapter, we do not intend to study the formation process and mechanism of the solar system structure, but only to discuss the possible effect of the distribution of the original nebula on the formation of the solar system structure.

1. PUZZLES ON THE ORIGIN OF THE SOLAR SYSTEM STRUCTURE

1.1. Titius-Bode law of planets

There is a famous law of planetary distances in the solar system, usually called Titius-Bode law (also simply called Bode's law). This law was first discovered in 1766 by the German astronomer Johann Daniel Titius (1729- 1796). Titius found that the mean distance of of each of the six known planets from the Sun approximately follows a sequence that can be expressed in a simple formula. He first inserted this finding into the main text of his German translation, and changed the insertion to a footnote in his second edition in 1772, which was read and popularized by another German astronomer Johann Elert Bode (1747–1826) [1]. Titius-Bode law was generally denoted as

$$a_n = 0.4 + 0.3 \times 2^n \ , \tag{1}$$

where a_n is the planetary orbital semi-major axis in astronomical units (AU), $n = -\infty, 0,1,2,4,5$ for Mercury, Venus, Earth, Mars Jupiter and Saturn, which were known planets at that time. The great success of formula (1) for these planets is shown in Table 1. Especially, Uranus discovered in 1781 by William Herschel (1738–1822) at orbital distance 19.2 AU, almost an exact distance for $n = 6$, was considered as one of the triumphs of the law.

However, there remained an anomalous gap between Mars and Jupiter, where there might exist a "missing" planet at 2.8 AU. On New Year's Day, 1801, Giuseppe Piazzi (1746–1826), an Italian astronomer fortuitously discovered a minor planet with orbit semi-major axis of the of 2.77 AU, which should be an excellent candidate for the "missing planet". Piazzi named his object Ceres after the patron goddess of Sicily [2]. But Ceres' diameter is only about 1000 km, far less than that of a full-size planet. Now we know that Ceres is the largest one of thousands of minor planets in Asteroids Belt, which is well suited to $n = 3$ in Titius-Bode law.

Neptune was found in 1846 by Johann Galle (1812–1910) on the basis of the calculations of Urbain le Verrier (1811–1877) by means of celestial mechanics. At the same time, John Couch Adams (1819 - 1892) had also done the same calculation, so the discovery of Neptune is attributed to both le Verrier and Adams [2]. Unfortunately, Neptune was discovered to follow an orbit with radius of 30.1 AU, while the Titius-Bode law predicts 38.8 AU for $n = 7$. Pluto was discovered in 1930 by American astronomer Clyde William Tombaugh (1906–1997). Its orbit semi-major axis of 39.5 AU was only one half of the Titius-Bode law's prediction of 77.2 AU. The Titius-Bode law was quickly losing its scientific reliability. So far no larger planet have been found around the places of 38.8 AU and 77.2 AU. It appears that Titius-Bode law is only a plausible rule.

Astronomers never give up, and continued to look for the more accurate formula to describe planetary distances. The expression mentioned frequently are

$$a_n = a_0 d^n \quad \text{or} \quad a_{n+1} / a_n = d \ , \tag{2}$$

where a_0 and d are constants given different values in different articles in order to fit in with the observations [3-12]. The formula $a_n = 0.285 \times 1.523^n$ published by Basano and Hughes in 1979 has attracted more attention [3-7], and was called B-H law, but was later pointed out that it was a repeat of the Armellini's law published in 1978 [4]. In addition, Peter Lynch thinks the parameters a_0=0.2139 and d = 1.706 for planets are "in broad agreement with the observed values" [7]. The results obtained from different parameters of the formula (2) and Titius-Bode law (T-B law) show in Table 1, here d is relative deviation, and the data in the last a_n column are recalculated according to the formula, which are a little smaller than thoset in original paper.

Table 1. Comparison of various planetary distance laws

Planets	Observed a (AU)	T-B Law $a_n = 0.4 + 0.3 \times 2^n$			B-H law (1) $a_n = 0.285 \times 1.523^n$			B-H law (2) $a_n = 0.2139 \times 1.706^n$		
		n	a_n	d %	n	a_n	d %	n	a_n	d %
Mercury	0.39	$-\infty$	0.4	3	1	0.43	15	1	0.37	5
Venus	0.72	0	0.7	3	2	0.66	8	2	0.62	14
Earth	1.00	1	1.0	0	3	1.01	1	3	1.06	6
Mars	1.52	2	1.6	5	4	1.53	1	4	1.81	19
(Ceres)	2.77	3	2.8	0	5	2.34	16	5	3.09	12
Jupiter	5.20	4	5.2	0	7	5.42	4	6	5.27	1
Saturn	9.54	5	10.0	5	8	8.25	13	7	9.00	6
Uranus	19.18	6	19.6	2	10	19.14	0	8	15.35	20
Neptune	30.06	7	38.8	29	11	29.14	3	9	26.18	13
(Pluto)	39.44	8	77.2	96	12	44.39	13	10	44.67	13

It is clear that B-H law dose not fundamentally solve the problem, the deviations of some planets are still larger. In particular, in order to eliminate the deviations of Uranus and Pluto in T-B law, B-H law (1) used the discontinuity of the integer sequence of n, the missing of n = 6 and 9, being difficult to understand. In fact, Pluto is unnecessarily to considered separately, because it is only a member in the Kuiper belt.

1.2. Distance law of satellite systems

The same method is used to investigate satellite distances [5-7]. The work result on satellites of Neuhaeuser and Feitzinger [6] is shown in Table 2, in which some observational data have been updated according to the book "Solar System Dynamics" [13], and the n = 5 for $r_n = a + b \cdot 2^n$ of Saturn is supplemented here. Taking into account the orbits instability of the small satellites, these listed in Table 2 are just some regular satellites with large mass, except Phoebe.

Clearly, the distance distribution of satellites appears to have some regular, but it is also not ideal to use the formulas similar to equations (1) and (2), especially for the Saturn system, it is almost a failure.

Table 2. Distance laws of satellite systems

Planet	Satellite	Semi-major axis r		$r_n = a + b \cdot 2^n$			$r_n = c \cdot d^n$		
		in 10^6 m	in r_{mark}	n	r_n	$d\%$	n	r_n	$d\%$
Jupiter	Amalthea	181	0.43	-1	0.44	2	1	0.39	7
$a = 0.3$	Thebe	222	0.53	0	0.58	9	2	0.62	17
$b = 0.28$	Io	422	1.00	1	0.86	14	3	0.97	3
$c = 0.25$	Europa	671	1.59	2	1.42	11	4	1.52	4
$d = 1.57$	Genymede	1070	2.54	3	2.54	0	5	2.38	6
	Callisto	1883	4.47	4	4.78	7	6	3.74	16
Saturn	Prometheus	139	0.98	/	/	/	/	/	/
$a = 0.25$	Pandora	142	1.00	0.5	0.91	9	3	0.91	9
$b = 0.465$	Janus	151	1.06	/	/	/	/	/	/
$c = 0.25$	Mimas	186	1.29	1	1.18	9	4	1.41	9
$d = 1.54$	Enceladus	238	1.67	/	/	/	/	/	/
	Tethys	295	2.08	2	2.11	1	5	2.17	4
	Dione	377	2.65	/	/	/	/	/	/
	Rhea	527	3.66	3	3.97	8	6	3.33	9
	Titan	1222	8.61	4	7.69	11	8	7.91	8
	Hyperion	1481	10.43	5	15.13	45	/	/	/
	Iapetus	3561	25.08	6	30.01	20	11	28.88	15
	*Phoebe	12952	91.21	7.5	84.42	7	14	105.49	16
Uranus	Puck	86	0.45	-2	0.41	9	1	0.49	9
$a = 0.23$	Miranda	130	0.68	-1	0.58	15	2	0.68	0
$b = 0.7$	Ariel	191	1.00	0	0.93	7	3	0.94	6
$c = 0.35$	Umbriel	266	1.39	1	1.63	17	4	1.31	6
$d = 1.39$	Titania	436	2.28	1.5	2.21	3	5	1.82	20
	Oberon	584	3.06	2	3.03	1	6	2.52	18

Note: Observational data are from the book "Solar System Dynamics"pp.532-535 [13]

In short, the distance problem of planets and satellites has attracted and baffled people for over two hundred years, some more complex formulas have been proposed, but just as the above, the purpose is just to fit the observational data, lacking physical meaning.

1.3. Planetary rings

The origin of planetary rings is one of the prominent unsolved problems of planetary science. Four giant planets (Jupiter, Saturn, Uranus and Neptune) have ring systems. The ring system of Saturn is most famous, which can be seen with ground-based telescopes. The main ring system of Saturn consists of four broad rings, which are bright A and B rings and optically thinner C and D rings in order from outer to inner. The A and B rings are separated by the Cassini division. Three new rings found by the space probes, narrow F ring and broader diffuse G and E rings, lie beyond the outer edge of the main rings. The distribution of these rings is listed in Table 3. Some narrow gaps exist in C ring, at the edge of the Cassini division, and in the outer part of the A ring. Gaps in C ring and Cassini division also contain many narrow, eccentric rings. Most of the structure in the A ring can be explained by resonances of the $p+1: p$ with the smaller satellites Prometheus and Pandora [13], which lie near the A ring. Cassini spacecraft got close to satellite Enceladus on 14 July 2005, registered micron-sized dust particles enveloping this satellite. So it is inferred that production or release of dust particles related to these processes may provide the dominant source of Saturn's E ring [14]. But the B ring and others are still poorly understood.

The ring system of Jupiter was discovered by Voyager 1 spacecraft in March 1979 and further images were obtained by Voyager 2 later that year [15,16]. The main ring of Jupiter is about 6,000 km wide with a sharp outer edge and a faint “gossamer” ring extending outwards beyond the orbit of the small satellite Thebe. At the inner edge of the main ring there is a toroidal-shaped halo, whose vertical structure extends for perhaps 10,000 km above and below the equatorial plane. Two small satellites, Metis and Adrastea, orbiting in outer the main ring, are thought to be the source of most of the ring material. A hypothesis on Lorentz force resonance has been suggested to explain some of the structure of the ring system of Jupiter [17], and shown that the $k = 3$ Lorentz resonance occurs close to the transition point between the main ring and halo [18].

The Uranus ring system was first discovered serendipitously in 1977 during observations of an occultation of a star by the planet. Subsequent ground-based experiments confirmed that there are 9 narrow rings (6, 5, 4, α, β, η, γ, δ and ε) around the Uranus [19]. The Voyager 2 encounters with Uranus in 1986. The images shown the presence of numerous dust rings in addition to the nine rings, and also identified a new λ ring between δ and ε rings [13]. The shepherding satellite model for narrow rings is considered a considerable success in explaining the confinement of the ε ring by the satellites Cordelia and Ophelia, which orbit interior and exterior to the ring. However, despite extensive searches, no other satellites have been found in the main ring system [13].

Prior to the Voyager reconnaissance of Neptune, the ground experiments on stellar occultation suggested that Neptune have a system of “arcs” of ring. The distinct images of rings obtained by Voyager 2 in 1989 show that Neptune has two distinct, narrow rings, the Le Verrier and Adams; two broad ring, the innermost Gelle and Lassell extending outwards from Le Verrier and bounded by the narrow Arago ring; and a unnamed ring that appears to share the same orbit as the small satellite Galatea (see Table 3)[13,20]. There are few research on the formation of the Neptune’s ring structure. Porco proposed that resonances with moon Galatea were responsible for the arcs in the Adams ring, because the Adams lies close to a 42:43 resonance with Galatea [21].

Many old problems on planetary rings have not yet been resolved, some new problems have arisen. Verbiscer et al. report that Saturn has an enormous ring associated with its outer moon Phoebe, extending from at least 128 R_S to 207 R_S (Saturn’s radius R_S is 60,330 km) [22].

2. A Quantum-like Model on the Solar System

The similarity between quantum mechanics and stochastic motion of particles has been noticed in some early papers. Comisar (1965) has introduced a Brownian motion model to explain the behavior of electron described in quantum mechanics [23]. Nelson (1966) has given a derivation of the Schrödinger equation for an electron motion from the theory of Brownian motion [24]. Following Nelson, Nottale et al. (1997) have applied a Schrödinger-like equation to planets in the solar system [25]. Since 1992, Nie et al. have begun to try to explore the origin of the solar system structure by simulating the quantum mechanics. Distances of the planets 26], the distribution of the Kuiper Belt objects [27,28], and the rotations of planets [29] have been explored. In a recently published article, the main issues of the solar system structure, including the problems on the planetary energy and angular momentum, have been interpreted [30]. The work in this book tries to give a systematic and comprehensive description of the solar system structure.

Table 3. The ring systems of planets

Ring		Inner edge (10^6 m)	Outer edge (10^6 m)	Satellite in rings (10^6 m)	
Jupiter	Halo	89.400	123.000	Metis	127.98
	main ring	123.000	128.940	Adrastea	128.98
	gossamer	128.940	242.000	Amalthea	181.30
				Thebe	221.90
Sature	D	66.900	74.658	Pan	133.583
	C	74.658	91.975	Atlas	137.640
	B	91.975	117.507	Prometheus	139.350
	Cassini			Pandora	141.700
	division	122.340	136.780	Epimetheus	151.422
	A	140.219 (core)		Janus	151.472
	F	166.000	173.200	Mimas	185.520
	G	180.000	408.000	Enceladus	238.020
	E				
Uranus	6	41.84			
	5	42.23			
	4	42.57			
	α	44.72			
	β	45.66			
	η	47.17			
	γ	47.63			
	δ	48.30			
	λ	50.02		Cordelia	49.752
	ε	51.14		Ophelia	53.764
Neptune	Galle	42.000-1000	42.000+1000		
	Le Verrier	53.200		Naiad	48.227
	Lassell	53.200	57.200	Thalassa	50.075
	Arago	57.200		Despina	52.526
	[unnamed]	61.953		Galatea	61.953
	Adams	62.932			

Note: Data are from the book "Solar System Dynamics"pp.532-535 [13]

The quantization character of planetary orbits has been expressed in teams of Bohr orbit $a_n = n^2 a_0$ by some researchers (e.g., Yang [31], Agnese & Festa [32]), where a_0 is the "Bohr radius" of planetary system with different values in various literatures, and n is the principal quantum number. The difficulty in applying Bohr orbit is that the values of n must be discrete integer in order to fit the planets and satellites.

If we divide the planets into two groups, the first group including the Mercury, Venus, Earth, Mars and asteroid belt, the second group including the Jupiter, Saturn, Uranus, Neptune and Kuiper belt, the planetary distances can approximately form two same successive sequences with the continuous integer $n = 2, 3, 4, 5, 6$. Similarly, the energy and angular momentum of unit mass can also constitute the two sequences respectively. It is important that the ratio of the mean "Bohr radius" of two groups of planetary orbits is approximately 16, i. e. $(\bar{a}_0)_1 / (\bar{a}_0)_2 \approx 1/16$.

Comparing the format between gravitational force and Coulomb force,

$$\frac{1}{4\pi\varepsilon_0}\frac{e^2}{r^2} \rightarrow G\frac{M\mu}{r^2}, \tag{3}$$

using the analogy $1/4\pi\,\varepsilon_0 \rightarrow G$, $e^2 \rightarrow M\mu$ and $\hbar \rightarrow \hbar_g$, the corresponding relationship between Coulomb field in a hydrogen atom and gravitational field in the solar system is

$$a_0 = \frac{4\pi\,\varepsilon_0 \hbar^2}{\mu e^2} \quad \rightarrow \quad a_0 = \frac{\hbar_g^2}{GM\mu^2}, \tag{4}$$

where a_0 is the Bohr radius, M the mass of the Sun, and μ the mass of the particle. From equation (4), we can suppose that the formations of the proto-planets in two groups are related to the particles with different masses μ_1 and μ_2, and $\mu_1/\mu_2 \approx 4$. This ratio happens to be the mass ratio of H atom and He atom, which are the most abundant elements in the universe.

Equation (4) seems to remind us of Bohr-Sommerfeld theory, which is an old quantum theory. But our numerical simulation, in which 28000 test particles in Keplerian motion are distributed on the 28 different Sommerfeld's elliptical orbits, shows that the radial distribution of the particles is no associated with the distances of planets. Many simulation tests indicate that the model of old quantum theory is not reliable. As the progress from the old quantum theory to the new theory, the application of Schrödinger equation in gravitational field can be naturally considered

According to the modern hypothesis on origin of solar system, the solar system formed from a rotary gas nebula. This chapter attempts to explore the slow shrinking phase of the nebula, which can be regarded as stable in a long time. The object studied is the large number of micro-particles. The motion tracks of the particles are chaotic due to the superposition of Brownian motion with Kepler motion. The Kepler orbit is only the probability orbit of the large number of particles. This behavior of gas particles is very similar to the electrons'.

Therefore, we attempt to borrow a time-independent Schrödinger equation in gravitational field to describe the motion of nebular gas particles. The equation can be written as

$$\frac{\hbar_g^2}{2\mu}\nabla^2\psi + (E - V)\psi = 0 \ , \tag{5}$$

where E and V are the total energy and potential energy of a particle respectively, μ is the mass of the particle, and $\hbar_g$ a constant on cosmic scale with the same meaning to the Planck constant. The density of the nebular gas is ψ^2.

The potential energy of the particle in gravitational field is $V = -GM\mu / r$, where r is the distance of the particle from the gravitational center. From quantum theory, on condition of $E < 0$, the normalized radial function of equation (5) in spherical coordinate is

$$R_{nl}(r) = -\{(\frac{2}{na_0})^3 \frac{(n-l-1)!}{2n[(n+l)!]^3}\}^{\frac{1}{2}} e^{-\frac{\rho}{2}} \rho^l L_{n+l}^{2l+1}(\rho),$$

$$\rho = \frac{2r}{na_0} \ , \ a_0 = \frac{\hbar_g^2}{GM\mu^2} \ , \tag{6}$$

where $L_{n+l}^{2l+1}(\rho)$ is associated Laguerre polynomial,

$$L_{n+l}^{2l+1}(\rho) = \sum_{k=0}^{n-l-1} (-1)^{k+1} \frac{[(l+n)!]^2}{(n-l-1-k)!(2l+1+k)!k!} \rho^k .$$

The allowed values of quantum numbers are $n = 1,2,3,\cdots$ and $l = 0,1,2\cdots n-1$. The "Bohr radius" a_0 is consistent with equation (4). According to the shape and the initial revolving angular momentum of observed proto-planetary nebula disks, most of the nebular particles should be in states of the azimuthal quantum number $m_l = l$, in which the particles revolve in the same direction around the equatorial plane of the system. The numerical simulation shows that all particles in these states form a gas disk similar to the observed nebular disk.

The key question is what particles have affected the formation of planets. Though the old quantum model is rough, it can give us some valuable information: the formation of the distances of inner and outer proto-planets might be the result of the different distributions of hydrogen and helium particles. It is known that the mass of the proto-solar nebula is mostly made of hydrogen (71%) and helium (27%) gases, with tiny traces of other chemical elements and tiny dust particles. Therefore, hydrogen and helium, the most abundant gas particles, will be given the special attention.

Based on the Boltzmann equiprobability principle, we can assume that the particles distribute in each of states described by above quantum number with equal probability. Since radial probability density of particles is $[rR_{nl}(r)]^2$, which can affect planetary distances, if the number of particles in every state is N , the radial mass density of some kind of particle with the mass μ in all these states can be written as

$$m(r) = N\mu \sum_{n.l} [rR_{nl}(r)]^2 . \tag{7}$$

Analytical expression of $m(r)$ is very complicated, so the numerical calculating is used here. It provides a clear nature of the radial mass density. In calculation, the unit of distance is a_0 of hydrogen atom, and N is an appropriate constant due to the unimportance of its absolute magnitude. Though n may be all integers, numerical experiments show that when n>20, the increase of n has almost no influence on the characteristic of $m(r)$ curve used in the solar system. Here the maximum of n is set at >30 for hydrogen and helium particles due to their large abundance. But the maximum of n is set at <30 for heavy element particles, such as carbon and oxygen. The numerical experiments also show that only when l is all the allowed values, the regular fluctuation of the $m(r)$ curve can appear. In this case, the spatial periods of the fluctuation correspond to the spacing of planets or satellites in the solar system.

3. RESEARCH ON PLANETARY DISTANCES

3.1. Correspondence between Planetary Distances and m(r)

This model is first applied to the planets in the solar system. Based on the phenomena and analyses in above sections, H, H_2 and He, the most abundant particles in nebular disk, are chosen as the investigated objects. The plots of $m(r)$ are shown in Figure 1, where the a_0 of the H atom is set as 1 AU, and the ratios of the $m(r)$ values for the three curves have no real significance. The curves in Figure 1 have the following properties: (1) The wave amplitudes of $m(r)$ values is decreasing with the distance *r*, while wavelengths increasing. (2) The wave cycles of H curve from the first to the sixth correspond to the terrestrial planets, Jupiter, Saturn, Uranus, and Kuiper belt, respectively. (3) Six cycles of the He curve are included within the first cycle of the H curve, and the wave cycles of He curve from the second to the fifth respectively correspond to the Mercury, Venus, Earth and Mars, while the mean distance of asteroids belt is just at the sixth He wave crest but H wave trough.

There is no good correspondence between the wave cycles of the H_2 curve and the planets, so we will not pay any more attention to it in the following discussions. But it should be noticed that the wave crests of H curve from first to fifth overlap with some crests of H_2 cure, and there are the large planets existing in the superposition. But the position of the sixth crest of H curve overlaps in the wave trough of the H_2 curve, and there is no large planet but the Kuiper belt .

It is noticeable that the positions of planets are not exactly consistent with the wave crests of the $m(r)$ curves of H or He, especially in terrestrial planets. These deviations are actually reasonable, which will be further specially explained.

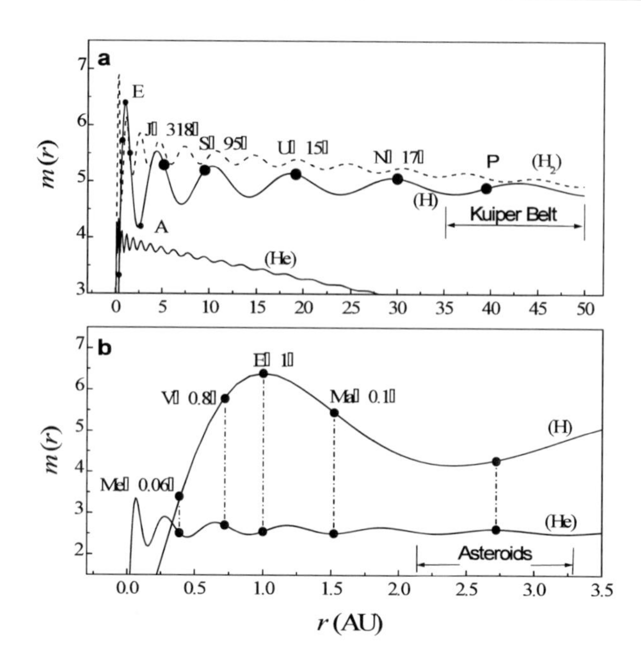

Figure 1. The $m(r)$ curves of particles and planetary distances.
The solid lines are $m(r)$curves of H and He particles, the dashed is one of H_2, the filled circles express the planetary positions, and the number in bracket is planetary mass.

3.2. Significance of the Correspondence

The above correspondences and deviations can be easily understood once the model is accepted. The fluctuation of mass density of nebula is an important initial condition for the proto-planet formation. Numerical integration can show that there are the gravitational potential wells at the wave crests of the $m(r)$ curve. This means that the formation of proto-planets is the easiest at these radial density peaks. In the terrestrial region, the oscillation of radial density of helium particles is comparatively obvious, i.e., the action of the potential wells of helium is stronger than that of other kinds of particles. Therefore, the particles can be aggregated more easily in these regions to form the proto-planets. This is just the reason why the terrestrial planets correspond to the $m(r)$ waves of He. But the influence of the hydrogen particles is also important on the formation of the terrestrial planets, because the largest gravity potential well caused by the first peak of the H curve lies in the middle of the terrestrial region, and it might cause the planetesimals or embryo in the region to move towards it during their growth. This can explain why the terrestrial planets all deviate from the He curve peaks towards the first peak of the H curve. In the giant planet region, the oscillation of the He curve becomes very small, and the potential wells corresponding to wave crests of the H curve act as the local gravity center. Therefore there exists a good correspondence between the giant planets and the $m(r)$ peak of H.

From modern hypothesis on planetary formation[33,34], giant planets are formed because of the gravitational instability of gas in the accretion disk around the young star, and terrestrial planets are formed by collisional accumulation of successively larger planetesimals and planetary embryos. If it is true, the gravitational instability could be caused by the gravitational potential wells in the gas disk, and larger planetesimals and planetary embryos should be first formed nearby the potential wells. Actually, the quantum-like model can give this planetary formation mechanism a convincing support. From Figure 1, it can be seen that there are higher frequency and greater intensity radial density wave of heavier elements in terrestrial region in comparison with giant planets region. Accordingly, it can be speculated that: (1) Formation of planetesimals is easier, and the number is larger in terrestrial region than in outer solar system. So the collision probability of planetesimals is larger, and the planetary embryos form easily and grow quickly at the wave crests of He. (2) In giant planets region, for the lack of heavier particles and the weakness of the fluctuation of their density, the formation of planetesimals is more difficult. But because of the strong fluctuates of the radial density of hydrogen gas, the proto-planets most possibly formed because of gravitational instability of gas.

3.3. Signification of the Deviation

The deviations of the planetary positions will be specially explained here. The locations of curve crests are about a_0, 4.4 a_0, 10.3 a_0, 18.6 a_0, 29.5 a_0, 42.9 a_0, etc, where a_0 is equal to 1 AU for H, and about 1/16 AU for He. Figure 1(b) displays the obvious deviation of the position of each terrestrial planet from the corresponding wave crest of He curve towards the first crest of H curve. It is worth noting that the planetary deviation degree is roughly related to the slope of the H curve at the planetary corresponding position, and larger deviation corresponds to larger slope in either side of the crest of H. For example, the deviation of the Mercury is larger than that of the Venus in inner side of the H wave crest, while the deviation of the Mars is larger than that of the Earth in outer side. Since the sixth crest of the He curve is near the minimum point of H curve, where the slope is the smallest, the deviation of the asteroids from the He crest is also the smallest. If the matter distribution of the proto-solar system is surely similar to that in Figure 1, these deviations are reasonable. As explained above, the formation of the terrestrial planetary positions depends not only on the distribution of helium, but also on that of hydrogen. If an embryo formed at the wave crest of He, it should shift from the crest towards the location with larger density of hydrogen in later accretion process. Besides, other mechanisms, such as their mutual gravitation, can also cause them to congregate. So we can speculate that if there is no strong tidal force from the Sun and Jupiter, a giant planet might be formed. Since the distributions of He atoms and other heavier particles lean to the interior of nebula disk, the position of this aborted giant planet should lie at the first crest of the H curve but deviate a little towards the Sun, i.e., its distance from the Sun should be less than 1 AU. Such giant planets are possible in extrasolar systems.

The deviations of the Jupiter and Saturn shown in Figure 1(a) can be explained by resonance mechanism. The resonance can cause the changes of some planetary positions in later evolvement [35]. Among all giant planets, since Jupiter and Saturn both have the largest masses and the shortest distance between them, their strong mutual attraction can easily draw themselves close to each other. It is known that the mean-motion resonance of 5:2 exists in the Jupiter-Saturn system. If the mass ratio of the Jupiter to Saturn remains a constant in evolvement, the ratio of 10.2/10.7 between the total angular momentum of initial to that of modern for the Jupiter-Saturn system can be obtained by using the initial distances of 4.4 AU for the Jupiter and 10.3 AU for the Saturn. Obviously, the ratio approximately obeys the conservation law of angular momentum. The increase of 5% of modern angular momentum may be explained by action of the solar wind or other mechanisms.

4. Planetary Mass and Radial Density of Material

Another interesting phenomenon in Figure 1 is the order correspondence between the mass of a planet and the value of $m(r)$ curve of H. This relationship is clearly shown in Figure 2, where the height of vertical bar represents the mass of the planet. Apparently, except the Neptune, two corresponding relations appear in two sets of planets respectively. This correlation can be understood easily, because hydrogen is the most abundant component of the solar system, and the final mass of planet should mainly depend on the radial density and the total mass of hydrogen in the region of planet formed. The larger radial density can cause the proto-planet to form earlier and grow more quickly, and so become a larger mass planet. The mass of the Neptune is slightly larger than Uranus, which is an intelligible exception. The Neptune is the outermost giant planet, and its region of provisioning material can be extended outwards into the Kuiper belt, so its mass can become larger.

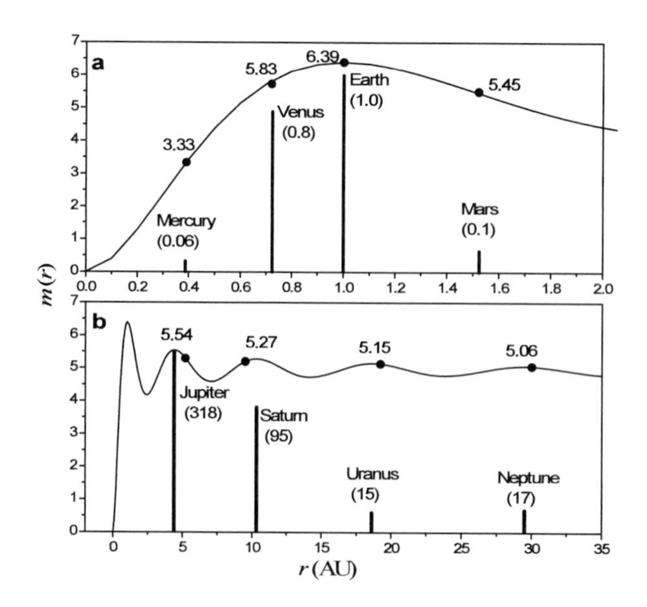

Figure 2. Relation between the planetary mass and the radial density of H.

5. Asteroid Belt and Kuiper Belt

An interesting coincidence between the asteroid belt and Kuiper belt can be seen from Figure 1. Asteroid belt and Kuiper belt respectively correspond to the sixth wave cycle of the He curve and the H curve, and the asteroid belt lies at the overlapping region of the crest of He with the trough of the H curve, while the Kuiper belt lies at the overlapping region of the crest of H with the trough of the H_2 curve. This means that the material and gas density in the two regions are smaller, resulting in the difficulty of the formation of the planets. This may be one of the reasons why larger planets have not formed at these positions.

In Figure 1(b), the distance between the two troughs of the sixth wave of He curve is 2.2-3.1 AU. It is close to the width of about 2.1-3.3 AU of the asteroid belt shown in Figure 3.

The position of the sixth crest of the He curve is 2.7 AU, which is just the radial geometrical center of the asteroid belt.

Similarly, from Figure 1(a), the distance between the two troughs of the sixth wave of the H curve is 36-50 AU, which is consistent with the current observed range of Transneptunian Objects (TNOs), also called as Kuiper belt objects (KBOs). Figure 4 is the statistics for population of 1163 TNOs promulgated by MPC in 2011 March [29]. Their orbital semi-major axis distribute between 35-50 AU, which is called as the main Kuiper belt [30]. This population pattern has not changed since 2002 [22], although the number of observed objects has increased twice. In addition, we do not worry about the effect of Centanurs and SKBOs (scattered Kuiper belt objects), which number is smaller, and the most of them distribute inside the Neptune orbit and outside 50AU respectively. Their properties are different from the TNOs, and their population is not innate. So they are not considered here.

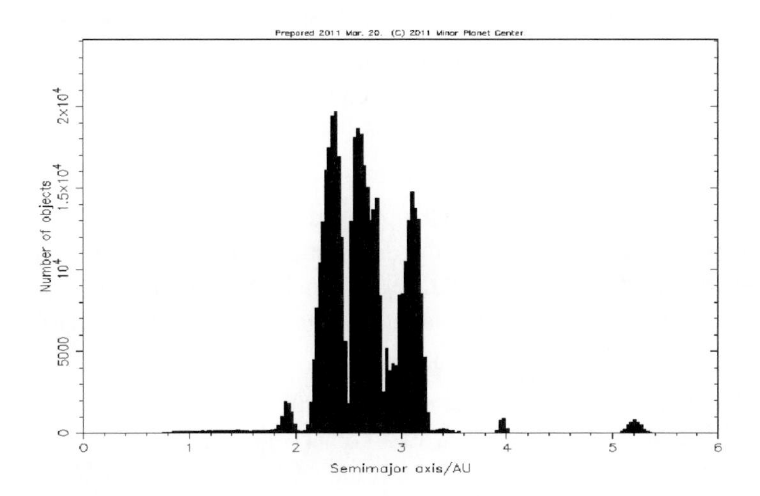

Figure 3. Distribution of the minor planets
Taken from the IAU Minor Planet Center (MPC), prepared on 2011 Mar. 20 [36].

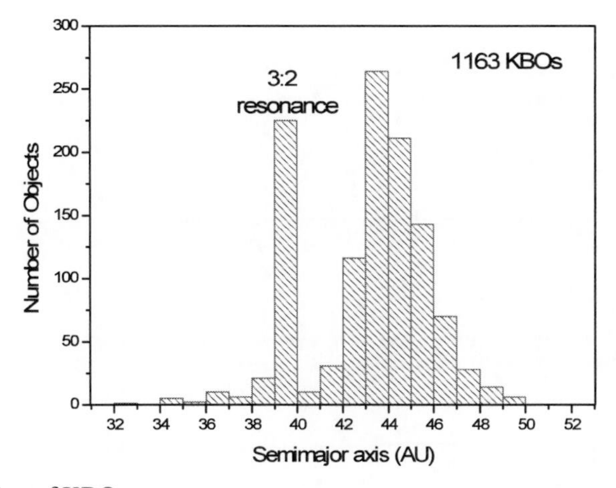

Figure 4. Distribution of KBOs.

For the origin of the Kuiper belt, a representative viewpoint holds that the Kuiper belt itself is a remnant of the solar nebula [36]. The quantum-like model supports this view. The major part of the KBOs possibly retains the characteristics of the radial distribution of the primeval material, and a fraction of them were turned into the 3:2 resonance region of Neptune or were scattered into the other space. There is an absence of objects between the resonance belt of 3:2 and 43 AU. The major part of these absent objects might have been removed. This characteristic of distribution of KBOs shown in Figure 4 cannot be completely illustrated by the mechanism of the Neptune resonance [37], unless the distribution of initial material is same as the $m(r)$ curve shown in Figure 1.

Zuo et al. have simulated the evolution of the Kuiper belt by the time regression method, using *N*-body dynamical model including the Sun, 8 planets, Pluto, UB313 and 551 massless KBOs with determined orbital parameters observed in 2007. The simulated results show that 10×10^8 years ago, more then 1/3 of the KBOs now observed had resided in the region of the Kuiper main belt, a few located inner Neptune orbit, and rest distributed outer 50 AU; 4.5×10^8 years ago, no large numbers of objects congregated in Neptune 3 : 2 resonance, and the distribution of objects in the Kuiper main belt is almost normal, which is almost same as the $m(r)$ curve [28]. This simulated result is more credible, because the initial condition of the simulation is current observational orbital parameters, which is different with other works. Figure 5 shows the evolution process of the distribution pattern of these KBOs, from top to bottom correspond to the patterns 4, 3, 2, 1 hundred million years ago respectively. Obviously, the distribution is gradually closer to a normal distribution, and the maximum is located between 43 and 45 AU, which is almost consistent with the sixth peak of $m(r)$ curve of H atoms. This result supports the guess of Nie in 2002[27].

It can be predictd that despite the increase of objects number the center of the Kuiper Belt should remain at about 43-45 AU. Since Pluto is only the largest object lying in 3:2 resonance region in the Kuiper belt, and have lost its large plane status, no attention will be paid it here.

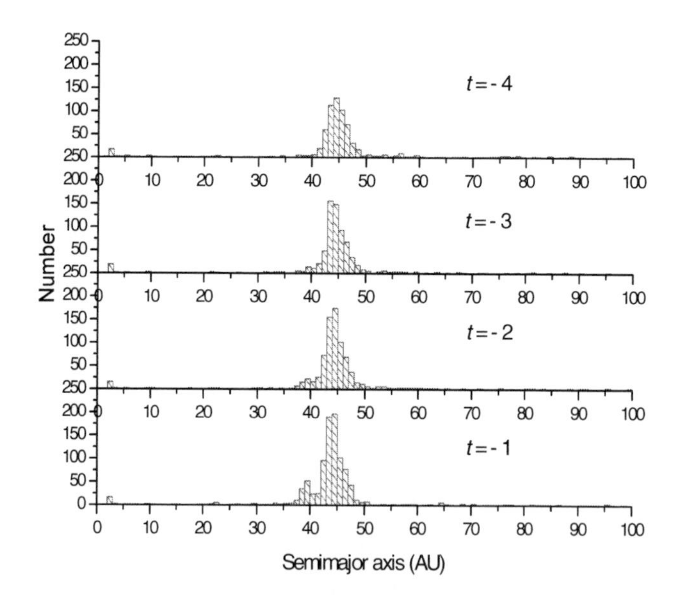

Figure 5. Evolution process of the distribution pattern of KBOs. The unit of *t* is 10^8 years.

6. STELLITES AND RING SYSTEMS

The model can also be applied to explain the regular satellites and ring systems. Numerical fitting shows that if the "Bohr radius" of H particle in planetary gravitational field is $a_J = 95.2\times10^3$ km for Jupiter, $a_S = 28.4\times10^3$ km for Saturn, and $a_U = 4.4\times10^3$ km for Uranus, many corresponding relations in Figure 5 can be found.

Figure 6(a) shows the relation between the regular satellites of Jupiter and the distribution of the materials in the model. The positions of the four Galilean satellites with large masses correspond to the four wave crests of H_2 curve respectively, and the largest three among them, Io (893×1020 kg), Ganymede (1482×1020 kg) and Callisto (1076×1020 kg), just locate at the superposition region of the wave crests of H and H_2 curves. The Europa with smaller mass (480×1020 kg) is at crest of H_2 curve but in trough of H curve. Besides, most of the inner small moons of Jupiter also lie near crests of H_2 curve, only the slightly larger Amalthea lies in the bottom of H_2 curve but just on the crest of He curve. This means that the role of the hydrogen molecules in the formation of satellites might be more important than that of hydrogen atoms. In addition, Jupiter has a ring system including halo (89.4-123.0×103 km), main (123.0-128.9×103 km) and gossamer (128.9 -242.0×103 km) . The brighter halo and main ring just occupy one wave of He curve, and the end of the gossamer ring is at the overlapping region of two trough of H and He waves.

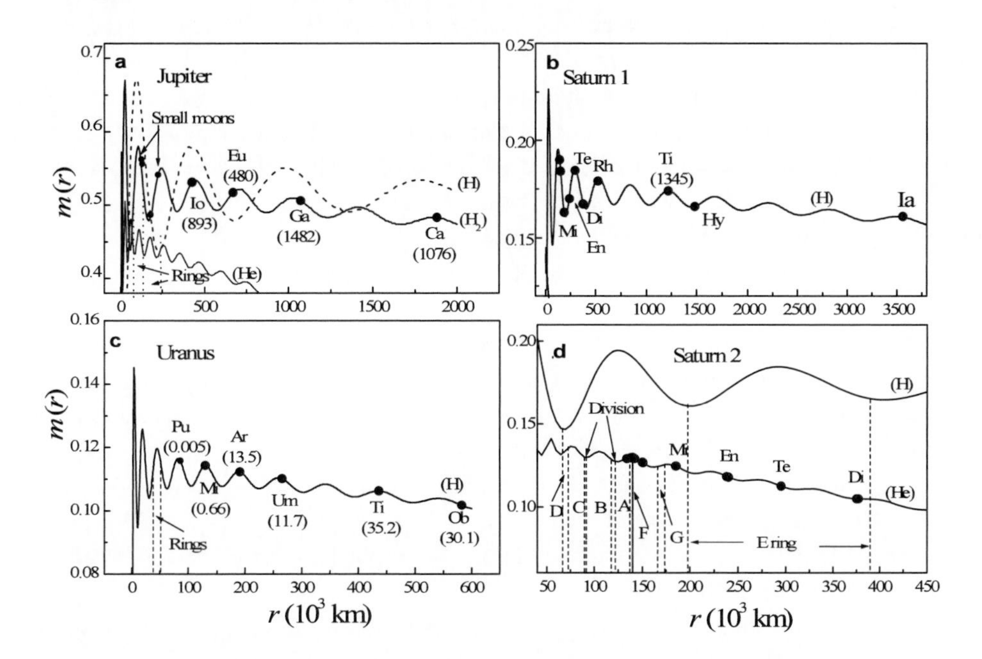

Figure 6. correspondences of $m(r)$ curves with the satellites and rings.

Here, every satellite is marked with the first two letters of its name. The dashed vertical lines give the locations of the rings. The numbers in brackets are the masses of the satellites (in unit 10^{20} kg).

Figure 6(b) reveals mainly the relation between the distribution of H atoms and positions of ten larger satellites of Saturn, which with their radii greater than 50 km. Among the five larger satellites (mass >1×1020 kg), Titan (1345.5×1020 kg), Rhea (23.1×1020 kg), Iapetus

(15.9×1020 kg), Dione (10.5×1020 kg), and Tethys (6.2×1020 kg), except for Dione, the four accurately lie at the wave crests of H curve. The distribution of the Dione and Tethys as well as smaller satellites Enceladus (0.73×1020 kg) and Mimas (0.38×1020 kg), is very similar to that of the terrestrial planets (see Figure 6(d)). Four satellites correspond to the four wave cycles of He curve respectively, and are included in a wave cycle of H curve. Mimas, Enceladus, and Tethys all slide down the wave crests of He curve and are close to the crests of H curve. Their mass order is also similar to Mercury, Venus and Earth. But Dione is special, which move away from the the crests of H. This can be understood because we can find an about 3:5 resonance between Dione and the next-door neighbor, Rhea with larger mass (see Figure 6(b)). The location of the small satellite Hyperion appears to be wrong , but it can be explained because there is a 4:3 resonance between the Hyperion and the largest satellite Titan. Their mass ratio is greater than 5000. Obviously, Hyperion can be attracted by Titan, and we should worry that Hyperion will be devoured because abound the Titan is empty.

Figure 6(d) shows clearly that the distributions of rings and the most of the moonlets of Saturn also accord with the fluctuation of He curve. The D ring and C ring occupy one wave cycle of He curve, B ring just takes up the next one, the narrow F ring and nearby five moonlets are at the He wave crest following B ring, while the French division (90-92×103 km) and Cassini division (118-122×103 km) are exactly in the two troughs of wave corresponding B ring respectively. Another coincidence is that if E ring is really defined observationally to lie in the region about 3.3-6.5 Saturn radii [39], which about is 60×103 km, it will be surprising again that E ring takes up a complete wave cycle of H curve, from about 200×103 km to 390×103 km. The outer edge is inconsistent with the Table 3.

Figure 6(c) displays that six larger regular satellites of Uranus (puck, Miranda, Ariel, Umbriel, Tiania and Oberon) correspond to the wave crests of H curve well, and the ring system of Uranus is just at a wave crest. The many narrow rings and small satellites (moons) inner of the satellite Puck will be shown in Table 4.

The satellite and ring system of Neptune is more special, it is difficult to plot their locations, which will be listed in Table 5.

From Figure 6 we can think that the formations of the three planet systems are related to the distributions of initial hydrogen. The molecule composition of hydrogen around the Jupiter is more than that of the Saturn and Uranus due to larger gas density caused by large mass of the Jupiter, so the correspondences of them with H_2 or H curve are reasonable respectively. In addition, there is an absent object in every curve in Figure 6, while the masses of satellites near these spaces are all the largest, such as Ganymede of Jupiter, Titan of Saturn, and Titania of Uranus. It can be assumed that the materials here have been captured by these largest satellites by some mechanism.

It can be found that the distributions of the small satellites and narrow rings around Neptune and Uranus are relative to the heavier elements, carbon and oxygen, which are the most abundant elements following the hydrogen and helium in the solar system. Table 4 and Table 5 list the locations of regular moons and rings, the corresponding locations of crests of $m(r)$ curves of particles, and the relative difference between them. In the two tables τ is optical depth of the ring, R is the mean radius of the moon, a is the semi-major axis of the orbit, a_n is the distance of the crest of the $m(r)$ curve from the planet, which can be calculated by the following experiential formula

$$a_n = [n + n(n-1) \cdot x + 0.01 \cdot \sum_{i=1}^{n} (i-1)(i-2)]a_1 \cdot \quad (8)$$

This formula is concluded from the $m(r)$ curves of H atom and heavy element particles, where x=1.23 for heavy element, and x=1.21 for H_2 and He; $a_1 = a_0$ is decided by equation (6), but they are different values in different planetary system for the same element particle. The a_0 of H atom is still used as a standard, which is 4.4 ×103 km for Uranus system and 3.96×103 km for Neptune system. The ratio of the a_0 of other particle to the H atom is inversely proportional the square of the mass, such as 1/122 and 1/162 for the carbon atom and oxygen atom respectively.

Table 4. The rings and moons of Uranus

Ring	τ	Moon	R (km)	a (10^6 m)	Crest of particle				a_n (10^6 m)	D (%)
6	0.3			41.84			C32		41.29	1.3
5	0.5			42.23						
4	0.3			42.57				O43	43.16	1.4
α	0.4			44.72			C33		44.03	1.5
β	0.3			45.66	H3			 O44	45.23 45.31	 0.8
η	≤0.4			47.17	$H_2$6		 C34		46.97 46.86	 0.6
γ	≥1.5			47.63				O45	47.51	0.2
δ	0.5			48.30		He12			48.43	0.3
		Cordelia	13	49.75				O46	49.77	0.0
λ	0.1			50.02			C35		49.79	0.4
ε	0.5-2.3			51.14				O47	52.09	1.8
							C36		52.81	
		Ophelia	16	53.76				O48	54.46	1.3
							C37		55.94	
								O49	56.90	
		Δ				He13			57.06	
		Bianca	22	59.16			C38	 O50	59.09 59.41	0.1 0.4
		Cressida	33	61.77				O51	61.96	0.3
		Desdemona	29	62.66			C39		62.48	0.3
		Juliet	42	64.36	 $H_2$7			O52	64.57 64.37	0.3
		Portia	55	66.10		 He14	C40		65.89 66.41	0.3 0.5
		Rosalind	29	69.93			C41		69.40	0.7
		Belinda	34	75.25		He15			76.50	1.7

In tables 4 and 5, the symbols C, O, He, and H_2 express the carbon atom, oxygen atom, helium atom, and hydrogen molecule respectively, and the number following the symbol is the ordinal of the radial density crest of the particle, for example, C32 expresses the 32nd crest of the carbon atom radial density $m(r)$ curves.

Tables 4 shows the results as follows:

(1) The locations of these moons and rings correspond to the crests of H6-H8, He12-H16, C32-C41, and O43-O52 respectively. The differences between of them are very small, and much less than that of ways of T-B law and B-H law in Table 1 and Table 2.

(2) The γ ring with the almost greatest optical depth lies on the overlapping region of the crests of $H_2$6 and He12. Similarly, the maximal moon Puck lies on the overlapping region of three crests of H4, $H_2$8 and He16; the next Portia correspond to the overlapping of He14 and C40, and closes the crest of $H_2$7; the third great Juliet is also close the crest of $H_2$7. Interestingly, the moons near $H_2$7 are greater than that near the He13. These phenomena mean that there are richer original material to form the moons and rings in these overlapping regions. This rule is similar to the mass distribution of terrestrial planet.

(3) There are three absent moons at the crests of the C36, C37, and O49, where is a trough between the He12 and He13. It is possible that smaller moons or fainter rings lie at these regions.

(4) The corresponding crests with brighter rings 5 and δ are not be found, which might be other mechanism.

Table 5. The rings and moons of Neptune

Ring	τ	Moon	R (km)	a (10^6 m)	Crest of particle				a_n (10^6 m)	D (%)
Galle	1×10⁻⁴			41.84			C34		42.17	0.8
						$H_2$6			42.27	1.0
							C35		44.81	
		Naiad	29	48.23			C36		47.54	1.4
								O48	49.02	1.6
		Thalassa	40	50.07			C37		50.34	0.5
		Despina	74	52.53				O49	51.21	2.5
Le Verrier	0.01			53.20			C38		53.24	0.0
								O50	53.45	0.5
Lassell	1×10⁻⁴			55.20			C39		56.22	1.8
								O51	55.75	1.0
Arago				57.20		$H_2$7			57.93	1.3
								O52	58.10	1.6
unnamed		Galatea	79	61.95			C40		59.30	4.3
								O53	60.50	2.3
Adams	0.01-0.1			62.93			C41		62.46	0.7
		Larissa	94	73.55	H4				73.66	0.1
		Proteus	209	117.65	H5				116.42	1.0

The distribution of the rings and moons of Neptune is similar to Uranus. The two greatst regular satellites, Proteus and Larissa, correspond to two wave crests of H atom accurately. Rings and small moons correspond to the crests of C and O. But the action of He is not found.

Moon Despina may correspond with crest O49, but the difference is relatively large. This can also be understood because there is a 3:5 resonance between Despina and Larissa.

7. Energy and Angular Momentum of Planetary Region

Whether the original nebular radial density variation is really as shown by $m(r)$ curves, we will give further evidence from comparison of the energy and angular momentum in planets districts. Considering interaction between planets during their formation and growth, the system is divided into four regions as given in Table 6, which adopts Dai method [35]. The inner boundary of the terrestrial region is the trough of the helium $m(r)$ curve. From quantum mechanics, the energy of a particle with mass μ in n state is $E_n = -GM\mu/2n^2 a_0$. From equation (7), the radial number density of a certain particle is $N\sum_{n,l}[rR_{nl}(r)]^2$. The total energy of particles in a thin spherical shell with thickness $\mathrm{d}r$ is $E = N\sum_{n,l} E_n[rR_{nl}(r)]^2 \mathrm{d}r$. If there are two kinds of particles, H and He atoms, the energy density in each of the regions can be expressed as $\varepsilon_{\mathrm{m}} = \sum E_i / \sum m_i$, where E_i and m_i are the total energy and the total mass of a certain kind of particles respectively, and they can be written as

$$E_i = \sum_{n,l} \frac{-GM\mu_i}{2n^2 a_{i0}} N_i \int_{r_1}^{r_2} [rR_{nl}^{(i)}(r)]^2 \mathrm{d}r \ , \tag{9}$$

$$m_i = N_i \mu_i \int_{r_1}^{r_2} \sum_{n,l} [rR_{nl}^{(i)}(r)]^2 \mathrm{d}r \ , \tag{10}$$

where r_1 and r_2 are the inner and the outer radii of the region respectively, and $R_{nl}^{(i)}(r)$ represents the radial wave function of a certain kind of particles. Based on the above two equations, the energy carried by the unit mass of the nebula (called energy density) in each of the four regions can be calculated by letting the ratio of total mass $\mu_i N_i$ of H and He in the solar system be 71: 27, and the unknown N_i can be eliminated. The calculated values of ε_{m} in unit $(-GM/2a_{0(\mathrm{H})})$ are given in Table 6, where the $a_{0(\mathrm{H})}$ is "Bohr radius" of H particle. The actual energy of planets can be calculated by $-GMm/2a$, where m and a are planetary mass and semi-major axis respectively. The planetary energy density ε_{p} in same unit is also given in Table 1 for comparison. It is clear that the data of the model and the observation agree well.

The similar method is used to calculate the angular momentum of the unit mass nebula (called angular momentum density). z axis component of the angular momentum of a particle in l states is written as $L_z = l\hbar_g$. In each region, the total angular momentum of a certain kind of particles can be written as

$$J_{zi} = N_i \hbar_g \int_{r_1}^{r_2} \sum_{n,l} l [rR_{nl}^{(i)}(r)]^2 \, \mathrm{d}r \ . \tag{11}$$

The angular momentum density in the each region is calculated by $j_{\mathrm{m}} = \sum J_{zi} / \sum m_i$. Since $\hbar_g$ is unknown, we take the value of the Uranus region as unit. The approximate equation calculating the angular momentum of planet is $J_p = m\sqrt{G(M+m)a(1-e^2)}$, where e is orbital eccentricity. The angular momentum density of a planet is $j_p = J_p / m$. The calculated values of nebular and the actual values of the planets in each region are listed in Table 6. It is also seen that the model is consistent with observation, except the terrestrial region.

Table 6. The comparison of the angular momentum and energy

Region	$r_1 - r_2$ (AU)	$\varepsilon_{\mathrm{m}} / \varepsilon_{\mathrm{p}}$	$j_{\mathrm{m}} / j_{\mathrm{p}}$
Terrestrial	0.15 - 1.7	1.075 / 1.182	0.074 / 0.214
Jupiter& Saturn	1.7 - 16	0.170 / 0.172	0.589 / 0.562
Uranus	16 - 25	0.052 / 0.052	1 / 1 (as unit)
Neptune	25 - 38	0.034 / 0.033	1.267 / 1.252

In conclusion, Table 6 shows a strong consistency between the theory and observation in three giant planets regions, but the deviations in the terrestrial region are larger, especially for the angular momentum. Actually, this is reasonable as a result when most of the original gas materials have escaped from the terrestrial region. There are two reasons for increase in the angular momentum of unit mass of planets: (1) planets had captured particles with large angular momentum, (2) planets formation region had lost the materials with smaller angular momentum. In the terrestrial region, the maximum amount of the primordial materials is H particles which are in $l = 0$ state with zero angular momentum, and next is smaller ratio of helium and other heavier particles being in any possible l state with larger angular momentum. At the early stage of the Sun formation, the solar wind is very weak, so that H particles with zero angular momentum can be easily captured by the Sun. In late stage, the strong solar wind took away the most of H particles again and left some angular momentum in the collision with planets materials. Both the mechanisms can make terrestrial region lose a lot of mass, while lose only small amount of angular momentum. Even the angular momentum taken away by solar wind might be smaller than brought in. As a result, the angular momentum of unit mass of planets in the terrestrial region is increased. We can make a simple estimation. Let the original total mass of terrestrial region be m, the lost mass with zero angular momentum be $x \cdot m$, then $(m - xm)$ is planetary mass. From the ratio of 0.074 / 0.214 (see Table 6), the total original angular momentum is $0.074m = 0.214(m - xm)$. The ratio of lost mass can be obtained as $x = 65.4\%$, which is reasonable. In addition, since the energy of the particle is proportion to mass, the energy will naturally decreases with the decrease of the mass, and the ratio of energy and mass will be a constant approximately. So there is only small deviation of energy of the unit mass in the terrestrial region.

In this section, the planetary rotation has not been mentioned because the rotation energy and angular momentum are very smaller than the orbital ones.

8. The Rotations of Planets

The understanding for rotations of planets in solar system is also a difficult problem. It is known that the rotation of Venus differ from other planets in the solar system, which is retrograde and long-period. To explain this observed phenomenon, same mechanisms such as core mantle friction inside planet [40], atmospheric tide [41-46], or twain actions together [47-50], and impact with a giant object [51,52] have been suggested. However, these works need specific initial conditions with a remote probability. In addition, the slow rotation of Mercury cannot be explained very well. Using the quantum-like hypothesis, the unusual rotations of the Venus and Mercury can be understood easily, which might be the naturally evolution by interaction with interplanetary matter. This conjecture has been proved by our initial work, in which we calculate the angular momentum obtained during the planet revolution and rotation[29,53].

8.1. The model

From Figure 1(b) we can see that the positions of Venus and Mercury on the H curve are different from other plants, the radial density of H atoms outside of their bodies are greater than the inside. This means when they run around the Sun, the collisions of the H atoms with outside of the bodies are more that the inside, so they can obtain the inverted angular momentum.

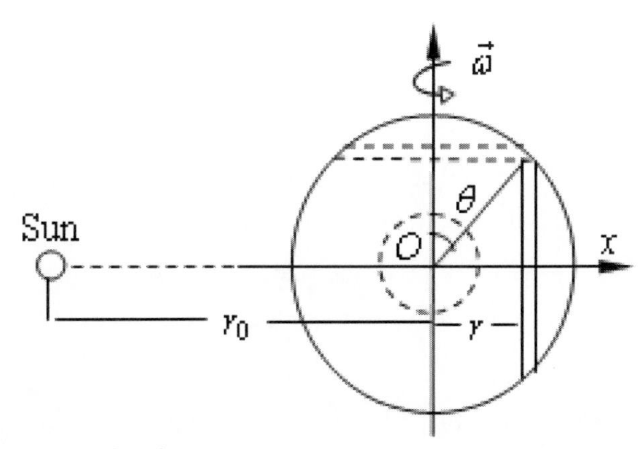

Figure 7. Revolution section of a planet.

To calculate the evolution of planetary rotation, two models for planets revolution and rotation should be established by means of the sketch in Figure 7, where the circle expresses a planetary Hill sphere with the radius R, r_0 is the distance of the planetary center from the

Sun. Assuming the planet spin axis is perpendicular to the orbital plane and the spin direction is same as revolution, we can draw a thin ring with the width dr on the surface of the sphere, the spin angular momentum obtained during one period of revolution by the ring in collision with interplanetary matter is

$$\mathrm{d}L = -\rho(r)[2\pi r \cdot 2\sqrt{R^2-(r-r_0)^2}\,\mathrm{d}r]\eta v(r-r_0), \tag{12}$$

where r is the distance of the ring from the center of the planet, $\rho(r)$ is the mass density at the r, v is the revolution velocity of the planet, and ηv is the relative velocity of planets to interplanetary gas. The theory on origin of the solar system suggests that solids and gas do not have the same angular velocities in solar nebula, and angular velocity of the gas is smaller than that of a large body [54], so $0<\eta<1$.

The key of equation (12) is mass density $\rho(r)$, which depend on the $m(r)$ value of H. The $m(r)$ given by equation (7) is the non-normalized radial probability density of interplanetary matter. Using the near-Earth mass density $\rho_\oplus$ and the radial density $m(r_\oplus)$, the relation between $\rho(r)$ and $m(r)$ can be given. To derive this relation, we first must consider the angular distribution of interplanetary matte. It is known that many younger stars have a circumstellar nebular disk, where planets will be formed. The thickness of a nebular disk is generally expressed as $h(r)\propto r^\beta$, where parameter β is a little larger than 1. For instance, β= 1.10-1.32 for HK Tau/c in the young binary system HK Tauri [55] , 9/8 for Orion 114-426 [56], 1.29 for IRAS 04302+2247 [57], and 1.25 for HH 30 IRS [58]. We take β=1 approximately in the region of $r<6$ AU. Due to $h_\oplus / h(r) = r_\oplus / r$ and $m(r)=\rho(r)\cdot 2\pi r\cdot h(r)$, the mass density can be expressed as

$$\rho(r) = \frac{\rho_\oplus r_\oplus^2}{m(r_\oplus)}\cdot\frac{m(r)}{r^2} \tag{13}$$

where $r_\oplus$ is the distance of Earth from the Sun, and the $\rho_\oplus$ can be determined by observation.

The difficulty to integrate equations (12) is that there is not a formulation of $m(r)$. But it can be written as $m(r)=m(r_0)+kx$ in the scale of the planet body, where $x=r-r_0$. If Ω denotes the angular velocity of revolution, the integral of equation (12) is

$$\begin{aligned}\Delta L_1 &= -4\pi\eta\Omega\frac{\rho_\oplus r_\oplus^2}{m(r_\oplus)}\int_{-R}^{R}[m(r_0)+kx]\sqrt{R^2-x^2}\,x\mathrm{d}x \\ &= -\frac{\rho_\oplus r_\oplus^2\pi^2\eta}{2m(r_\oplus)}\Omega R^4 k\,. \end{aligned}\tag{14}$$

This is the angular momentum obtained by the planet during one period of revolution, the positive and negative of which depends on the slope k of the $m(r)$ curve of H. If $k>0$, the spin angular momentum is decreasing, which is the case of the Venus and Mercury. Assuming that the planet is a sphere with uniform density, the moment of inertia is $J = 2mR_0^2/5$, where m is the planet mass, R_0 is the radius of the planetary body. Using $\Delta L_1 = J\Delta\omega_1$, the increment of angular velocity due to the revolution of the planet is

$$\Delta\omega_1 = -\frac{5\pi^2 \rho_\oplus r_\oplus^2 \eta}{4m(r_\oplus)} \cdot \frac{\Omega R^4 k}{mR_0^2} \tag{15}$$

In addition, the resistance of interplanetary matter during the planet rotation could also slow down its spin velocity. During one period of revolution, the mass of interplanetary matter interacting with the Hill sphere of the planet is approximately $\Delta m = \rho(r_0)\cdot\pi R^2 \cdot 2\pi r_0$. Based on the conservation of angular momentum , we can give

$$(2mR_0^2/5)\cdot\omega = [2(m+\Delta m)R_0^2/5]\cdot(\omega+\Delta\omega_2) \tag{16}$$

Ignoring the second order small quantity, then $\Delta\omega_2 = -\Delta m\omega/m$, where ω is spin angular velocity. Using equation (13) again, the increment of angular velocity due to the rotation of the planet is

$$\Delta\omega_2 = -\frac{2\pi^2 \rho_\oplus r_\oplus^2}{m(r_\oplus)} \cdot \frac{m(r_0)R^2}{mr_0}\omega \tag{17}$$

During one period of revolution, the increment of the spin angular velocity is $\Delta\omega = \Delta\omega_1 + \Delta\omega_2$, and the planet spin angular acceleration is

$$\frac{\mathrm{d}\omega}{\mathrm{d}t} = -\frac{\pi\rho_\oplus r_\oplus^2}{m(r_\oplus)} \cdot \frac{R^2}{m}\Omega[\frac{5}{8}(\frac{R}{R_0})^2 \eta k\Omega + \frac{m(r_0)}{r_0}\omega] \tag{18}$$

Assuming that the revolution angular velocity Ω, the planetary mass m, and the Hill sphere radius R are constants, the mass density $\rho_\oplus$ is a unique variable in equation (18), which should decrease with time. The rule of the mass decrease of the dust disks around sun-like stars, $\mathrm{d}m_{disk}/\mathrm{d}t \propto -m_{disk}$, was given by Wyatt et al [59]. The evolution equation $\rho_\oplus = \rho_0 \mathrm{e}^{-at} = \rho_0 \mathrm{b}^{-t}$ can be obtained by solving equation $\mathrm{d}\rho_\oplus/\mathrm{d}t = -a\rho_\oplus$, where $b > 1$. Finally, the planet spin angular velocity, which is function of time, is obtained from the integral of equation (18)

$$\omega = (Q+\omega_0)\exp[P\rho_0(b^{-t}-1)/\ln b] - Q \tag{18}$$

where $P = \frac{\pi r_\oplus^2}{m(r_\oplus)} \cdot \frac{m(r_0)R^2}{mr_0}\Omega$, and $Q = \frac{5\eta}{8m(r_0)}(\frac{R}{R_0})^2 kr_0\Omega$.

8.2. The results

Taking current time $t = 0$, then ω_0 and ρ_0 are the current spin angular velocity and mass density in near-Earth space, respectively. The ρ_0 and b should first be clarified. The mass density of interplanetary dust have been estimated at 10^{-19} kg/m3 [60], the ratio of dust to nebula is about 2%, then the density of interplanetary matter is about 5×10^{-18} kg/m3. Dwek et al. have proposed that the number density of hydrogen atoms is ~104 cm-3 [61], so the mass density should be ~10-17 kg/m3. Here we take ρ_0=5×10^{-18} kg/m3 =1.7×10^{16} kg/AU3. Taking ρ_B as the mass density at 4.5 Gyr (the Earth age) ago, then we can obtain the relation $\rho_B / \rho_0 = b^{45}$. The parameter b in equation (18) can be estimate by comparison and selection (see Table 7). It is found that changing mass density ρ_B can bring different effects on the planetary rotation. The effect is obvious for Venus and Mercury, but unobvious for other planets. When $\rho_B / \rho_0 = 1.0\times10^9$, corresponding to b = 1.585, the periods of the terrestrial planets are all close to one day, which are typical expected values. So we take the b = 1.585, η = 0.5, and $r_\oplus$ =1 AU in calculation of equation (18). The Hill radius $R = (m/3M)^{1/3} r_0$ is used, where the mass of the Sun $M = 1.989\times10^{30}$ kg, the radial density near Earth is taken as $m(r_\oplus) = 0.64$, which does not have practical significance. real meaning. Other parameters are derived and shown in Table 8.

The evolution of spin periods of the Venus and Mercury are shown in Figure 8. It is found that the spin of the Venus changes naturally from positive to retrograde around 3.5 Gyr ago, the period is from the initial about 3 day to the current – 243 day; the spin of the Mercury changes from quick to slow, the period is from the initial about 1 day to the current 58.6 day. Another fact should be noticed that the spins of the two planets remain almost steady from about 3 Gyr ago until the foreseeable future.

Table 7. Comparison of the original rotation periods for different densities

ρ_B / ρ_0		10^{11}	10^{10}	10^9	10^8	10^7	10^6	10^5	10^4	10^3	10^2
b		1.756	1.668	1.585	1.506	1.431	1.359	1.292	1.227	1.166	1.108
T_B day	Mer	0.001	0.088	0.963	7.732	33.50	53.93	58.04	58.57	58.64	58.65
	Ven	0.011	0.330	3.361	33.99	-3471	-272	-246	-243	-243	-243
	Ear	0.153	0.811	0.974	0.995	0.997	0.997	0.997	0.997	0.997	0.997
	Mars	-0.004	-0.21	2.356	1.094	1.033	1.027	1.026	1.026	1.026	1.026
	Jup	0.627	0.425	0.412	0.410	0.410	0.410	0.410	0.410	0.410	0.410

Note: The first three letters represent the names of the planets.

Table 8. The data for computing

planets	m 10^{24}kg	R_0 10^{-5}AU	r_0 AU	R AU	ω_0 yr^{-1}	T yr	$m(r_0)$	k AU^{-1}	τ_B day	τ_0 day
Merc	0.33	1.63	0.387	0.0015	39.11	0.241	0.33	1.03	1.054	58.6

Ven	4.87	4.04	0.723	0.0068	-9.44	0.615	0.59	0.43	3.465	-243
Ear	5.97	4.25	1.000	0.0100	2210	1.000	0.64	0	0.985	0.997
Mars	0.64	2.26	1.524	0.0073	2236	1.881	0.54	-0.24	2.314	1.026
Jup	1900	47.6	5.203	0.3551	5594	11.86	0.53	-0.05	0.412	0.41

Note: Here m and R_0 are the planet mass and radius respectively. R is the Hill radius, r_0 is the distance of the planet center from the Sun , ω_0 is the current spin angular velocity, and T is the revolution period, τ_{B} and τ_0 are 4.5 Gyr ago and current spin periods.

Appling equation (18) to other planets, we can find that the changes their spin periods are very small. In detail, the period of the Earth decreases 0.01 days, the Mars increases 1.3 days, the Jupiter decreases 0.02 days, the changes of Saturn and Neptune are much smaller than these. However, equation (18) cannot be applied to the Uranus due to the particularity of its spin axis.

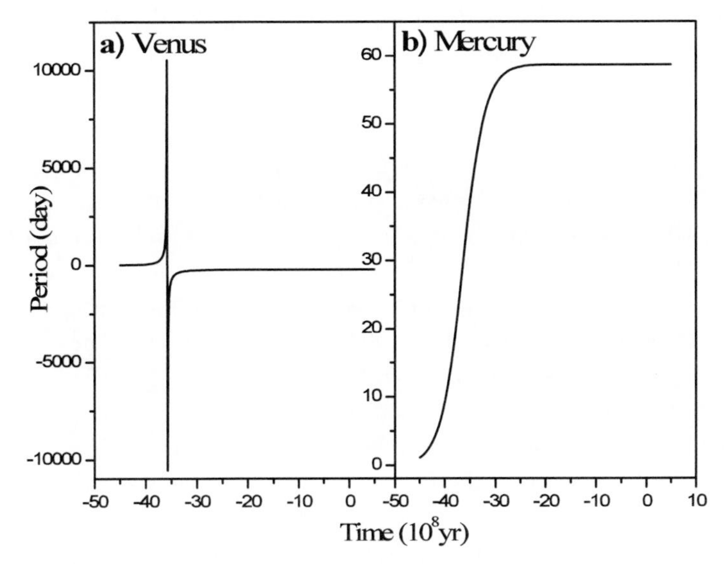

Figure 8. Evolution of the rotation periods of the Venus and Mercury in 4.5 billion years.

9. THE SUPPOSITION ON ORIGIN OF THE SOLAR SYSTEM STRUCTURE

In conclusion, in the study of the basic problems of solar system structure origin, the quantum-like model described in this chapter would be more efficient than other ones. Many correspondences mentioned above are neither far-fetched and obscure nor haphazard coincidence. It might reveal an unknown truth. These researches can inspire us to make following hypotheses:

(1) The chaos behavior of nebular particles in gravitational field can be described by the wave function satisfying the Schrödinger equation when we replace the $\hbar$ with $\hbar_g$.

(2) The distributions of H and He particles might play the important roles in formation of the planets. Waves of radial density of H atoms formed a series of the largest rings, and the terrestrial planets, Jupiter, Saturn, Uranus, Neptune, as well as the Kuiper

belt formed in these rings respectively. The rings of He atoms are narrower than those of H. The first H ring covers four He rings, where the original terrestrial planets formed respectively, but due to the influence of H ring, they have shifted toward the center of H ring.

(3) The distribution of heavier elements is closer to the Sun than hydrogen and helium. The rings made from heavier particles are narrower and denser than those of He. One He ring contains tens or hundreds of thin rings of heavier particles, and the thin ring contains the thinner rings, which is like the form of the observed rings of Saturn. With the decrease of the distance from the Sun, rings become denser gradually. They might be important for the formation of planetesimals.The heavy elements and planetesimals in the terrestrial region are more than in other regions, so the formation of terrestrial planets is easier. In the giant planets region, due to the lack of heavy elements and planetesimals, the formation of proto-planets can only depend on the gravitational instability of gas.

(4) The formation of the proto-planet was earlier on the higher crest of the particle radial density wave. In the giant planet region, the order of planets formation is Jupiter, Saturn, Uranus and Neptune. Since Jupiter forms early and grows quickly, it can capture a large amount of materials of the terrestrial region and Saturn, and becomes the largest planet. Though Neptune forms later, because a large amount of materials in the widest region outside its exterior boundary can contribute to it, its mass is still larger than that of Uranus. Due to strong solar wind and the strong tidal forces from the Sun and Jupiter, the terrestrial region cannot form a giant planet. But if the mass of a star and nebula density around it satisfy the condition to cause the accretion of a planet, the giant can also be form in terrestrial region, such as the extrasolar planets.

(5) The formation mechanism of satellites differs from that of planets. The regular satellites might form in a circumplanetary accretion disk produced by a slow inflow of gas and solids during the end of the planet formation [62]. The gas should form an envelope, in which the gas particles distribution is the same as shown in Figure 6. When the planetesimals in exterior space fall into the envelope, they should be likely to stay nearby the radial density crest of gas by the stickiness of gas, and finally form satellites. In this way, the mass distribution of satellites is naturally larger in the outer than in the inner.

(6) Due to the special position in H wave, the rotation speeds of Mercury and Venus gradually slow down by asymmetric viscous resistance. The resistance is so effective that the rotation direction of Venus is reversed.

(7) This chapter shows a perfect structure of the solar system, which can be divided into two symmetrical quantum groups: (1) the Jupiter, Saturn, Uranus, Neptune and the Kuiper belt; (2) the Mercury, Venus, Earth, Mars and the asteroid belt. Their quantum orders are all described as n = 2, 3, 4, 5, 6, and the width of the planetary districts of the giant planets group is 16 times larger than that of the terrestrial group. This ratio is just the square of the atom mass ratio of helium to hydrogen, which are the main particles in primitive nebula of the solar system. According to these facts, it can be predicted that the structure of the solar system has been basically fixed, and it is impossible to find major planets inside the orbit of Mercury and outside the Kuiper belt, but the minor planets or dwarf planets similar to Pluto may be found.

(8) This chapter provides an initial distribution state of the solar nebula. Using the initial conditions, the formations of planets and satellites can be simulated, although the workload is very large.

10. THE QUESTION ON"QUANTUM THEORY OF THE UNIVERSE"

Many literates have raised doubts about whether the quantum law can only exist at the microcosmic fields.This chapter gives some clues on quantum phenomena in cosmic scale. The quantum phenomena exist generally in various cosmic scales. The quantum law might be a fundamental law of the universe, but there are different quantum constants $\hbar$ in different cosmic scales. From the clues in this chapter, we can find that the quantum gravitational constant $\hbar_g$ is related to the mass of the celestial system, and is a possible proportional relation. However, $\hbar_g$ is also very small, so only the microscopic particles such as atoms and molecules can be observed the quantum law, while the macroscopic objects such as planets and satellites cannot.

One century ago, Rutherford proposed “the solar system model” of the atom structure. Today, we explore the origin of the solar system structure by referencing the quantum law in the atom. What we doing now is just based on the reverse thinking of Rutherford's model. Thereby, I’d like to express my sincere honor to the great scientist, Rutherford, at the time of the 100th anniversary of the atom model of Rutherford.

ACKNOWLEDGMENTS

The author would like to thank Prof. Chuanfu Cheng and Prof. Hongmin Geng for their help in English writing, and also to thank David Jewitt for the use of data from his *Internet Station*. At last, the author would express her thanks to *The SAO/NASA Astrophysics Data System* for the consulting on references.

REFERENCES

[1] M. M. Nieto, "Conclusions about the Titius Bode Law of Planetary Distances," *Astronomy and Astrophysics*, Vol. 8, 1970, pp. 105-111.

[2] David M. F. Chapman, “Reflections: The Titius-Bode Rule, Part 1: Discovering the Asteroids” *Journal of the Royal Astronomical Society of Canada* , Vol. 95, 2001, pp.135-136.

[3] L.Basano and D. W. Hughes, "A Modified Titius-Bode Law for Planetary Orbits," *Nuovo Cimento C*, Vol. 2C, 1979, pp. 505-510.

[4] E. Badolati, "A Supposed New Law for Planetary Distances," *Moon and the Planets*, Vol. 26, 1982, pp. 339-341.

[5] V. Pletser, "Exponential Distance Laws for Satellite Systems," *Earth, Moon, and Planets*, Vol. 36, 1986, pp. 193-210.

[6] R. Neuhaeuser and J. V. Feitzinger, "A Generalized Distance Formula for Planetary and Satellite Systems, " *Astronomy and Astrophysics*, Vol. 170, 1986, pp. 174-178.

[7] Peter Lynch, "On the Significance of the Titius-Bode Law for the Distribution of the Planets" *Monthly Notice of the Royal Astronomical Society*, Vol. 341, 2003, pp. 1174-1178.

[8] L. Neslušan, "The Significance of the Titius-Bode Law and the Peculiar Location of the Earth's Orbit" *Monthly Notice of the Royal Astronomical Society*, Vol. 351, 2004, pp. 133-136.

[9] G. Gladyshev, "The Physicochemical Mechanisms of the Formation of Planetary Systems" *Moon and the Planets*, Vol. 18, 1978, pp. 217-222.

[10] J. J. Rawal, "Contraction of the Solar Nebula" *Earth, Moon, and Planets*, vol. 31, 1984, pp. 175-182.

[11] X. Q. Li, Q. B. Li and H. Zhang, "Self-similar Collapse in Nebular Disk and the Titius-Bode Law" *Astronomy and Astrophysics*, Vol. 304, 1995, pp.617-621.

[12] F. Graner and B. Dubrulle, "Titius-Bode Laws in the Solar System. 1: Scale Invariance Explains Everything, " *Astronomy and Astrophysics*, Vol. 282, 1994, pp. 262-268.

[13] C. D. Murray and S. F. Dermott, Solar System Dynamics. Cambridge University press, Cambridge, (1999).

[14] Frank Spahn, Jürgen Schmidt, Nicole Albers, et al., "Cassini Dust Measurements at Enceladus and Implications for the Origin of thee Ring" Science, Vol. 311, 2006, pp. 1416-1418.

[15] B. A. Smith, L. A. Soderblom, and T. V. Johnson, "The Jupiter system through the eyes of Voyager 1," *Science*, Vol. 204, 1979, pp. 951-957, 960-972.

[16] B. A. Smith, L. A. Soderblom, R. Beebe, "The Galilean satellites and Jupiter - Voyager 2 imaging science results," *Science*, Vol. 206, 1979, pp. 927-950.

[17] G. J. Consolmagno, "Lorentz forces on the dust in Jupiter's ring" *Journal of Geophysical Research*, vol. 88, 1983, pp. 5607-5612.

[18] L. Schaffer and J. A. Burns, "The dynamics of weakly charged dust - Motion through Jupiter's gravitational and magnetic fields," Journal of Geophysical Research, vol. 92, 1987, pp. 2264-2280.

[19] P. D.Nicholson, S. E. Persson, ; K. Matthews, et al., "The rings of Uranus - Results of the 10 April 1978 occultation" *Astronomical Journal*, vol. 83, 1978, pp. 1240-1248

[20] B. A. Smith, L. A. Soderblom, and D. Banfield, "Voyager 2 at Neptune - Imaging science results," *Science*, vol. 246, 1989, pp. 1422-1449.

[21] Carolyn C. Porco, "An explanation for Neptune's ring arcs,"Science, vol. 253, Aug. 30, 1991, p. 995-1001.

[22] Anne J Verbiscer, Michael F. Skrutskie, and Douglas P. Hamilton, "Saturn's largest ring," Nature, Vol. 461, 2009, pp. 1098-1100.

[23] G. G. Comisar, "Brownian-Motion Model of Nonrelativistic Quantum Mechanics" *Physical Review*, Vol. 138, 1965, pp. 1332-1337.

[24] E. Nelson, "Derivation of the Schrödinger Equation from Newtonian Mechanics" *Physical Review*, Vol. 150, 1966, pp.1079-1085.

[25] L. Nottale, G. Schumacher and J. Gay, "Scale Relativity and Quantization of the Solar System," *Astronomy and Astrophysics,* Vol. 322, 1997, pp. 1018–1025 .

[26] Qingxiang Nie, "Simulated Quantum Theory for Seeking the Mystery of Regularity of Planetary Distances," *Acta Astronomica. Sinica.*, Vol. 34, 1993, pp. 333 – 340.

[27] Qing-xiang Nie, "The Characteristics of Orbital Distribution of Kuiper Belt Objects," *Chinese Astronomy and Astrophysics,* Vol. 27, 2003, pp. 94-98.

[28] Qing-lin Zuo, Qing-xiang Nie, Yuanling Yang, et al.,"Simulations for Original Distribution of KBOs ," *Acta Astronomica. Sinica.*, Vol. 49, 2008, pp.413-418.

[29] Qingxiang Nie, Cuan Li, and Feng-shou Liu, "Effect of Interplanetary Matter on the Spin Evolutions of Venus and Mercury," *International Journal of Astronomy and Astrophysics*, Vol. 1, No.1, 2011, pp.1-5.

[30] Qingxiang Nie, "Comprehensive Research on the Origin of the Solar System Structure by Quantum-like Model," *International Journal of Astronomy and Astrophysics*, Vol. 1, No. 2, 2011, pp. 51-60.

[31] Bu-en Yang (1996) Introduction to Quantum theory on Planets and Satellites（in Chinese）. *Dalian University of Technology Press*, Dalian, pp. 27-32.

[32] A. G. Agnese and R. Festa, "Clues to Discretization on the Cosmic Scale," *Physics Letters A*, Vol. 227, 1997, pp.165-171.

[33] Alan P. Boss, "Astrometric Signatures of Giant-planet Formation," *Nature,* Vol. 393, 1998, pp. 141-143.

[34] Alan P. Boss, "Rapid Formation of Outer Giant Planets by Disk Instability," *The Astrophysical Journal*, Vol. 599, 2003, pp. 577-581.

[35] N. Murray and M. Holman, "The Origin of Chaos in the Outer Solar System" *Science*, Vol. 283, 1999, pp.1877.

[36] David Jewitt, (2011) Kuiper Belt *http://www2.ess.ucla.edu/~jewitt/kb.html.*

[37] Jane X. Luu and David C. Jewitt, ",Kuiper Belt Objects: Relics from the Accretion Disk of the Sun," *Annual Review of Astronomy and Astrophysics*, Vol. 40, 2002, pp. 63-101.

[38] Renu Malhotra, "The Origin of Pluto's Orbit: Implications for the Solar System beyond Neptune" *Astronomical Journal,* Vol. 110, 1995, pp.420-429.

[39] J. A. van Allen et al., "Saturn's magnetosphere, rings, and inner satellites" *Science*, Vol. 207, 1980, pp. 415-421.

[40] A. C. M. Correia, "The Core Mantle Friction Effect on the Secular Spin Evolution of Terrestrial Planets," *Earth and Planetary Science Letters*, Vol. 252, 2006, pp. 398-412.

[41] W. Kundt, "Spin and Atmospheric Tides of Venus," *Astronomy & Astrophysics*, Vol. 60, 1977, pp. 85-91.

[42] A. R. Dobrovolskis, "Atmospheric Tides and the Rotation of Venus. II-Spin Evolution," *Icarus*, Vol. 41, 1980, pp. 18-35.

[43] J. McCue, J. R. Dormand and A. M. Gadian, "Estimates of Venusian Atmospheric Torque," *Earth, Moon, and Planets*, Vol. 57, 1992, pp. 1-11.

[44] J. McCue and J. R. Dormand, "Evolution of the Spin of Venus," *Earth, Moon, and Planets*, Vol. 63, 1993, pp. 209-225.

[45] J. Laskar and P. Robutel, "The Chaotic Obliquity of the Planets," *Nature*, Vol. 361, 1993, pp. 608-612.

[46] A. C. M. Correia, *et al*., "Long-Term Evolution of the Spin of VenusI. Theory," *Icarus*, Vol. 163, 2003, pp. 1-23.

[47] A. C. M. Correia and J. Laskar, "The Four Final Rotation States of Venus," *Nature*, Vol. 411, 2001, pp. 767-770.

[48] P. Goldreich and S. J. Peale, "The Obliquity of Venus," *Astronomical Journal*, Vol. 75, 1970, pp. 273-285.

[49] B. Lago and A. Cazenave, "Possible Dynamical Evolution of the Rotation of Venus since Formation," *Moon and the Planets*, Vol. 21, 1979, pp. 127-154.

[50] M. Shen and C. Z. Zhang, "Dynamical Evolution of the Rotation of Venus," *Earth, Moon, and Planets* , Vol. 43, 1988, pp. 275-287.

[51] S. F. Singer, "How Did Venus Lose Its Angular Momentum," *Science*, Vol. 170, 1970, pp. 1196-1198.

[52] S. Tremaine and L. Dones, "On the Statistical Distribution of Massive Impactors," *Icarus*, Vol. 106, 1993, pp. 335-341.

[53] Feng-shou LIU and Qing-xiang NIE, "A Possible Formation Mechanism of Venus's Abnormal Rotation" *ACTA ASTRONOMICA SINICA,* Vol. 51, 2010, pp.173-183.

[54] A. G. W. Cameron, "Origin of the Solar System," *Annual Review of Astronomy and Astrophysics*, Vol. 26, 1988, pp. 441-472.

[55] K. R. Stapelfeldt, *et al*., "An Edge-On Circumstellar Disk in the Young Binary System HK Tauri," *The Astrophysical Journal*, Vol. 502, 1998, pp. L65-L69.

[56] M. J. McCaughrean, *et al*., "High-Resolution Near-Infrared Imaging of the Orion 114-426 Silhouette Disk," *The Astrophysical Journal*, Vol. 492, 1998, L157-L161.

[57] S. Wolf, D. L. Padgett and K. R. Stapelfeldt, "The Circumstellar Disk of the Butterfly Star in Taurus," *The Astrophysical Journal*, Vol. 588, 2003, pp. 373-386.

[58] K. W. Wood, J. Michael, J. E. Bjorkman and B. Whitney, "The Spectral Energy Distribution of HH 30 IRS: Con-straining the Circumstellar Dust Size Distribution," *The Astrophysical Journal*, Vol. 564, 2002, pp. 887-895.

[59] M. C. Wyatt, R. Smith, J. S. Greaves, *et al*., "Transience of Hot Dust around Sun-Like Stars," *The Astrophysical Journal*, Vol. 658, 2007, pp. 569-583.

[60] P. R. Weissman, L. A. McFadden and T. V. Johnson, "Encyclopedia of the Solar System," Academic Press, San Diego, 2006.

[61] E. Dwek, R. G. Arendt and F. Krennrich, "The Near-Infrared Background: Interplanetary Dust or Primordial Stars," *The Astrophysical Journal*, Vol. 635, 2005, pp. 784-794.

[62] R. M. Canup and W. R. Ward, "Formation of the Galilean Satellites: Conditions of Accretion" *The Astronomical Journal,* Vol. 124, 2002, pp. 3404-3423.

In: Solar System: Structure, Formation and Exploration ISBN: 978-1-62100-057-0
Editor: Matteo de Rossi

Chapter 2

CLOSE BINARIES, ECCENTRIC EXOJUPITERS AND THE SOLAR SYSTEM

Edward M. Drobyshevski*

Ioffe Physical-Technical Institute of RAS, St. Petersburg, Russia

ABSTRACT

Most of the stars are members of multiple systems, which can be traced to multi-stage fragmentation of turbulent protostellar clouds bearing a standard excess of angular momentum. Close binaries are products of the last fragmentation of a rapidly collapsed, differentially rotating protostar (PS). It breaks down as a result of rotational-exchange instability, when the angular momentum is carried out by turbulent viscosity as the outer Hayashi convection sets in in the PS. This culminates in a massive gas ring being detached from the PS, with ensuing immediate azimuthal fragmentation. Interaction of these ~3-10 fragments of mass ~1-10 M_J (M_J is a mass of Jupiter) with the PS and with one another leads to part of them becoming ejected out of the system, with some of them merging with the PS, and the others entering strongly eccentric orbits around the PS. These are exo-Jupiters. The convection-embraced matter outflows rapidly onto the most "lucky" fragment, which at the perigee of its low-eccentricity orbit comes close to the PS, so that the first Lagrangian point falls into its convective zone. In the close binary thus formed, the mass ratio of the components is determined by that of the Hayashi convective zone and the stable core of the PS.

If convection extends over all of the PS (the $M \leq 1.5\ M_{\odot}$ case), overflow of all convective mass results in mass reversal of the components (a process well known for the Algol-type binaries) strong enough to produce a system with a mass ratio $q \sim 1{:}1000$. Here mass transfer from the fully convective PS stops after its mass has dropped to ~M_J because of the onset of water condensation in its gas that became adiabatically cooled in expansion. Eventually, a rapidly rotating PS converts into a Jupiter-like planet, while the "lucky" fragment that obtained almost all the initial PS mass becomes the "Sun".

Cooling of matter in such a rapidly rotating and extremely dense *proto-Jupiter* (PJ) will initiate, starting from $q \sim 0.1$-0.02, condensation of nonvolatile compounds and formation of numerous (~10^4-10^5) Pluto-like and larger bodies moving in it along

* E-mail address: emdrob@mail.ioffe.ru

unwinding orbits and, hence, leaving it. Thus, PJ may be considered here as a very dense analog of a traditional protoplanetary disk. This is how a Solar-type planetary system forms. The latter four bodies survived in our system as Galilean satellites, the most massive ones retained heliocentric planetary orbits, the larger part of these 10^4-10^5 bodies were ejected out of the system, and a small part of them (~10^3) formed, primarily at a distance of 50÷300 AU, a trans-Neptunian cloud of interacting dwarf planets moving in disordered orbits. This prediction finds convincing support in the discovery made in the recent decade of planets of the type of Sedna, Xena etc.

To sum up, the above approach is essentially a synthesis of the gas-dynamic scenario of the formation and evolution of binary stars with the classical pattern of the planets' formation in circumstellar disks. In contrast to the latter, however, it is capable of accounting for the totality of available data and offering predictions borne out by observations.

1. Introduction. Problems with Planets

Studies of planets of the Solar system (SS) during half a century of the cosmic era and the discovery of more than half a thousand exoplanets in the last one-and-a-half decades [e.g., Schneider 2011; Baraffe et al 2010; Marcy et al 2005; Udry and Santos 2007] have shown that, despite the great scientific and technical progress, the present-day concepts of the origin of planetary systems remain almost at the level of the late XVIII century. These concepts are based on the nebular Kant–Laplace cosmogony, which assumes that planets (and comets) originate in vast gas-dust disk surrounding the central star. Surprisingly, these concepts, put forward at times when the caloric theory was predominant and nearly nothing was known about the structure and evolution of stars, have survived in an almost unchanged form until now, although incessant and persistent attempts to create on their basis (and using modern computational mathematics) scenarios that would be consistent with at least the basic facts cannot be considered as very successful (any specialist is aware of these attempts, some believe in having recipes for overcoming the difficulties involved, but, seemingly, nobody ventures to solve the problem as a whole; see, e.g., Flaig et al [2010]). Not going into details, we briefly list the main problems encountered by disk-type planetary cosmogonies.

(1) The Reynolds number in Keplerian gas disks reaches values Re ≈ 1011. Owing to the existence of a great velocity shear (V_φ ~ R–1/2), this inevitably gives rise to a turbulence leading to dissipation of a disk in ~102 revolutions only. Judging from the whole body of available data (for meteorites, lunar regolith, etc.) obtained in last decades, the bodies of the SS had been formed on the time scale of ~104 yr [e.g., Cameron & Pine 1973], which was a great surprise for advocates of disk cosmogonies. The reason is that the time t of planet formation in classical disks with $M_{\rm disk} \approx 0.05 M_\odot$ amounts to ~108 yr (and more, according to Safronov [1972]). This time t depends on the surface density of the disk σ as ~σ–2.5 [Drobyshevski 2011]. That is why Cameron & Pine [1973] considered a disk with $M_{\rm disk} = 2M_\odot$ without any central condensation (here σ = 105 g/cm3, and $t \approx$ 104 yr); the potentialities of disks with σ = 5×104 g/cm3 were recently studied in [Desch 2007]. It is noteworthy that just this circumstance causes perplexity when planetary systems are discovered at apparently young stars, say, at β Pic (see below point 5).

(2) The turbulent velocity dispersion in a disk is comparable with the Keplerian velocity itself, so that the forming rock-type and other conglomerates with sizes, say, of ~1÷10 m

collide with each other at $V \approx 3 \div 10$ km/s, which inevitably leads to their disintegration. As Blum and Wurm [2008] noted, 'the formation of planetesimals, the kilometer-sized planetary precursors, is still a puzzling process'.

(3) It is unclear why almost all planets have satellites. From the nebular standpoint, this must be an exotic occasion. To explain the origin of satellites, one is forced to use *ad hoc* hypotheses (of the type of hyperimpact for the Moon whose high-temperature initial composition markedly differs from that of the Earth). At the same time, we have to state that, even for dwarf planets, multiple systems are a rule, rather than an exception. (This suggests to believe that the process of planet formation goes just via stages of permanent successive capture of satellites which is realizable under conditions of an exceedingly low relative velocity and giant (of gas origin?) drag.)

(4) The reason for the slow rotation of the Sun is unclear.

(5) The discovery of exoplanets, many of which have unexpectedly large eccentricities, was a surprise for many of those who believe in the nebular cosmogony. In fact, this cosmogony occasionally becomes a hindrance to understanding and finding of relationships between the observational facts. (A good example in this sense is an interesting paper by Lagrange et al (2010) about β Pic b planet, - see Sec. 3. Until recently, the eccentricities of Uranus and Neptune were preferred not to be mentioned (see below, Sec. 4)).

(6) The origin and properties of small bodies in the SS (comets, asteroids, Trojans, rings, and irregular planetary satellites) remain unclear. (The Nice model of the SS evolution, while claiming explanation of Trojans, irregular satellites and similar bodies' origin [Morbidelli et al 2005; Nesvorný et al 2007], is circumventing a discussion of causes of absence of metallic- or stone-type bodies among them or, say, the well-documented excess of Greeks over Trojans, etc.)

(7) Disk cosmogonies cannot boast of any confirmed prediction.[1] All new discoveries are adapted to their concepts *a posteriori*.

The disk planetary cosmogony has been put forward (and still nearly remains so) solely for explaining the origin of our SS. It could not take into account, from the very beginning, the existence of multiple star systems, which include the majority of stars.

Below, we make an attempt to demonstrate that all the problems inherent in this theory are automatically eliminated if we proceed from, and are based on (gas-dynamic) processes leading to appearance of binary stars and accompanying their evolution. Planetary systems are either the limiting case or an inevitable by-product of the formation and evolution of multiple stellar systems. This approach, we formulated based on an analysis of the statistics of binary stars and have been developing since 1974, proved to be rather self-consistent and fruitful. It enabled us to explain from a common standpoint (*i*) the origin and the known properties and statistics of binary stars and (*ii*) specific features of the SS and other planetary systems with (*iii*) making a number of predictions that were confirmed by new discoveries and observations and recent theoretical works (e.g., by the Nice model). Moreover, the confidence in its validity forced us to be insistent in a search for concrete physical mechanisms leading to origination and observed manifestations of various groups of minor bodies, and this search has also been crowned with success [Drobyshevski 2011].

[1] Recent comparison of oxygen and nitrogen isotopic compositions of the Solar wind and planets based on data of the Genesis mission [McKeegan et al 2011; Marty et al 2011] gave results opposite to that predicted by the disk cosmogonies [Crida 2009].

2. Close Binaries: Their Origin and Evolution

The first physically binary stars were discovered in the XVIII century and originally seemed to be exotic objects. It is clear now that rather single stars belong to untypical objects [e.g., Abt 1983].

2.1. ABC of the Evolution of Stars and, Especially, Binary Stars

It is presently believed that as the energy source of stars serve nuclear fusion reactions, first involving hydrogen in the Main Sequence (MS) stars. After hydrogen burns out, a helium core appears at the center of a star, with burning of hydrogen continued within a layer at its surface. In the process, the star swells to become a red giant with a thick convective envelope and with a radius of up to several AU. On average, the specific entropy of the envelope matter permanently agitated by convection is nearly constant and independent of radius (depth). Therefore, the envelope structure can be approximated by an equation of state $p = K\rho\gamma = K\rho^{((n+1)/n)}$ (where p is pressure, ρ is density and K is a constant) with adiabatic exponent $\gamma = 5/3$ or polytrope index $n = 3/2$. For a self-gravitating gas sphere with $n = 3/2$ at a fixed specific entropy, the relation $RM1/3$ = const is valid, so that a decrease in the mass M of the sphere leads to an increase in its size R [Chandrasekhar 1939].

Therefore, if a more massive component in a close binary (CB) star system becomes a red giant and its envelope overfills the critical Roche lobe whose surface includes the first Lagrange point L_1, then loss of a part of its matter through the region L_1 into the other component's sphere of action leads to further swelling of the primary component's envelope. In addition, the components' mass equalization draws them closer (their separation gets minimum shortly after their masses become equal; after the mass reversal, the separation is increasing). As a result, the process of loss of the red giant mass is self-accelerated. Almost the whole matter of the convective envelope of the red giant is lost in several years [Paczyński & Sienkiewicz 1972] and even days [Drobyshevski & Reznikov 1974]. Mostly flowing to the secondary component, this matter creates around it a short-living disk and a common disk-like gaseous envelope around the system as a whole (part of matter is lost, together with the angular momentum, from the system at all). Here, the characteristic time scale t is set by the dynamic time scale $\tau_{dyn} = (R_*3/GM_*)1/2$, in fact, $\sim 102\tau_{dyn}$ (in the subsequent stages, the overflow may also proceed on the Kelvin and/or nuclear time scales). Only a helium white dwarf remains from the primary component, red giant. This dwarf retains ~1–2% of the initial convective mass of the red giant. This mass reversal solves the problem of the 'Algol paradox', when in a CB system, the less massive component is, contrary to simple expectations, a more evolved star.

2.2. On the Origin of Stars and Their Multiple Systems

The evolution of stars in the stage of residence in the MS and in the stage of a red giant, when first central and then layered burning of hydrogen occurs in the stars, is rather well understood and developed. However, this cannot be said about initial stages of their origin.

It is believed that stars appear as a result of collapse of a prestellar gas cloud whose size exceeds the Jeans radius, i.e., in the case when self-gravitation overcomes the gas pressure in the cloud. The fall of the gas toward the center of such a cloud and details of appearance of a protostar (PS) widely differ, depending on initial conditions (e.g., Girihidis et al [2011]).

At an initial density of 10–11÷10–13 g/cm3, the collapse is very fast [Hayashi 1966; Narita et al 1970]. The fall energy has not enough time to be emitted, the gas in the PS is strongly heated, and convection arises in the collapsed PS, which propagates from the outside inwards in 10–1 to 102 year (for $M_* \approx 1M_\odot$) and spreads over a significant part of the star. The radius of a fully convective PS was estimated by Hayashi [1966] to be $R_{n=3/2} \approx 50R_\odot M_*/M_\odot$. The convection does not necessarily involve the whole PS. In more massive clouds, the collapse is longer, part of energy has enough time to be emitted, and the convection (at $M_* \approx 3M_\odot$) covers, roughly speaking, only a half of the PS mass.

A similar situation occurs if the initial density is low ($\rho \approx$ 10–16–10–20 g/cm3). Here, the fall time is also long, the energy has enough time to be emitted, and the PS size is small (~$2R_\odot$ at $M_* \approx 1M_\odot$), and convection does not appear in the PS at all, even at $M_* < 1.5\ M_\odot$ [Larson 1972].

During the last four decades, details of how stars appear have become significantly clearer in part [see reviews by Bergin & Tafalla 2007; Larson 2003; McKee & Ostriker 2007; Zinnecker & Yorke 2007], new aspects have been revealed (e.g., the important role of turbulence and, possibly, magnetic fields in prestellar clouds has been discovered [Bergin & Tafalla 2007; Zinnecker & Yorke 2007]), but no ultimate answers to numerous questions have been obtained. In particular, it can be believed that the truth, as usual, lies somewhere between the two extreme cases mentioned above. In this sense, consideration of processes in which binary stars, and especially close binaries, appear can shed light on the nature of real processes of star formation [see Mathieu 1994; Zinnecker & Yorke 2007].

As for processes in which multiple systems appear, the following three broad categories are presently distinguished [e.g., by Tohline 2002].

(1) Capture: independent collapses of adjacent regions (turbulent elements or filaments) of a cloud.
(2) "Prompt fragmentation": already in the course of a started inhomogeneous collapse, new collapse centers appear.
(3) "Delayed breakup": "The central-most regions of a rotating gas cloud may collapse to form an equilibrium configuration that is initially stable against fragmentation." Then, subsequently contracting to the MS, "the core (or its surrounding accretion disk) may encounter an instability that leads to the formation of a binary system."

The above processes, with possible exception of the last of these, lead to wide systems. To explain the appearance of CBs, Zinnecker & Yorke [2007] added to this list the possibility of capture of stars by each other (mergers) in interaction with near-star disks and capture by a single star, or, inversely, ejection of separate members in dense star aggregations, with third bodies involved.

Numerous attempts to numerically simulate the 3D gas-dynamics of appearance of binary systems encounter numerous difficulties that cast doubt on the validity of the results obtained (for details, see Tohline [2002]). These are the following: strong influence naturally exerted in nonlinear problems by fine nuances of initial conditions [Girihidis et al 2011]; calculation accuracy; huge differences of pressures and properties (transparency, molecular and phase composition, ultrasonic turbulence, etc.) of the gas, when it is *a priori* unclear which are important for the result and which are not; existence of numerous time scales; problem of adequacy of the algorithms employed in calculations; and so forth.[2]

Seemingly, the present-day result of numerous efforts is (see, e.g., Zinnecker & Yorke [2007]) the doubt in (1) successive multistage fragmentation in collapse of a rotating prestellar cloud and (2) appearance of planar rotating structures (disks), which, generally speaking, can also undergo fragmentation. In contrast to previous concepts addressing the general-galactic rotation, it is believed now that turbulent motions in the prestellar cloud are the source of the excessive (per unit mass) angular momentum. As a rule, the collapsing turbulent elements (eddies) bear an angular momentum markedly exceeding the possible angular momentum of a single star. Numerous studies (see, e.g., Machida et al 2008) have been devoted to the effect of a magnetic field on the gas dynamics and redistribution of the angular momentum at the multiple systems' origin. However, as far as it can be judged from publications, in this case scientists consider an idealized situation with a large-scale magnetic field fully or partly frozen in the gas. It is unlikely that anybody has considered the influence exerted on the fragmentation of molecular clouds by MHD turbulence and instabilities, the processes that give no way, during already half a century, of implementing controlled fusion on the Earth and limit the lifetime of sunspots to days, whereas these spots could seemingly have lived for thousands of years if their decay is governed by the Ohmic dissipation mainly.

2.3. Some Statistical Features of CBs

A consideration of the part played by duplicity in the appearance of the phenomenon of magnetic Ap and Am stars [Drobyshevski 1973] and of the related problem of non-conservative evolution of a CB system with an outflow of (magnetized) gas from the red giant to the secondary component, with partial loss of matter together with the angular momentum [Drobyshevski & Reznikov 1974], favored a deeper understanding of the role of the latter in the evolution of stars. We started [Drobyshevski 1973] from an analysis of the fundamental statistical aspects inherent in real detached CBs with F, A, and B primary components. Figure 2.1 presents an outline of distribution of systems with spectroscopically defined mass ratio q, depending on their period and the primary component spectral type (B, A, or F). SB systems with two spectra (SB2) and high quality (5, 4, or 3) orbits are taken from 'SB9: The ninth

[2] Generally speaking, a situation of this kind is typical of numerical simulations of more or less complex processes. The point is that the overwhelming majority of publications devoted to numerical simulation of various situations are mostly aimed to overcome purely technical or mathematical calculation difficulties, so that due attention is not always given to the initial physically correct formulation of the problem. As a result, situations that are in principle impossible physically are frequently simulated. I recollect my talk with A.G.W. Cameron at our banquet table at 23rd LPS Conference in Houston in 1992. We discussed problems related to the origin of the SS. AGWC said that (in the framework of the disk cosmogony) he would make an attempt to go over different basic variants and simulate processes leading to planets' formation. I described my concept of the Jupiter–Sun system as a limiting case of a close binary (see below Sec. 4) and then asked which type of computer he believes to be the most powerful. AGWC proudly said that this is a multiprocessor Cray and he computes his problems just on this device. In response, I noted that, still, the most powerful computer is human brain, which not only can assess the validity of particular results without making complicated calculations, but, moreover, is capable of putting forward some basic ideas and choosing the most promising among these. I emphasized that a valid basic idea is necessary before starting calculations. AGWC found nothing to answer.

catalogue of spectroscopic binary orbits' by Pourbaix et al [2004]. There were no indications of any deviation of the components from the mass-luminosity curve for the MS (spectral types IV and V), while Roche lobe radius exceeds the star radius by more than 20%. Ap stars were omitted from the study. Orbital periods for systems with eccentric orbits were recalculated to correspond to the periastron distance, as $P(1-e)3/2$.

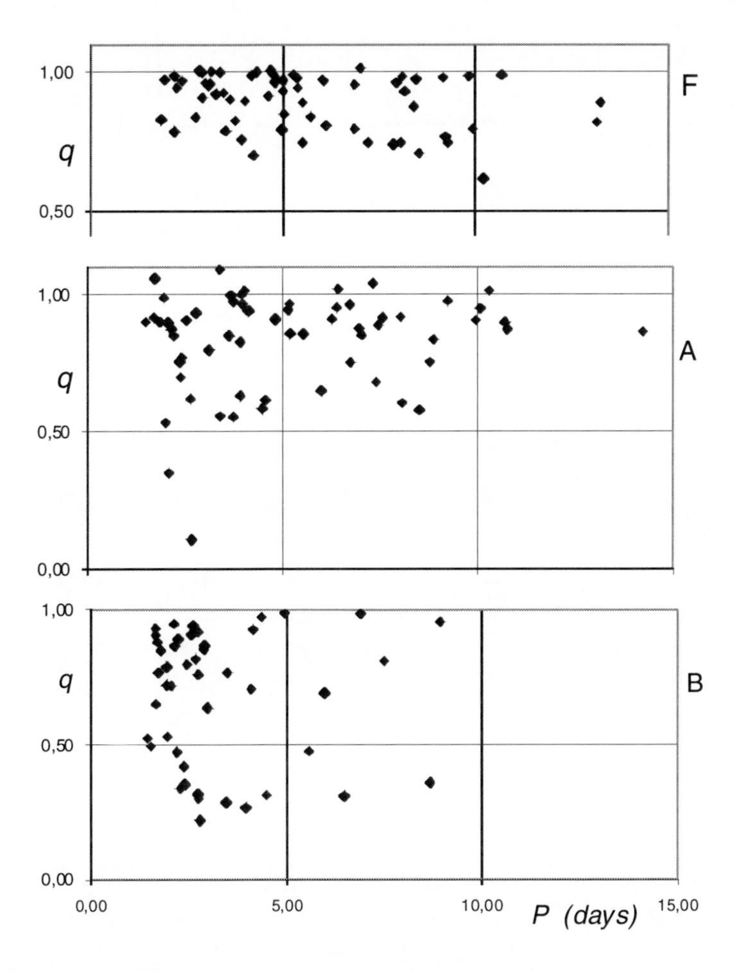

Figure 2.1. Period P – mass ratio q diagram for SB2 with primaries of different spectral type (B, A, and F). Data of good quality (5, 4, and 3) by Pourbaix et al [2004].

If one assumes that there is no correlation between q and P, then effects of observational selection in the detection of SB2s are such that one should expect a monotonous decrease of the number of known binaries with increasing P and decreasing q, no matter what is the spectral class (i.e., mass) of the primary component. However, a glance onto the Figure 2.1 reveals two features: (*i*) a pronounced decrease in number of low-q systems when one shifts from B to F systems, and (*ii*) a rather sharp drop of the systems' number with the period growth (at $P \geq$ 7-9d for B systems and $P \geq$10-11d for A and F systems). If we set, on the average, the mass of an F star as 1.5$M_{\odot}$, an A star - as 2$M_{\odot}$, and that of a B star, as 5$M_{\odot}$, then we obtain about 25-30$R_\odot$ for maximum value of the corresponding separation between components of the non-evolved (young) systems.

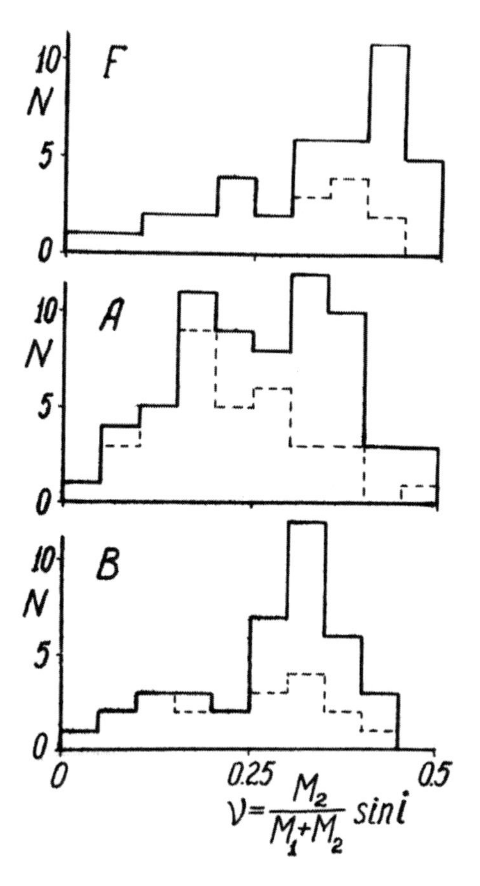

Figure 2.2. $N(v)$ distributions for SBs (of a good quality a, b, and c) with $2^d \le P \le 200^d$ and with bright ($m_V \le 6.5^m$) primaries of F, A, and B spectral classes [by Drobyshevski 1973, 1976b] (with making use of catalogues of Batten [1967] and Svechnikov [1969]). The dashed line confines the SB1 systems. A common feature of all distributions is an existence of two maxima. The maxima at $0.30 < v < 0.45$ belong to the initial distribution, i.e., to younger systems. They are somewhat shifted toward smaller v ($v = q \times \sin i/(1+q)$) when one goes from F to B systems. This fact confirms the similar feature for the SB2 distributions (see Figure 2.1) and should be taken into account (and explained) by any formation scenario of CBs. The weaker maxima at $v < 0.25$ are due to the already evolved systems (for more detail, see Drobyshevski [1973, 1976b]).

Before proceeding further, we note that, if matter is accreted from the residual disk surrounding a newly born CB, this leads, as shown in [Drobyshevski 1974a], to equalization of the component masses, i.e., $q \to 1$. Indeed, at $q \approx 0.004$–0.2, the less massive component moving at a larger distance from the center of masses of the system, intercepts the matter coming to the system from the disk (this effect was also noted by Kraicheva et al [1978], Artymovicz [1983], and Bate & Bonnel [1997]).

2.4. Scenario of the Last Fragmentation

Analysis of Figs. 2.1 and 2.2 shows that the original value of q for binary systems with A and B components falls within the range 0.3–0.8, whereas the characteristic distance between the components is about $(10\div25)R_{\odot}$. The last values suggest that the initial size of a disintegrated PS lies closer to old estimates by Hayashi [1966], rather than to those by Larson [1972]. This means that it can be believed that the final collapse begins at a density of ≥10–13 g/cm3 and the primary outer convection covers a significant part of the mass of the collapsed PS.

It would be expected that the collapse of a prestellar cloud can be divided into the following two stages (see, e.g., Bergin & Tafalla [2007], McKee & Ostriker [2007], and refs. therein).

(1) At $\rho < 10^{-13}$ g/cm^3, the matter is optically thin for the IR radiation and the process is isothermal (it is commonly taken that $T \approx 10$ K).
(2) At $\rho \approx 10^{-13}\div10^{-12}$ g/cm^3, the radiation is locked-in and the gas is compressed quasi-adiabatically. Interestingly, the boundary between close and wide multiple systems ($P \approx 10^2$ yr) rather well corresponds to the average density of turbulent elements $\rho \approx 10^{-13}\div10^{-12}$ g/cm^3.

Presumably, the only CB formation scenario that takes into account details of physical and gas-dynamic processes accompanying the appearance of stars themselves is that suggested by the author nearly 40 years ago [Drobyshevski 1974a] (by our opinion, the closest to this scenario approach was considered by Hubber & Whitworth [2005]). To say directly, this scenario is not a result of a numerical solution of a problem with unclearly selected initial conditions or problems that, as noted above, cannot be accurately solved because of the existence of turbulence and the difficultly imaginable (occasionally by tens of orders of magnitude) scatter of the simultaneously existing spatial, temporal, and parametric scales. The scenario was a result of revealing and simple physical analysis of main manifestations of CB stars, and a system search for possible ways of understanding of quite a number of their specific features (including the statistical ones, see above). Therefore, it stands to reason that these observed patterns find simple explanation in terms of our hypothesis.

Briefly, the main stages of appearance of close binaries can be described as follows [Drobyshevski 1974a] (a popular description, together with basic statistical evidence of binaries, can be found in [Drobyshevski 1976b]).

After the collapse, turbulent agitation immediately develops in a protostar, possibly as a result of convection. As a consequence of its rapid contraction, the rotation of a PS is originally differentiated over the volume: the more dense inner regions rotate faster than the outer regions (possibly, even not coaxially), with the rotation rate such that the balance of forces in the volume of a star is close to the rotation stability limit. The turbulent convective agitation tends to make the rotation uniform throughout the volume and thereby transfers the angular momentum outwards. The turbulence may also be of non-convective nature, appearing as a result of the instability of the shear flow itself, especially if the angular velocity falls outwards. Therefore, it can, in principle, be even suggested that the region of the angular momentum carrying out goes after (but with a certain lag) the shock wave that decelerates the gas falling onto a collapsing PS.

In any case, the carry-over of the angular momentum from inside to the outside leads to fast centrifugal detachment of outer layers of a protostar and to formation of a massive rotating ring around it. As is known (see above), the loss of mass by a star leads to expansion of its convective (turbulent) envelope, so that the mass transfer to the ring or to its remnants occurs on a dynamic time scale until complete loss of the matter subjected to convection.

The massive gas ring is unstable against azimuthal perturbations and, in a certain stage, the ring disintegrates into several fragments (see also, e.g., Bonnell & Bate 1994; Hubber & Whitworth [2005]). The fragments interact with each other and with the PS. Part of these fragments fall back onto the protostar, part collide with each other and dissipate or merge, and a part is at all ejected from the system by mutual gravitational perturbations. The final outflow of turbulized matter from the PS occurs to the "luckiest" of the fragments (see Figure 2.3 by Drobyshevski [1996]), if such a fragment appears at all (and does not move along, say, a too eccentric orbit; see below).

As a result of the "rotational-exchange" PS disintegration described above, a close binary system is formed, in which the component mass ratio q is determined by the ratio between the masses of the outer convective zone and the stable core of the starting configuration. The minimal distance between the components is defined by the size of the PS. It is clear that the both values have, in turn, to depend on the initial conditions of the collapse.

It can be easily seen that the effect of the following two factors is necessary for a rotational-exchange disintegration of a PS.

(1) Sufficiently fast collapse of a PS, so that the protostar (*i*) would have a large size and (*ii*) a large angular momentum would be reserved in its differential rotation.
(2) Rapid actuation of the turbulent viscosity for the angular momentum to be so rapidly transferred to outer layers that they would be suspended as a sufficiently massive self-gravitating formation and have not enough time to scatter as jets or an extended disk.

From the standpoint of the suggested scenario, it can be seen in Figs. 2.1 and 2.2 that (*i*) the convection in PSs with $M_* \approx 3M_\odot$, which give birth to close binaries with F components (and with $q \approx 1$), covers about a half of mass of the PS (*ii*) whose radius is up to $(10\div25)R_\odot$ (see above, Sec. 2.2). It is reasonable to believe that, in PSs with $M_* < 3M_\odot$, the external convection covers, as a rule, more than a half of their mass, so that the newly born component receives more than a half of the initial mass and becomes more massive(!) member of a CB. In this case, in the P–q diagram for F and later stars an excess of systems with $q < 0.8$ and $P >$ 8÷10d may appear. However, its discovery is apparently complicated by the observational selection (running ahead, we can say that CBs with giant planets must belong here, too; see below Sec. 4.1).

Thus, Hayashi's collapse scenario with outer convection is operative to a considerable extent, but not fully: according to Hayashi, at $M_* \approx 3M_\odot$, the radius of a PS with developed convection would be as large as $150R_\odot$, i.e., approximately an order of magnitude larger than the value following from the statistics of binaries.

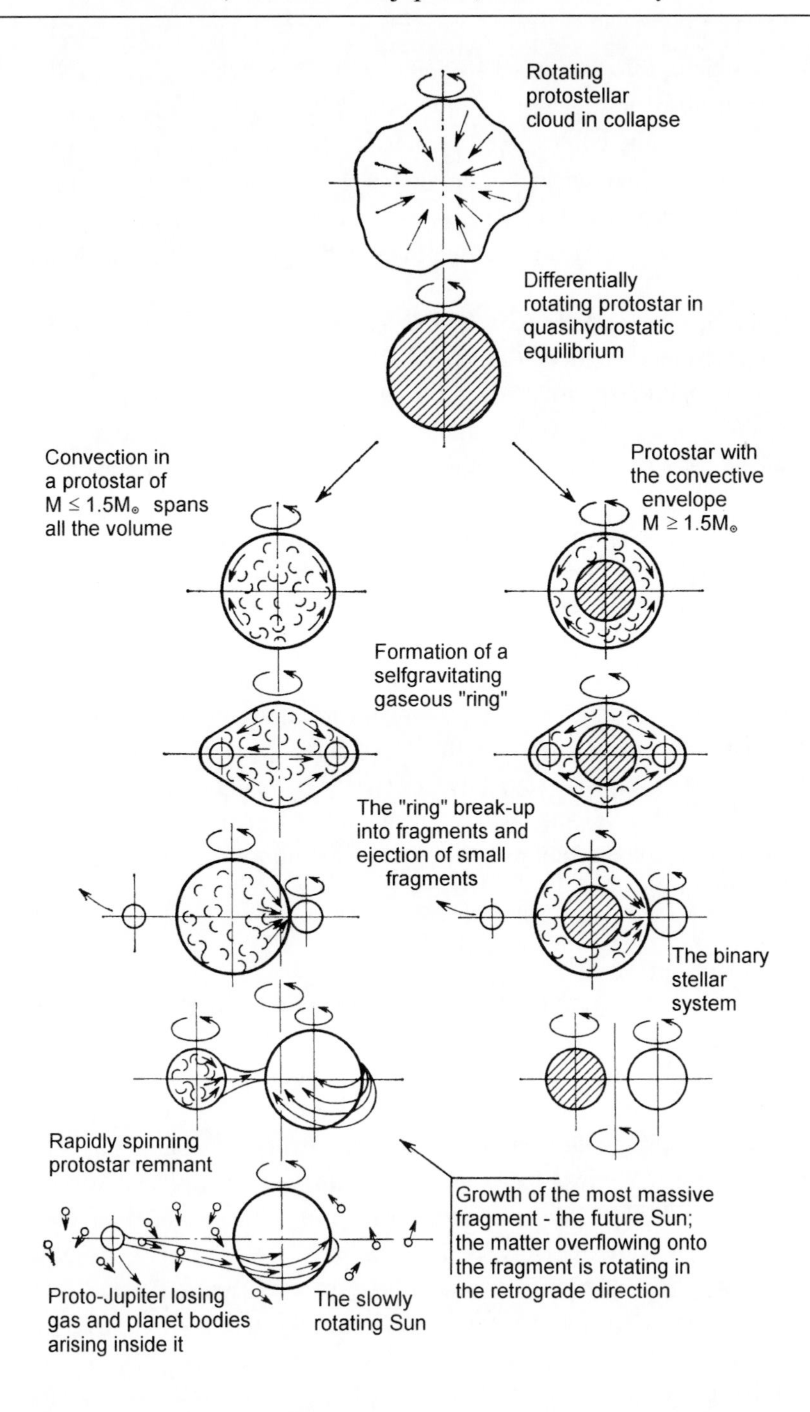

Figure 2.3. Origin of a close binary stellar system (right) owing to the rotational-exchange fission of the differentially rotating dense protostar (PS).

In the case $M_* \leq 1.5M_\odot$ (left), when the Hayashi convection involves all the PS mass, the original PS remnant (PJ) finally plays subordinate role of a giant planet. Note that due to great equatorial PJ velocity which can exceed its orbital velocity, the matter flowing from PJ onto the anew-forming component - the future Sun - rotates in the retrograde direction, thus giving rise to the slow rotation of the Sun.

At the same time, calculations of an isothermal collapse (e.g., Larson 1972, 2003) yielded no indications at all of any external convection at $M_* = 3M_\odot$. Thus, to all appearance, the truth does lie between the two extremities indeed. Here, it would be expected as this follows from Figs. 2.1 and 2.2 that, depending on the mass of a collapsing formation, the convection will cover different fractions of the PS mass: e.g., at $M_* \approx 5M_\odot$, this will only be about a third of its mass, which will give rise to a close binary with a primary component of the A type, but with $q \sim 0.5$.

Highly provocative observations of Cohen & Kuhi [1979] of some 500 young stars demonstrated that most of them "lie on convective tracks in the H-R diagram, and their distribution does not correspond to any published dynamical evolutionary racks".

On the other hand, it is worth noting once more that up to now nobody considered a possibility of a mass exchange (and the mass reversal even) between the just split apart components.

Unfortunately, calculations of the collapse and further evolution of differentially rotating PSs are far beyond the present-day computation possibilities (the modern situation is illustrated, e.g., by paper of Hubber & Whitworth [2005]). This refers, to an even greater extent, to the carry-over of the angular momentum by turbulent convective motions. Here, parameters of the turbulent viscosity should be set manually, as it is done, e.g., when analyzing the evolution of the so-called protoplanetary disks. In the last case, however, for the lifetime of a disk to be sufficiently long for planets to appear in the disk, efforts are made to diminish the turbulent viscosity (say, to $\nu_t \approx (0.01\text{–}0.1)\lambda_t v_t$, where ν, λ, and v are the viscosity, characteristic space and velocity scales of the turbulence), whereas calculations of the rotational-exchange disintegration of a PS requires that more realistic values of the parameter are to be taken (up to $\nu_t \approx 1\lambda_t v_t$).

3. Eccentric Exojupiters

The discovery of extrasolar planets was an important break-through in observational astronomy. In fact, the era commenced in which the old idea of multiplicity of (inhabited) worlds became practical, the idea for which Giordano Bruno was burned in 1600.

However, these discoveries (461 planetary systems with 561 planets were known by May 17, 2011 [Schneider 2011]; statistics and properties of the discovered exoplanets are given, e.g., by Baraffe et al [2010], Udry & Santos [2007], Marcy et al [2005]) immediately produced a multitude of puzzling questions because they contradict the old Kant-Laplace concepts of how planets are formed in extended (and presumably long-lived) protoplanetary disks surrounding the central star [Safronov 1972].

To everybody's surprise, it was found that a considerable part of the planets are exojupiters: they have a mass of $\sim(1\div10)M_J$ and move along strongly elongated orbits (paradoxically, the less massive Neptune-like exoplanets have, as a rule, the modest eccentricity orbits [Udry & Santos 2007]; that is understandable as these objects belong to the next generation, see below Sec.4). A study characteristic in this regard is an impressive paper by Lagrange et al [2011], in which results of visual observations of a massive planet β Pic b revolving with a period of ~17 yr around a young ($\sim 12{+}8_{-4}$ Myr) β Pic star with a mass of $\sim 1.75M_\odot$ (A6 V) are presented. The system has a wide (several hundreds of AUs) tenuous circumstellar disk containing dust, gas, and, possibly, km-sized solid bodies and planets. The three pages of the paper contain several notes of the type: 'processes that are not completely understood', 'available models do not offer detailed description', 'it is not clear whether the basic assumption' applies to β Pic b or other planet, the 'hot start' or 'cold start' models are still the matter of debate', 'this mechanism does not straightforwardly explain', etc. Actually, if one believes the disk cosmogony, it is not understandable how this planet appeared in such a young system.

From our standpoint, planets are not, indeed, formed in circumstellar disks because, as it was said in the Introduction, these are short-lived and have low (surface) density, and the wide dispersion of Keplerian velocities disintegrates colliding planetesimals and thereby precludes their aggregation and growth.

Above, in Sec.2, we mentioned that a massive gas ring is separated from a rapidly rotating collapsed PS as a result of the turbulent (convective) carrying of the angular momentum outwards, and this ring is immediately fragmented azimuthally. This process is shown schematically in Figure 2.3 [Drobyshevski 1996]. It was also mentioned that the interaction of the fragments with each other differently predetermines the fate of each fragment. Some are at all ejected from the system by mutual perturbations (Figure 2.3). If the matter from the convective envelope of a PS does not outflow to the "luckiest" of these, then the preserved fragments, which have ejected other fragments (outwards or back to the PS remnant) and themselves go to disordered noncircular orbits, will, naturally, continue to move around the PS along eccentric orbits of this kind.

The mass of the ring that was originally separated from the PS hardly reached ~0.1 of its mass. In all probability, this was $\sim$10–2$M_\odot$. Therefore, as regards their mass, the ring fragments are just those eccentric exojupiters that could not arise in an extended disk, but are products of vigorous evolution and rapid fragmentation of a rather massive ring that separated from the PS and then disintegrated into gaseous jupiters, without being transformed into a disk. These 'exojupiters', in contrast to our Jupiter, have a low density, as they cannot contain a rocky core exceeding ~0.5% of their mass (see below Sec. 4).

This scenario is an offshoot of the scenario in which close binaries are formed. It certainly needs a detailed elaboration, but it can be immediately seen that it simply eliminates numerous questions that cause bewilderment as regards unexpected properties of the numerous class of exojupiters of the β Pic b type.

It is possible that, in the case of a strongly eccentric fragment of the disintegrated ring, a potential nucleus of a newly formed secondary component, the effective flow of the convective mass from the initial rotating PS is simply suppressed. Indeed, even though L_1 point enters the convective zone when a fragment moving along an eccentric orbit approaches this point, so that the initial PS seemingly overfills its critical Roche lobe, its tidal deformation during this short time is such that the tidal hump containing the matter lags behind the fragment. As a result, there is only a small amount of matter or no matter at all at the L_1 point region (the orbital velocity of the fragment at the perihelion of its elongated orbit in any case exceeds the equatorial rotation velocity of the PS, even if it touches the PS surface, but the orbital revolution period may substantially exceed the period of motion along the minimum circular orbit).

As a result, in many passes of this kind, the PS may have enough time to radiate the excess energy and shrink below the Roche lobe surface, so that there is no well developed outflow of the PS convective matter onto the strongly eccentric fragments. In the end, the potentially binary system remains an underdeveloped planetary system with an eccentric exojupiter, instead of a massive stellar component that was never formed. Nevertheless, such a scenario elucidates the reasons why the closer exojupiters turn out to be more massive, as it was noted by Pont et al [2011] with a certain perplexity, or why the exojupiters, while having the greater mass, have lower density than our Jupiter nevertheless [Baraffe et al 2010].

Unfortunately the effect of eccentricity on the process of mass exchange in CBs has not been properly considered so far. Only Sepinsky et al [2009], who have studied the situation, noted that, depending on particular conditions, the eccentricity may not only decrease in the mass-exchange process, but also increase. Owing to specific features of the gas flow, a transfer of the angular momentum from the axial rotation of components to that of their orbital revolution and back can proceed (here, it will be recalled that the slow rotation of the Sun can be explained from this standpoint [Drobyshevski 1974b]; see Figure 2.3 and below, Sec. 4). Thus, a detailed development of the concept of appearance of exojupiters as a certain subsidiary branch of the genesis of CBs seems to be rather promising.

4. Origin of Developed Planetary Systems

4.1. Jupiter as a Core of the Proto-Sun

In Sec. 2, it followed from the statistics of CBs that, on average, the initial, Hayashi's convection involves half the mass of PSs with M $\approx$ 3M$_{\odot}$, with the rotational-exchange instability leading to disintegration of a PS into two approximately identical components (each with $M_{*} \approx 1.5$M$_{\odot}$) of the spectral class F. As a result, F-systems with $q < 0.8$ hardly exist. In the same Sec. 2, we suggested that, in a rotational-exchange disintegration of less massive PSs in which the primary convection covers more than a half of their mass, the secondary component newly born from a fragment of a disintegrated ring will be more massive than the remnant of the starting PS, because the former uptakes nearly the whole amount of the convective matter of the latter (if you wish, there occurs 'mass reversal', so typical of close binaries at their late stages of evolution). It was also noted that in the $P(q)$ diagram, CBs with later than the F-type primaries and with longer than ~10d period but small q must appear. Discovery of systems of this kind is complicated by the observational selection.

Of particular interest is the rotational-exchange disintegration of PSs with $M_{*} \leq 1.5$M$_{\odot}$ in which the initial convection covers the entire PS by penetrating as far as its center (it is noteworthy that objects of this kind are products of a fast non-isothermal collapse of very dense ($\rho \geq$ 10–13 g/cm3) gas clouds, or, more precisely, of their separate rapidly rotating turbulent elements; it seems that, after Hayashi [1966], nobody has performed detailed calculations of so dense objects). In this case, the convective matter of the PS starts to flow to the luckiest fragment of the massive (~10–2M$_{\odot}$) ring that was separated from the PS as a result of a fast turbulent carrying out of angular momentum from its interior outwards and was immediately broken down into several (2–5 and even more) fragments. That fragment can be named lucky which survived (was not disintegrated or ejected from the system) in interactions with other fragments of the disintegrated ring and occupied a nearly circular orbit that is sufficiently close to the PS for its first Lagrange point L_1 to be within the (convective) remnant of the PS. Then, as known from the evolution theory of binary stars (see Sec. 2.1), the whole amount of the convective matter must overflow in $\sim 102\tau_{dyn}$ onto this fragment. (If a fragment is "unlucky" because of, say, moving along a strongly eccentric orbit, then, as shown in the preceding section, the mass exchange conditions are complicated, so that the scenario described here may be inoperative.)

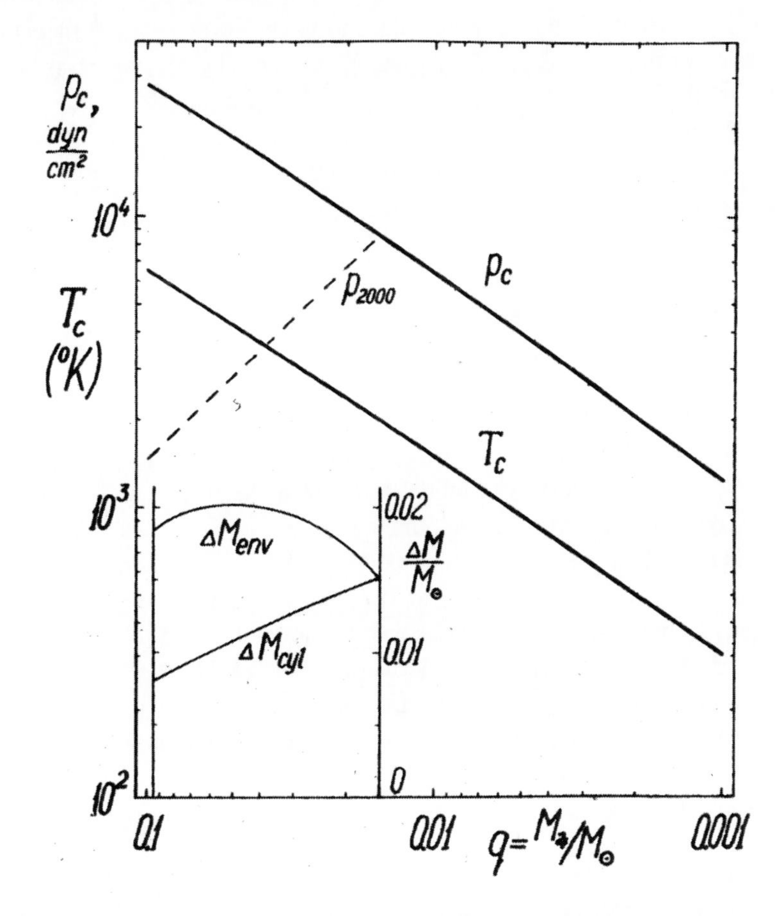

Figure 4.1. Parameters of matter in proto-Jupiter depending on q. T_c, p_c are temperature and pressure in the center; p_{2000} is a pressure at the level where $T = 2000$ K; M_{env} is a mass of the envelope where $T <$ 2000 K; M_{cyl} is a mass outside the 'hot' cylinder adjoining the core with $T > 2000$ K.

Actually, the isoentropy of expansion of the convective matter is disturbed in the envelope in the course of the outflow both because of the partial energy emission by the PS and due to its cooling if only as a consequence of a seemingly purely adiabatic expansion. Cooling of the matter of solar composition leads, beginning at $T_c \approx 3000$ K at $q \approx 0.03$, to precipitation of a condensate in the form of flakes and drops of silicates, iron, etc. (see Figure 4.1 by Drobyshevski [1975b]). The precipitation of this condensate leads to formation of a rocky core in the primary (see Sec. 4.3). At a mass ratio $q \approx 0.001$ between the PS remnant and the anew-born component, the temperature at the center of the remnant falls to $T_c \approx 300$ K, water starts to condense, and the object sinks into its Roche lobe. The outflow of matter from the PS remnant to the anew born component with $M_* \approx 1M_\odot$ terminates.

Here, three points should be noted right away.

(1) The process described results in a system of the Jupiter/Sun type with $q \approx 0.001$.
(2) The slow rotation of the newly born Sun can be understood if the equatorial velocity of the rapidly rotating proto-Jupiter (PJ) exceeded its orbital velocity, so that matter transferred to the Sun could occasionally form around it a counterrotating disk transferring a negative angular momentum to the proto-Sun [Drobyshevski 1974b,c]. A deviation of the rotation axis of the Sun from normal to the Jovian orbit plane may be due to the initial non-axial rotation of different layers in the spatially non-uniformly rotating PS.
(3) As a result of matter condensation at $q < 0.02$, a great rocky core with a mass of ~25M⊕ [Zharkov et al 1974] appears in Jupiter finally.

4.2. On Conditions of the Planet Origin

In the course of matter exchange and mass reversal of components in the proto-Jupiter/proto-Sun system, when the ratio of their masses lowered from $q \approx 102$ to $q \approx 10$–3, gas disks appeared both in the space between components around the accreting component and outside the system. Therefore, it would be believed that the rest of the planets were formed in these disks, together with their satellites.

However, this "classical" nebular approach encounters a multitude of equally "classical" difficulties, which are frequently avoided to be mentioned (it is, e.g., believed that the cosmogonic Nice model consistently describes the evolution of our planetary system [Tsiganis et al 2005; Morbidelli et al 2005; Hernández-Mena & Benet 2011], and we have to agree with this conclusion if only we disregard the problems related to the origin of the planets themselves with their satellites and of the cloud of planetoids with a necessary total mass of ~(10÷30)M⊕).

Some of these difficulties were listed in Introduction and we are not going to discuss them again.

Therefore, there is no point in discussing the hypothesis of planet origination in the gas that flowed from PJ to the proto-Sun and partly left the system at all. Let us, instead, consider another, more natural opportunity.

4.3. PJ as the Place of Formation of Planets with Satellites

The rapidly rotating PJ, overfilling its Roche lobe and losing its mass flowing to the proto-Sun with a characteristic time of some 104 yr, might well be the place of planet origination.

Indeed, the spacious Jupiter (see Table 4.1 by Drobyshevski [1978]) is a certain analog of a very dense (the surface density $\sigma \approx 105$ g/cm3) protoplanetary disk, with the only difference that the gravitation of gas in PJ is mostly balanced by the pressure gradient, rather than by the Keplerian rotation. Outer regions of PJ always (initially even at q ~ 10÷1) contain a zone with $T <$ 2000÷3000 K, in which precipitation of a condensate from the gas occurs and planetoid formation is possible (Figure 4.1) [Drobyshevski 1975b]. At $0.015 < q < 0.1$, its mass is

(0.015÷0.020)$M_⊙$. At $q \leq 0.015$ the condensation conditions are already satisfied throughout the PJ volume. At the beginning, when iron still remains in the gas phase, these are mostly Ca, Al, and Ti compounds of the type of perovskite $CaTiO_3$ and melitite, a solution of helenite $Ca_2Al_2SiO_7$ and akermanite $Ca_2MgSi_2O_7$. The crystal structure of these compounds is such that it provides space for capture of large ions of lithophilic elements of the type of Sr, Ba, U, Th and rare earths (REEs), with the resulting extraction of these elements from the gas phase [Drobyshevski 1975b]. Moreover, the presence of Ca++ ions creates certain advantages for capture of Sr++, Ba++, and Eu++ ions, and this circumstance can be used to make an attempt to explain the minima of concentrations of the last elements in "standard" abundances (which also reflect the abundances in the convective envelope of the Sun, whose mass is ≤0.01$M_⊙$). Hence follows that the photosphere abundances of the Sun may be not correspondent to its average composition. In this stage, a "rocky" core with a mass of ~25$M_⊕$ is formed in PJ [Drobyshevski 1975a,b], in quite a good agreement with the modern concept of the structure of Jupiter [Zharkov et al 1974]. So the refractory minerals were possibly precipitated into the core, together with the lithophilic elements they captured, whereas outer layers of the Sun were formed from a gas deficient in these elements.

It is pertinent to note here that the more late than the Sun stars have deeper outer convective zones (they can even extend down to levels where the nuclear reactions begin). That is why their observed surface metal abundances could better, than the solar photosphere abundances, reflect their total content in the star. Moving along this way, one could search for causes of correlation of the stars' metallicity with their having planets pointed out by Barbiery et al [2002] (on the other hand this may concern the observed anticorrelation of Li abundance with a presence of planets).

As PJ (or the remnant of the protostar) rotates, the condensed drops precipitate, generally speaking, to the equatorial plane, rather than towards the center directly. Some of them which condense outside the 'hot' cylinder will rotate within the hotter part of the cold envelope, and so will not reach the evaporation zone. (From this one can envisage the secondary heating of the condensate, resulting in chondrule formation.) Repeated sublimation and condensation, involving accumulation of the condensation products within a small doughnut-shaped volume near the equatorial plane in the cold envelope make higher their concentration in this zone, that favors formation of large bodies from the fractionated condensate. The relative motion velocities of planetesimals formed in the gas do not exceed hundreds of m/s (for reference, it should be noted that the outfall velocity of a body from the periphery to the PJ center amounts to only several km/s, see Table 4.1 (Drobyshevski 1978); at the same time, if the gas drag is disregarded, a body dropping out of the periphery and moving in the equatorial plane never finds its way to central regions of PJ because of the rotation of components of a close binary, if only about the center of masses). The larger the body, the weaker its Keplerian motion is impeded by the gas drag.

Bodies originating on the periphery of the rotating PJ and falling inside it move along irregular open rosette-type trajectories, which have large eccentricities, but, as a rule, bypass the central regions of PJ. As a result, these bodies are many times heated and cooled. At low relative velocities in the presence of a gas drag and liquid phase, condensation products stick together (form agglomerates) and reach sizes of planetary bodies with satellites (apparently, capture of satellites in gas-drag conditions at low relative velocities is not an exotic event, but, by contrast, is an inevitable stage of coagulation and formation of increasingly large bodies).

Table 4.1. Parameters of PJ at various stages of its evolution (σ = MJ/πRJ2 is average surface density, ρc and Tc are density and temperature at the centre, vc = 1.64×(GMJ/RJ)1/2 is velocity of free fall at the center, τdyn = (GMJ/RJ3)-1/2 ≈ 1.8×107 s). MJ + M☉ = 1.993×1033 g = const; R = 5.2 AU = const

$q = M_J/M_\odot$	R_J/R	σ (105 g/cm2)	ρ_c (10-8 g/cm3)	T_c (K)	v_c (105 cm/s)
0.1	0.201	2.35	6.73	6500	14.4
0.05	0.162	1.91	6.81	4240	11.6
0.02	0.120	1.42	6.83	2350	8.65
0.01	0.096	1.12	6.74	1480	6.86
0.005	0.077	0.88	6.61	930	5.43
0.001	0.046	0.50	6.26	314	3.15

If a body has a sufficiently large mass, so that the decelerating effect of the gas drag is small, then, falling from the periphery into PJ and again going outside along its unwinding trajectory it can reach, because of the *very rapid decrease of the PJ mass*, a higher surface level of the latter. As a result, such a large body can leave the PJ at all, despite the decelerating effect of the gas.

Simple estimates [Drobyshevski 1978] have shown that, if dt/dlnM$_{PJ}$ ≈ 104 yr, a body with $M \approx 1M_☾$ can leave PJ. This defines a time-scale for such a body formation. Then the total duration of the planetogenesis stage, corresponding to lowering of q from ~10 to 10-3, could reach ~105-106 years, - in agreement with isotopic data [Crida 2009]. Bodies with a considerably smaller mass, formed in outer layers of PJ, experience a strong gas drag and, falling into PJ, move because of the gas drag along contracting trajectories and thereby come to the center of PJ and form its rocky core (if $T_c < 3000$ K already).

Along with a hot environment in PJ, the rapid formation (in ~104 yr) of planet-sized bodies provides their initial melting due to accretion (not to slow radioactive heating). That caused the early differentiation evidenced by numerous facts (see, e.g., refs. in Crida [2009]).

Such a scenario makes clear the origin of Galilean satellites. These bodies, formed in the already cold PJ, are half composed of water ice and abiogenic organics with inclusions of both high- and low-melting minerals.[3] They could not leave PJ because it went below the surface of its Roche lobe and the mass of PJ ceased to decrease. Galilean satellites (moon-like bodies, MLBs) determine the lower characteristic scale for the mass of bodies capable of leaving PJ, whose mass is rapidly decreasing (larger mass bodies, of the type of the Earth with the Moon, or rocky cores of Uranus or Neptune have more opportunities to leave PJ; it is not improbable, either, that more massive objects are formed in different zones of the outer layers of PJ, compared with small-mass MLBs; a certain role is possibly played by gas-dynamic suspension and other factors of this kind [Drobyshevski 1978]). The Galilean satellites were presumably formed in the final stage of the PJ evolution, when the rate of its loss of mass started to become slower and $q \to 0.001$. It can be believed that earlier, when the PJ mass loss rate was greater, somewhat less massive bodies (with $M \approx 0.1M_☾$, i.e., of the type of Pluto) could have left PJ. Judging from the mass of the rocky core of Jupiter, which started

[3] For processes responsible for the present-day specific features and distinctions of Galilean satellites, see [Drobyshevski 2011]. Here we could add that at the bottom of their original water ocean, at the surface of their rocky core, a layer of heavy organics (carbonaceous matter) and chondrules could exist.

to be formed from the refractory condensate at $q < 0.015$, the total mass of bodies that left PJ could reach a value of ~102M⊕, so that their number achieved ~104–105.

4.4. Evolution of Ensemble of Planets and Formation of a Trans-Neptunian Planetary Cloud

Thus, the process of fast gas-dynamic evolution of a semi-detached CB PJ/proto-Sun system culminates in the appearance of a large, with, possibly, 104–105 members, ensemble of MLBs with a characteristic mass ~$M_☾$ (including giant planets and Saturn, which, as judged from its composition and low density (ρ = 0.7 g/cm3) is, possibly, a fragment of the original ring, similar to the eccentric exojupiters, see Sec.3). These bodies interact with each other, mostly via gravitation (Coulomb) forces, to form a nearly collisionless, in a sense, gas. Most part of its 'molecules', small-mass MLBs, were rather rapidly (in $t \approx (102 \div 103) P_J$) at all ejected outside the SS by gravitational perturbations from Saturn, Uranus, Neptune and, principally, by the slowly cooling PJ, which went below the surface of its Roche lobe (when modeling this stage of evolution, it is necessary to keep in mind that PJ cannot as yet be approximated by a point mass: MLBs may fly through its volume with a distributed mass, experiencing a gas drag there [Drobyshevski 2008]).

As follows from simple estimates [Drobyshevski 1978], the remaining part of MLBs (103–104, including giant planets beyond Jupiter) could relax in 4.5 Gyr to a distance not exceeding ~102 AU. To this distance, the planetary bodies could come into mutually quasi-resonance low (but not zero as that followed from the disk cosmogonies) eccentric and tilt orbits etc. owing to the 'viscosity' of the planetary 'gas'. It can explain [Drobyshevski 1978] that the "orbital period ratio of Pluto and Neptune being 3/2 rather than a more probable value of (9/4)-(8/3)". Beyond this ~102 AU limit lies a zone in which the quasi-stationary dynamic state of the objects has not been attained so far. "Here, the period ratios of adjacent planets are arbitrary, the orbits being of different tilt and intersecting one another" [Drobyshevski 1978]. We take the density of bodies "in such a planetary cloud to be maximal within 50 AU $< R <$ 300 AU".

This scenario of origination of planetary bodies in PJ solves a large number of problems. Let us mention some of these.

(1) This is a "fast" and highly dynamic scenario, as required by discoveries of the last decades.
(2) Because of the low relative velocities of particles in the primary condensate, of their agglomerates, and, further, of larger bodies, no difficulties caused by their disintegration in collisions are encountered.
(3) Simultaneously, various bodies appear in different adjacent and overlapping zones. These bodies can strongly differ in composition and contain products of varied thermodynamic origin (ices along with melitite ≈ CaAl-rich inclusions, etc.).
(4) The gas drag combined with the high volume concentration of bodies being formed (planetesimals etc.) make origination of multiple planets of both small and large masses a standard process (which also refers to systems constituted by bodies with different compositions: Earth–Moon, Venus–Mercury [Drobyshevski 1995], etc.), with sole bodies being exotic rather.

(5) The rocky cores of Uranus and Neptune (together with their satellites) are not exception, compared with the planets of the Earth group (as, possibly, also Saturn) or MLBs (and/or dwarf planets, Pluto etc.). The problem of the origin of their gas envelopes is also solved automatically (they emerged from PJ with these envelopes).
(6) Galilean satellites not only perfectly fit the given scenario, but even underlie it by determining the mass scale for MLBs formed within PJ and the characteristic time of the mass loss by PJ [Drobyshevski 1978].

Just the physical aspect of origination of planetary bodies has been the weak point of disk cosmogonies. The further evolution of a planetary system [again, with the interaction with a vast (up to 104–105 members) cloud of gravitationally interacting MLBs, including Uranus and Neptune, taken into account] is more or less apparent. It led us to the concept of an outward (relative PJ) drift of these planets and the resulting formation of a trans-Neptunian joint planeto-cometary cloud [Drobyshevski 1976a, 1977, 1978, 2011] just from which the massive Neptune separated out its family of MLBs. The problem of the final accretive and orbital evolution of Uranus and Neptune as a result of their interaction with the planetesimal disk has been considered later by Fernandez & Ip [1984] and is now developed in details with some impressive inferences in the framework of Nice model [Tsiganis et al 2005; Hernández-Mena & Benet 2011]. True, in its context, the origin of a sufficiently massive disk of planetesimals is not quite clear, and its mass ($\sim$20–50$M_\oplus$), like as of its individual members, is here a free parameter to be chosen to obtain a required result. In our case, its total mass scale is set by the more or less known mass of the rocky core of Jupiter, while the individual members are the MLBs mainly. We believe the most significant (and surprise, at the first glance) result of the Nice model is the initial compact configuration of the 4 outer planets [Crida 2009]. Such acompactness in no way follows from disk cosmogonies but it is inevitable if planets were born in the PJ.

5. Possibility of Appearance of Rocky Planets in CBs. Some Chemical Consequences and Am Phenomenon

Consideration of the Jupiter/Sun system as a limiting case of a CB and, further, of the entire SS as the product of evolution of this binary system, enables us to raise a question as to whether MLBs appear in CBs of other types.

And it seems that the answer to this question will be positive.

The point is that, both in early stages of formation of rather massive CBs of spectral classes F, A, or B and in later stages of their evolution, we have an overflow of matter from one component to the other. In the process, the mass-loosing component, or, more precisely, its large gas envelope involved in convection, overfills its critical Roche lobe and, losing its mass and expanding, is adiabatically cooled to temperatures that enable condensation (freezing-out) and separation of different components of its gas.

Here, however, two variants may be possible.

(1) In the initial appearance and separation of two components of a CB system, the rotational-exchange instability results in that the remnant of the PS rotates, by

definition, rapidly. And even though it retains the stellar mass, this enables formation of numerous MLBs in its outer cold layers and their exit therefrom.

(2) In later stages of evolution, the red giant component rotates slowly and, therefore, appearance of MLBs in its convective envelope, and the more so their exit from the envelope becomes more problematic. Here, it would be expected (see above, Sec. 4.3) that, in some stages of evolution of the envelope, its temperature drops to such an extent that a high-temperature condensate of the melitite type, enriched in Sr, Ba, and REE, starts to appear in the gas. The precipitation of this condensate toward the center leads to separation of the content of volatile and refractive elements in the outflowing matter of the envelope. This results in that, first, a gas rich in volatile components (elements) flows, and only then, that enriched with refractory components (here, surely, it should be remembered that (*i*) the matter of discussion is a turbulized gas, but (*ii*) the velocity of turbulent motions falls as the gas expands).

Both these opportunities can shed light on the origin of A-type stars with metallic lines in spectrum, the so-called Am stars.

5.1. Causes of the Am phenomenon

Stars of A-type (or, more precisely, F0–B5 stars) differ from later stars in that they have very thin (~10–10–10–8$M_{\odot}$) convective envelopes, and from earlier stars (with $M_* \geq$ 3-5$M_{\odot}$, in which there is no outer convection at all) in that their outer matter is not shed off by the light pressure.

It follows from the aforesaid that mass exchange in evolved systems leads just to such an enrichment of surface layers of the secondary (if they are stable), which will give rise to the Am phenomenon.

As for the detached young binary systems, here formation of metallicity may also be favored by MLBs appearing in the remnants of a rapidly rotating protostar, just as it occurred in PJ. The only difference is that ices are not condensed in this case and we deal with rocky MLBs mainly (of the type of our high-temperature condensate Moon and/or with MLBs that were formed, say, at slightly lower temperatures). The primitive igneous differentiation that follows their rapid formation, in which, being cooled, molten rocks gradually solidify under gravitation from below upwards, leads to formation of a solid surface anorthosite crust enriched in Sr, Ba, REE (Figure 5.1 by Drobyshevski [1975b]). This phenomenon is well documented with data on highland lunar rocks and terrestrial intrusions, volumes of magma that filled hollows in Earth' crust and solidified there (e.g., Skaergaard intrusion [Wager & Brown 1968] in Greenland).

The outfall of MLBs of this kind or their collisional (surface) fragments onto A-type stars must also lead to the observed enrichment of stable surface layers of these stars with elements that give rise to the Am phenomenon (for more details, see Drobyshevski [1975b]). Thus, it would be expected that binaries with Am components have Moon-like planets. The same may refer to some types of Ap stars [Drobyshevski 1985, 1986]. One cannot exclude that the dust disks observed just at A and later stars and which are thought to be remnants of protoplanetary disks [Wyatt 2008] are really created by (collisions) of such MLBs.

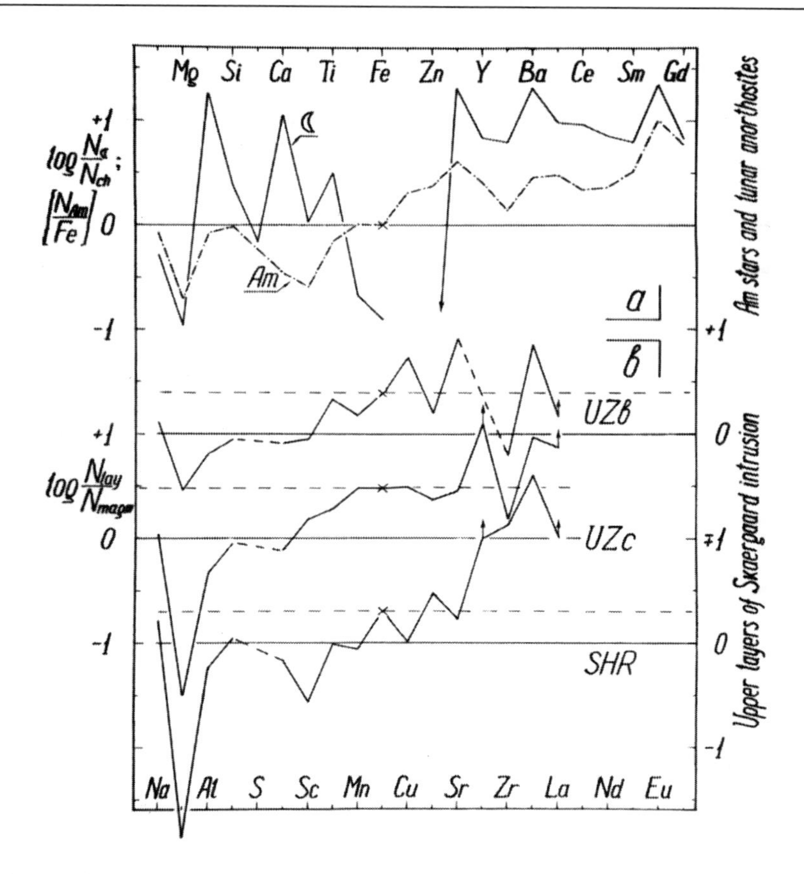

Figure 5.1. Comparison of anomalies in the chemical composition of Am stars with (a) the composition of the Moon's crust (excesses of Ca and Al are caused here by the overall deficiency of Fe in the Moon which is the high-temperature condensation product), and (b) of the upper layers of the Skaergaard intrusion (Greenland). Abundances for the Am stars are normalized to 'normal' stars and iron, for the Moon – to C1 chondrites, for the Skaergaard intrusion – to the composition of the original magma (to obtain figures normalized to iron, reckon from the horizontal dashed lines). The sandwich horizon rocks (SHR) representing the uppermost layer of the intrusion are considered to have resulted from crystallization of the latest liquid.

Conclusion

The modern science commenced when it became clear that its goal is to describe the maximum number of known phenomena with the minimum number of basic postulates (Occam's razor), without using *ad hoc* hypotheses.

We hope that our approach well conforms to this principle.

Indeed, the idea of a rotational-exchange break-up of a rapidly, but non-uniformly rotating PS, which originated from statistical data on close binary systems and is based on modern concepts of the gas dynamics of late stages of their evolution, proved to be rather fruitful in many respects.

This idea enabled understanding of the origin and properties of not only multiple systems with stellar components. It demonstrated that the Jupiter/Sun system is also a close binary system, its limiting case. It also accounts for the origin of eccentric exojupiters, on the one hand, and of SS planets on quasi-resonant, nearly circular orbits, on the other. The numerous difficulties encountered in attempts to construct disk-type planetary cosmogonies simply disappear in the given case. The CB cosmogony of the SS not only explains consistently all the known facts, many of which have baffled researchers (including the origin of the Moon and other numerous satellites), but also makes predictions of new phenomena, which are later confirmed (and this, according to Ch. Darwin, is the principal indication of the validity of any hypothesis, no matter how unbelievable it seems to be at the beginning). In particular, our prediction of existence of a trans-Neptunian ($R \approx 50 \div 300$ AU) cloud of Moon-(Pluto-)like planets (but not comets only(!) as expected by Oort or Kuiper) on irregular orbits has been confirmed in the last two decades by the discovery of objects of the type of Sedna, Eris, Xena, and many others (and thereby the prediction of the modern origin and nature of many long-period comets is also confirmed [Drobyshevski 2008]). Acceptance is being gained by the idea that numerous free (not bound any more to their parent stars) Jupiter- and Moon-like bodies (planets) exist in a ~1:104 ratio. It becomes clearer what are reasons for the correlation between the metallicity of stars and existence of rocky planets near these stars (hence follows the problem of the "standard-abundance" of the photosphere of the Sun, because its convective envelope is composed of a substance depleted of refractories, from which the rocky core of Jupiter is formed [Drobyshevski 1975b]) and, further on, what is the origin of the Am phenomenon (MLBs in multiple systems with Am components await their discovery).

Obviously, details of the new approach require thorough elaboration based on observational (including statistical) data, which must dictate selection of conditions for physically correct and practically realizable situations for numerical simulations. It seems that the priority challenging problems are the possibility of the rotational-exchange break-up of the initially rather dense PSs and ways in which planets are formed in proto-Jupiters (it is not improbable that this occurs also in mass-losing envelopes of red giants in close binary systems).

REFERENCES

Abt, H.A. Annu. Rev. Astron. Astrophys. 1983, 21, 343-372.

Artymowicz, P. *Acta Astron.* 1983, *33*, 223-241.

Baraffe, I.; Chabrier, G.; Barman, T. *Rept. Prog. Phys.* 2010, *73* (016901); arXiv:1001.3577.

Barbieri, M.; Marzari, F.; Scholl, H. *MNRAS* 2002, *396*, 219-224; arXiv:astro-ph/0209118.

Bate, M.R.; Bonnel, I.A. *MNRAS* 1997, 285, 33-43.

Batten, A.H. *Publ. Dominion Astrophys. Obs.Victoria* 1967, *13*, 119-251.

Bergin, E.A.; Tafalla, M. Annu. Rev. Astron. Astrophys. 2007, 45, 339-396.

Blum, J.; Wurm, G. Annu. Rev. Astron. Astrophys. 2008, 46, 21-56.

Bonnell, I.A.; Bate, M.R. *MNRAS* 1994, *271*, 999-1004.

Cameron, A.G.W.; Pine, M.R. Icarus 1973, 18, 377-406.

Chandrasekhar, S. An Introduction to the Study of Stellar Structure; Univ. Chicago Press, 1939.

Cohen, M.; Kuhi, L.V. *Ap. J. Suppl.* 1979, *41*, 743-843.

Crida, A. In *Reviews in Modern Astronimy21*; Rűser, S.; Ed.; WILEY-VCH Verlag GmbH &Co, Weinheim, 2009; pp.215-228.

Desch, S.J. *Ap. J.* 2007, *671*, 878-893.
Drobyshevski, E.M. Genesis and classification of magnetic stars. II. Am stars and duplicity of early stars. Preprint No446, Ioffe Inst., L-d; 1973.
Drobyshevski, E.M. Astron. Astrophys. 1974a, 36, 409-413.
Drobyshevski, E.M. Nature 1974b, 250, 35-36.
Drobyshevski, E.M. Astron. Tsirk. 1974c, No831, 5-7.
Drobyshevski, E.M. Astron. Tsirk. 1975a, No854, 3-5.
Drobyshevski, E.M. Earth Planet. Sci. Lett. 1975b, 25, 368-378.
Drobyshevski, E.M. How many planets may the Solar system contain? Preprint No508, Ioffe Inst., L-d; 1976a.
Drobyshevski, E.M. Zemlya i Vselennaya (The Earth and Universe) 1976b, No3, 70-76 (in Russian).
Drobyshevski, E.M. Astron. Tsirk. 1977, No942, 3-5.
Drobyshevski, E.M. Moon Planets 1978, 18, 145-194.
Drobyshevski, E.M. Planetoidal hypothesis of CP F, A, and B star formation: Possibilities and prospects. Preprint No942, Ioffe Inst., L-d; 1985.
Drobyshevski, E.M. In Upper Main Sequence Stars with Anomalous Abundances; IAU Coll. No90; Cowley, C.R. ; Dworetsky, M.M.; Megessier, C.; Eds.; D. Reidel Publ. Co, 1986; pp.473-476.
Drobyshevski, E.M. Int. J. Impact Engng. 1995, 17, 275-283.
Drobyshevski, E.M. Astron. Astrophys. Trans. 1996, 10, 211-217.
Drobyshevski, E.M. Icarus 2008, 197, 203-210; arXiv:astro-ph/0702601.
Drobyshevski, E.M. In Comets: Characteristics, Composition and Orbits; Melark, P.G.; Ed.; Nova Science Publ., 2011; Chapter 1.
Drobyshevski, E.M.; Reznikov, B.I. Acta Astron. 1974, 24, 29-43.
Fernandez J.A.; Ip W.H. Icarus 1984, 58, 109-120.
Flaig, M.; Kley, W.; Kissmann R. MNRAS 2010, 409, 1297-1306.
Girichidis, P.; Federrath, C.; Banerjee, R.; Klessen, R.S. MNRAS 2011, 413, 2741–2759.
Hayashi, C. Annu. Rev. Astron. Astrophys. 1966, 4, 171-192.
Hernández-Mena, C.; Benet, L. MNRAS 2011, 412, 95-106.
Hubber, D.A.; Whitworth, A.P. Astron. Astrophys. 2005, 437, 113-125; arXiv:astro-ph/0503412.
Kraicheva, Z.T.; Popova,E.I.; Tutukov, A.V.; Yungelson, L.R. Astron. Zh.1978, 55, 1176-1189.
Lagrange, A.-M.; Bonnefoy, M.; Chauvin, G.; Apai, D.; Ehrenreich, D.; Boccaletti, A.; Gratadour, D.; Rouan, D.; Mouillet, D.; Lacour, S.; Kasper, M. *Science* 2010, 329, 57-59; arXiv:1006.3314v1.
Larson, R.B. MNRAS 1972, 157, 121-145.
Larson, R.B. Rept. Prog. Phys. 2003, 66, 1651-1697; arXiv:astro-ph/0306595.
Machida, M.N.; Tomisaka, K.; Matsumoto, T.; Inutsuka, S.-I. Ap. J. 2008, 677, 327-347.
Marcy, G.; Butler, R.P.; Fischer, D.; Vogt, S.; Wright, J.T.; Tinney, C.G.; Jones, H.R.A. Prog. Theor. Phys. Suppl. 2005, No.158, 24-42.
Marty, B.; Chaussidon, M.; Wiens, R. C.; Jurewicz, A. J. G.; Burnett, D.S. *Science* 2011, *332*, 1533-1536.
Mathieu, R.D. Annu. Rev. Astron. Astrophys. 1994, 32, 465-530.
McKee, C.F.; Ostriker, E.C. Annu. Rev. Astron. Astrophys. 2007, 45, 565-687.

McKeegan, K.D.; Kallio, A.P.A.; Heber, V.S.; Jarzebinski, G.; Mao, P.H.; Coath, C.D.; Kunihiro, T.; Wiens, R.C.; Nordholt, J.E.; Moses Jr., R.W.; Reisenfeld, D.B.; Jurewicz, A.J.G. *Science* 2011, *332*, 1528-1532.

Morbidelli, A.; Levison, H.F.; Tsiganis, K.; Gomes, R. *Nature* 2005, *435*, 462–465.

Narita, S.; Nakano, T.; Hayashi, C. Prog. Theor. Phys. 1970, 43, 942-964.

Nesvorný, D.; Vokrouhlický, D.; Morbidelli, A. *Astron. J.* 2007, *133*, 1962–1976.

Paczyński, B.; Sienkiewicz, R. Acta Astron. 1972, 22, 73-91.

Pont, F.; Husnoo, N.; Mazeh, T.; Fabrycky, D. MNRAS 2011, 414, 1278–1284.

Pourbaix D.; Tokovinin A.A.; Batten A.H.; Fekel F.C.; Hartkopf W.I.; Levato H.; Morrell N.I.; Torres G.; Udry S. Astron. Astrophys. 2004, 424, 727-732; arXiv:astro-ph/0406573; SB9: The ninth catalogue of spectroscopic binary orbits, http://sb9.astro.ulb.ac.be/mainform.cgi.

Safronov, V.S. Evolution of Protoplanetary Cloud and Formation of the Earth and Planets; Keter Publ. House, Jerusalem (Israel), 1972.

Schneider, J. Interactive Extra-solar Planets Catalog; http://exoplanet.eu/catalog.php, 2011

Sepinsky, J.F.; Willems, B.; Kalogera, V.; Rasio, F.A. Ap. J. 2009, 702, 1387-1392; arXiv:0903.0621.

Svechnikov, M.A. Catalogue of Orbital Elements, Masses and Luminosities of Close Binary Stars; Publ. Ural State University, Sverdlovsk, 1969.

Tohline, J.E. Annu. Rev. Astron. Astrophys. 2002, 40, 349-385.

Udry, S.; Santos, N.C. Annu. Rev. Astron. Astrophys. 2007, 45, 397-439.

Wager, L.R.; Brown,G.M. Layered Igneous Rocks, Oliver & Boyd, Edinburg & London, 1968.

Wyatt, M.C. Annu. Rev. Astron. Astrophys. 2008, 46, 339-383.

Zharkov,V.N.; Makalkin,A.B.; Trubitsyn, V.P. Astron. Tsirk. 1974, No812, 3-5.

Zinnecker, H.; Yorke, H.V. Annu. Rev. Astron. Astrophys. 2007, 45, 481-564.

In: Solar System: Structure, Formation and Exploration
Editor: Matteo de Rossi
ISBN: 978-1-62100-057-0

Chapter 3

HARNESS ENERGY OF THE SUN BY SPLITTING WATER USING MN-OXO OR CO-BASED CATALYTIC SYSTEMS TO MIMIC PHOTOSYNTHESIS

Harvey J.M. Hou*
Department of Physical Sciences, Alabama State University,
Montgomery, Alabama, U. S.

ABSTRACT

The sun, one member of the solar system, provides all of the energy to support the life on our planet over several billion years, and the solar energy is harvested and storied via photosynthesis by green plants, cyanobacteria, and algae on the large scale at room temperature and neutral pH. Use of solar energy is completely carbon-free and is able to eliminate current concerns on the energy crisis and global climate change. A deep understanding of photosynthesis is the key to provide a solid foundation to facilitate the transformation from carbon-based energy source to sustainably solar fuels. The energy from the sun can be stored via water splitting, which is a chemical reaction through chemical bond rearrangement to convert the energy-deficient water molecule to energy-rich oxygen and hydrogen molecules. Water splitting chemistry is driven by sunlight in the reaction center of photosystem II located in the thylakoid membranes of plant leaves. The three-dimensional structures of photosystem II with oxygen-evolving activity have been determined at an atomic and a molecular level in the past ten years. To mimic the water splitting of photosystem II oxygen evolving complex, appealing systems including earth abundant element catalytic materials were discovered. In this chapter, recent progress in solar fuel production emphasizing on the development of Mn-oxo complexes and Co-phosphate catalytic systems were summarized and discussed. These systems, including Mn-oxo tetramer/Nafion, Mn-oxo dimer/TiO_2, Mn-oxo oligomer/WO_3, Co-Pi/Fe_2O_3, and Co-Pi/ZnO were, show a compelling working principle by combing the active Mn-oxo and Co-based catalysts in water splitting with semiconductor hetero-nanostructures for effective solar energy harnessing. The protocols are suit for preparing earth-abundant metal/semiconductor catalysts and highly likely open a new area of

* E-mail address: hhou@alasu.edu

fabricating next generation of highly efficient water splitting catalysts to store the energy from the sun. Grand challenges include the discovery of inexpensive, robust, and efficient water oxidation catalysts. It is particularly important forthe improvement in efficiency and durability of the water-splitting catalytic systemsfor their practical application as well as theutilization of visible and infrared light.

INTRODUCTION

Energy is increasingly becoming the top priority of national and international issues. This is because that global energy need is expected to double by midcentury and triple by the end of the century (Lewis and Nocera, 2006; Cook et al., 2010), largely due to the growing world population. As documented in the literature, current the energy sources are insufficient to keep pace with the global energy demand. The main energy source, fossil fuels, is nonrenewable and produces enormous amount of net greenhouse gases, which have substantial negative impact on the environment, as well as has limited source supply on earth. To address these issues, novel renewable carbon-free or carbon-neutral energy sources must be identified and generated in next 10 to 50 years.Nuclear energy is problematic to build fast and has been a concern in public safety. The wind energy is too low in producing enough energy density. Compared with all other energy options, solar energy is the most promising and the only source of truly renewable, plentiful, and secure energy (Lewis and Nocera, 2006; Cook et al., 2010).

The sun, one member of the solar system, provides all of the energy and generates oxygen molecules by water splitting reaction to support the life on our planet over several billion years. The solar spectrum of the sun and absorption spectra of photosynthetic pigments are shown in Figure 1. Through a variety of pigment and their protein complexes, sunlight energy is harvested and storied via photosynthesis on the large scale (Figure 2). The process can be performed at room temperature and neutral pHby green plants, cyanobacteria, and algae. Sunlight is far exceeds what is necessary to support the society. The ability of solar to meet the global energy need of the future is well documented. The major challenge for the development of solar energy on a large scale is its storage. The solar energy storage has been successfully accomplished by the water splitting reaction via rearranging the chemical bonds in photosynthesis (Figure 2). In the chemical reaction, the breaking of four O-H covalent bonds and forming of two H-H bond and one O-O bond result in the conversion of energy-deficient water molecule into the energy-rich hydrogen and oxygen molecules. The water splitting reaction (Figure 2) stores significant amount of energy (ΔH=570 kJ) in chemical form of hydrogen and oxygen, which may be released in the form of electric energy via a fuel cell through the reverse reaction.

In addition to photosynthesis, the water splitting reaction can be achieved by electrolysis. A key determinant of energy storage in artificial photosynthesis is the efficiency of the water splitting catalysts. These catalysts must operate close to the Nernstian potential (E, Figure 2) for the half-cell reaction. In general, an extra potential in addition to E, designated overpotential, limits the efficiency of the conversion of light to catalytic current. The water oxidation reaction is more complex as it requires a four-electron oxidation of two water molecules coupled to the removal of four protons. In addition, a catalyst must to tolerate prolonged exposure to highly oxidized conditions, which is able to cause most chemical functional groups to degrade. In artificial photosynthesis, water oxidation is considered a substantial challenging task. A deeper understanding of solar energy conversion, such as photosynthesis, is the key. The fundamental investigations of water splitting chemistry will provide a firm foundation for facilitating this transformation. Solar PV panel, solar energy cell, and fuel cell working together will promise to transform solar energy into affordable mainstream energy.

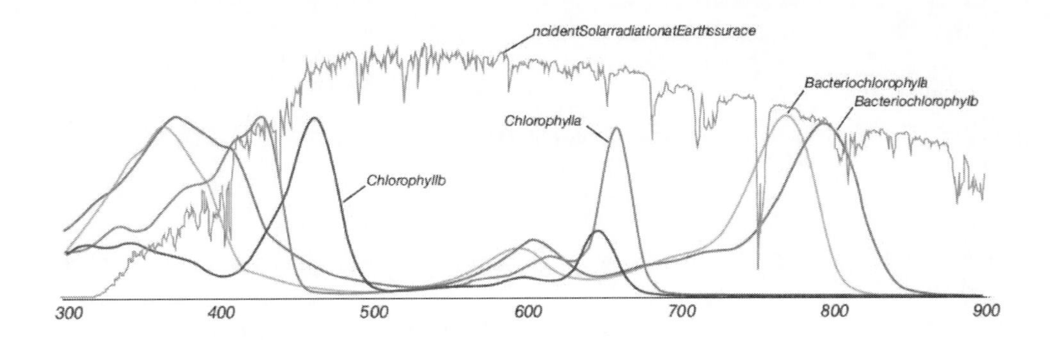

Figure 1. Solar Spectrum utilization in photosynthesis. The solar spectrum incident on the earth's surface (air mass 1.5, NREL) is in gray. Absorption spectra of chlorophyll a (green), chlorophyll b (dark green), bacteriochlorophyll a (yellow), and bacteriochlorophyll b (blue) are taken in methanol or ethanol (McConnell et al 2010). (Reproduced with permission from Elsevier).

$$O_2 + 4H^+ + 4e^- \rightleftharpoons 2H_2O \qquad E_{anodic} = 1.23\ V - 0.059\ (pH)\ V \text{ vs NHE}$$

$$4H^+ + 4e^- \rightleftharpoons 2H_2 \qquad E_{cathodic} = 0\ V - 0.059\ (pH)\ V \text{ vs NHE}$$

$$2H_2O \longrightarrow 2H_2 + O_2 \qquad E_{rxn} = -1.23\ V$$

Figure 2. Half-cell reactions and redox potentials for the water splitting reaction.

OVERVIEW AND MECHANISMS OF PHOTOSYNTHETIC WATER SPLITTING

In nature, photosynthetic organisms collect sunlight efficiently from the sun and convert the solar energy into organic molecules. At the heart of the photosynthetic process, is the splitting of water by sunlight into oxygen and 'hydrogen' (NADPH). The oxygen is released into the atmosphere for us to breathe and for burning fuels to drive our technologies. The 'hydrogen' is combined with greenhouse gas, carbon dioxide, to make sugars and other organic molecules. As shown in Figure 3, to extract one electron from water and transfer into carbon dioxide, two photons of light are required by two separated photosystems (PS I and PS II). One photon is absorbed by PS II to generate a strong oxidizing species ($P_{680}+$), which is able to drive the water splitting reaction. The other photon is used by PS I to produce a strong reducing species, NADPH, and a weak oxidant $P_{700}+$.

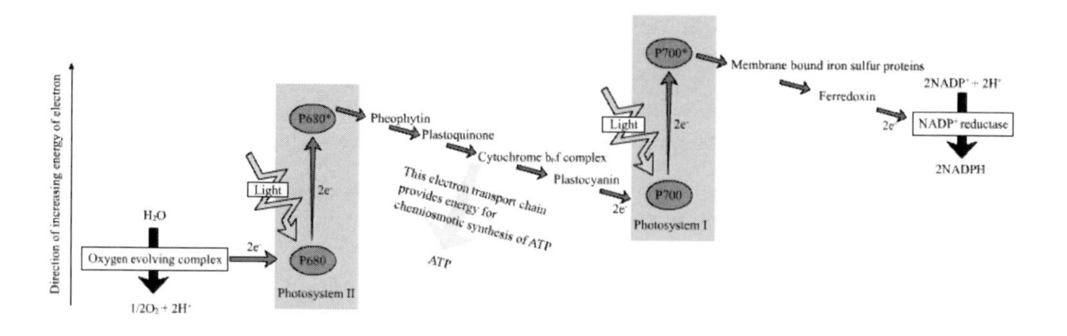

Figure 3. Z-scheme of the light reaction of photosynthesis (from http://en.wikipedia.org/wiki/photosynthesis).

Water splitting chemistry driven by sunlight for solar energy conversion occurs in the reaction center of PS II, which is located in the thylakoid membranes of green plants, cyanobacteria, and algae (Nanba and Satoh, 1987; Diner and Rappaport, 2002; Barber, 2009). PS II is the water-platoquinonephotooxidoreductase or oxygen-evolving enzyme. It performs a series of light-induced electron transfer reactions leading to the splitting of water into protons and molecular oxygen. The products of PS II, namely chemical energy and oxygen, are vital for sustaining life on earth. The three-dimensional structures of PS II with oxygen-evolving activity were determined in the past six years (Ferreira et al., 2004; Loll et al., 2005; Yano et al., 2006) and have laid solid foundation for mechanistic study of solar energy conversion at the molecular level (Figure 4, upper). When the primary donor P_{680} is excited by light, charge separation (2-20 ps) takes place mainly between the chlorophyll, Chl_{D1}, and the pheophytin, Ph_{D1} (Step 1). The cation is stabilized mainly in chlorophyll P_{D1}, designated P_{680}+. Ph_{D1}- species transfers an electron (~400 ps) to the quinone, Q_A (Step 2). P_{680}+ is able to oxidize (~20 ns) the tyrosine-161 of D1 protein, Tyr_Z, which loses a proton to the neighboring histidine (Step 3). Tyr_Z• oxidizes (~30 μs) the Mn cluster (S_1 to S_2) (Step 4). Q_A- transfers an electron (~100 μs) to the second quinone, Q_B (Step 5). Subsequent turnovers give similar reactions but with kinetic differences at steps affected by charge accumulation on the Mn cluster and on Q_B. The second electron on Q_B triggers the uptake of two protons and replaces a plastoquinone (PQ) from the pool in the membrane. The enzyme accumulates four positive charge-equivalents and releasing O_2. The valence of the Mn ions increases on the S_0 to S_1 to S_2 steps. However, it is unknown for the S_3 and S_4 states (Figure 4, lower right).

The model of Mn_4Ca cluster (Figure 4, lower left) in the PS II oxygen evolving complex (OEC) is proposed but remains to be confirmed(Yano et al., 2006; Sproviero et al., 2008), due to the photosensitivity of PS II to X-ray radiation and current resolution of X-ray crystallographic data.Recently, to suppress the possible radiation damage to a minimum level, using a slide-oscillation method, afull data set of oxygen-evolving photosystem II was collected and process to a resolution of 1.9 Å(Umena et al., 2011). The 1.9 Åcrystal structure reveals the geometric arrangement of the Mn_4CaO_5 cluster including its oxo bridges and ligands (Figure 5). Three manganese, one calcium and four oxygen atoms form a cubane-like structure, but the Mn_3CaO_4 is not an ideal, symmetric one. The fourth manganese (Mn5) is located outside the cubane and is liked to two manganeses (Mn1 and Mn3). The calcium is linked to all four manganeses by oxo bridges. In addition to the five oxygen, four water molecules (W1 to W4) were found to be associated with the Mn_4CaO_5 cluster. Two waters are coordinated to the Mn4 and two to the calcium. The direct ligands of the Mn_4CaO_5 cluster are identified: D1-Glu 189, D1-Asp 342, D1-Glu 333, D1-Asp 342, D1-Ala 344, CA43-Glu 354, and D1-His 332. The second coordination sphere includes D1-Asp 61, D1-His 337, and CP43-Arg 357. The O5 is likely a hydroxide ion in the S_1 state. The O-O bond formation may occurs in two of the three species O5, W2 and W3. The high-resolution structure of PS II at 1.9 Å resolution provides a basis for unraveling the mechanism of water splitting and O-O bond formation, one of the most fascinating and important reactins in nature.

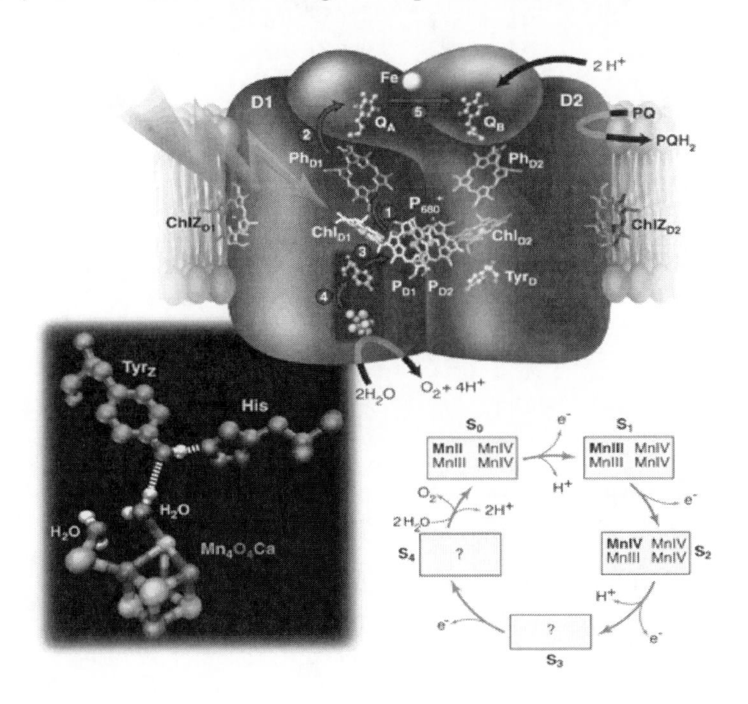

Figure 4. Structure and Mechanism of PS II OEC (Rutherford and Boussac, 2004).Model of PS II (upper), Mn_4Ca center (lower left), and S-state model (lower right) for oxygen evolution. (Reproduced with permission from the American Association for the Advancement of Science).

PS II water splitting chemistry involves four-oxidation steps with five intermediates known as S-states (Kok et al., 1970). However, limited information at the molecular level due the complexity of PS IIand its sensitivity to envionment. There were several mechanistic proposals in probing PS II water splitting chemistry (Hoganson and Babcock, 1997; Ruettinger et al., 2000; McEvoy et al., 2005). A body of evidence provides strong support for binding of the substrate water molecules as terminal ligands to manganese and calcium and for a direct role of calcium in the water-oxidation chemistry as a Lewis acid to activate a substrate water molecule as a nucleophile. Mn model chemistry also supports the possibility that water is activated for O-O bond formation in the OEC by binding to a high-valent manganese ion. It is generally established that the active catalytic species is Mn(V)=O or Mn(V)-oxo radical, which is capable of releasing oxygen and closes the S-state cycle (Brudvig, 2008).

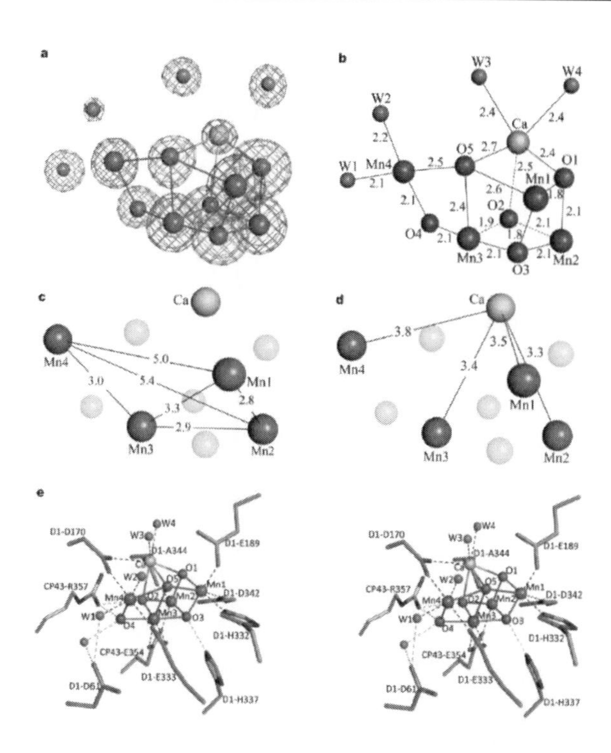

Figure 5. Structure of the Mn_4CaO_5 cluster in the oxygen-evolving photosystem II at a resolution of 1.9 Å(Umena et al., 2011). (Reproduced with permission from Macmillan Publisher).

The pioneery work in artificial water splitting chemistry is on Ru-based catalysts. The first water splitting Ru-based catalyst, blue dimer, was reported with a moderate number of turnover, and its mechanisms are greatly elucidated (Gersten et al., 1982; Liu et al., 2008). The mechanism involves Ru(V)-oxo active intermediate. Recently one new Iridium-based family of catalysts was reported (McDaniel et al., 2008; Meyer, 2008). In addition, two all-inorganic catalysts by two independent groups were synthesized (Geletii et al., 2008; Sartorel et al., 2008). However, the low abundance and high expense of Ru- and Ir-based catalysts are problematic for large scale solar energy conversion. It is urgent to develop earth abundant metal catalysts, such as Mn, Fe, Co, Ni, and Cu-based catalysts.The reason is obvious for practical purpose. The invention of earth-abundant metal-oxo catalysts is extremely important for transforming solar energy to affordable energy source in the next ten to fifty years.

Syntheses and Mechanisms of Mn-oxo Complexes Catalysts Mimicking Photosynthetic Water Splitting

In the field of artificial photosynthesis, the first functional mimic of Mn_4Ca center in PS II is a Mn-oxo tetramer complex (Ruettinger et al., 1997; Ruettinger et al., 2000), thereafter designated as Mn-oxo tetramer in the chapter as shown in Figure 6. The compound is synthesized and contains a cubical $[Mn_4O_4]^{n+}$ core with six bidentateligands chelating to the manganese ions, $(dpp)_6Mn_4O_4$ (dpp-=diphenylphosphinate anion). UV light absorption by the Mn ion produces aMn-O charge-transfer excited state, which efficiently release one dioxygen molecule. The development of the Mn-oxotetrameric model offer novel insights into the possible nature of PS II oxygen evolving complex in water splitting and play a vital role in illustrating photosynthetic oxygen evolving mechanism. However, the oxygen evolution is not continuous due to the light-induced decomposition of Mn-oxo tetramer cubane core.

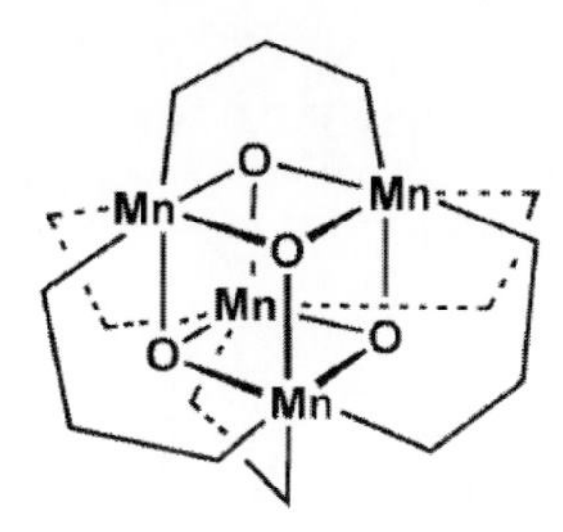

Figure 6. Schematic represention of manganese cubane, $Mn_4O_4[(RPh)_2PO_2]_6$. (Ruettinger et al., 2000; Brimblecombe et al., 2009). (Reproduced with permission from the Royal Society of Chemistry).

To probe the active site of Mn_4Ca center in PS II, Brudvig and co-workers discovered a dimericMn-oxo complex, $[H_2O(terpy)Mn(O)_2MnH_2O(terpy)](NO_3)_3$ (terpy is 2,2':6',2''-terpyridine), with continuous oxygen evolution activity in the presence of chemical oxidant such as oxone(Limburg et al., 1999). The compound (Figure 7), thereafter designated as Mn-oxodimer in the chapter, was characterized by Raman spectroscopy, EPR, MS, and enzymatic kinetics in the presence of variety of oxidants, such as, oxone, Ce4+, and hyperchrite(Limburg et al., 2001; Chen et al., 2005; Tagore et al., 2006; Chen et al., 2007; Tagore et al., 2007; Tagore et al., 2007; Brudvig, 2008; Cady et al., 2008; Tagore et al., 2008). The key feature of the Mn-oxodimer is Mn(III)/Mn(IV) mix-valence and the presence of one terminal water molecule on each Mn ion (Figure 7). The catalytic mechanism of Mn-oxodimer involves the valence change of Mn(III/IV) to Mn(IV/V) by the oxidant and followed by the molecular oxygen release from water splitting step as shown in Figure 8. The oxygen release step is associated with the reduction of Mn(IV/V) to Mn(II/III).

A recent study argued that PS II photoinhibition is triggered by a direct absorption of UV light in the Mn_4Ca cluster (Hakala et al., 2005). We investigated the UV effect on the functional PS II model, Mn-oxodimer, and PS II membranes. The Mn-oxodimer was unstable to UV light, as judged by the measurement of increasing absorption at 400 nm, which is assigned to the Mn(IV)/Mn(IV) species (Limburg et al., 2001). We suggested that the photodamage of the Mn-oxodimer may be associated with a valence change from Mn(III) to Mn(IV)(Wei et al., 2011). The oxygen-evolution activity of the Mn-oxodimer was decreased upon UV treatment, supporting the occurrence of photodamage. The action spectrum of Mn(III/IV)-oxodimer under strong light at six wavelengths (254, 312, 365, 452, 555, and 655 nm) revealed the presence of a stable species peaking at 440 nm. Fluorescence spectrometry showed that the UV-induced product has an intense fluorescence peak at 513 nm, confirming the formation of a novel stable species. The photodamage induced by UV radiation showed strong pH dependence, indicating that protons play a role in the photodamage reaction.

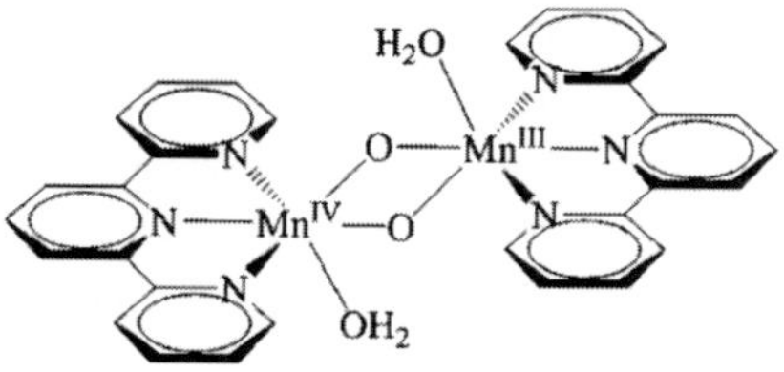

Figure 7. Structure of a dinucleardi-μ-oxoMn(III,IV) water-oxidation catalyst, which is a functional model for oxygen-evolving complex of PS II (McConnell et al. 2010; Limburg et al., 1999). (Reproduced with permission from Elsevier).

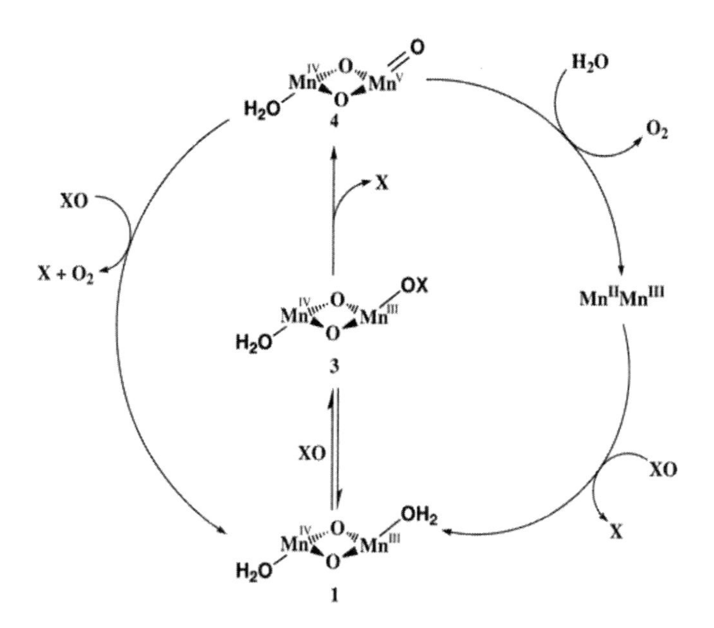

Figure 8. Catalytic mechanism of Mn-oxodimer for water oxidation (Limburg et al., 2001).Mn(III/IV) is the reactive species. The electron transfer species XO initially binds reversibly to 1 and form 3. Once formed 3 may react with XO to form an intermediate Mn(V)=O (4), which rapidly oxidize water to evolve O_2. (Reproduced with permission from the American Chemical Society).

The thermal stability of Mn-oxodimer showed the dissociation temperature of 60°C. The reaction was accompanied with formation of new products, judged by the formation of brown precipitates in solution and the observation of the colored Mn-oxodimer from green to colorless. Unexpectedly, the oxygen evolution measurements showed an activity increase after the decomposition reaction was completed (Figure 9). We concluded that one solid water-splitting material with higher activity, thereafter designated Mn-oxooligomer, is formed in the solution(Zhang et al., 2011).

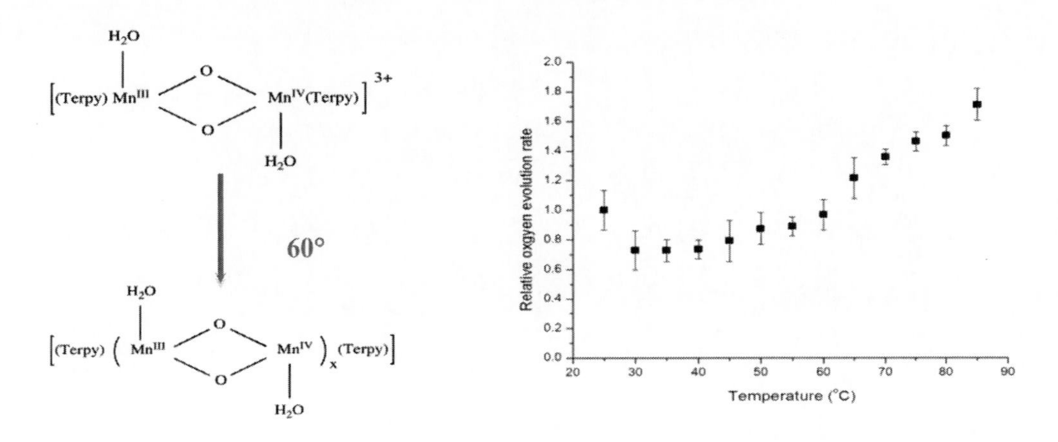

Figure 9. Formation of a Mn-oxooligomer water splitting catalyst by heating the Mn-oxodimer in aqueous solution at 60°C (Zhang et al., 2011). (Reproduced with permission from Elsevier).

The Mn-oxooligomer was characterized by FTIR, EPR, elemental analysis, XANES and EXAFS. FTIR data showed that the solid product has a different IR spectrum than MnO_2, suggesting the Mn-oxooligmer is not MnO_2. The EPR signal confirmed that the Mn-oxooligomer is different from Mn(III/IV)-oxodimer. The elemental analysis showed that the Mn-oxooligomer contains terpyridineligand. The TEM data indicated Mn-oxooligomer is amorphous on the nanometer scale. The XANES and EXAFS data suggested that the rising edge energy of the Mnoligomer is slightly shifted to higher energy compared to the MnTerpy sample, likely indicating an increased fraction of Mn(IV). However, this should still be a mixture of Mn(III) and Mn(IV) oxidation state. The EXAFS data indicated that the Mn-Mn distances are increased from ~2.7Å to 2.9Å (average) after the oligomerization. These lines of evidence suggested that the Mn-oxooligomer has unique new structural feature with bounded terpyligands. This material is thermal stable in nanoscale size and highly active in photosynthetic water splitting, which may be unique for fabricating novel catalysts in solar fuel production.

SYNTHESES AND MECHANISMS OF CO-BASEDCATALYSTS MIMICKING PHOTOSYNTHETIC WATER SPLITTING

A Co-based catalytic material that forms electrochemically on an ITO electrode in phosphate buffered water containing cobalt (II) ions was reported to operate in neutral water under room temperature (Kanan et al 2009, Kanan et al 2008). This type of Co-base catalyst was able to oxidize water in aqueous solutions containing 0.5 M NaCl (Surendranath et al 2009). The catalytic mechanism was shown in Figure 10. The active species is proposed to be the Co-oxocubane-like structure, which oxidizes water to produce O_2 by forming a Co(IV) intermediate via a proton-coupled electron transfer step. Phosphate ion may be the key player for the proton transfer reaction (Kanan et al., 2009; Lutterman et al., 2009). Further analysis revealed that the Co-Pi material is a robust heterogeneous water splitting catalyst and able to self-repair by self-assembly.

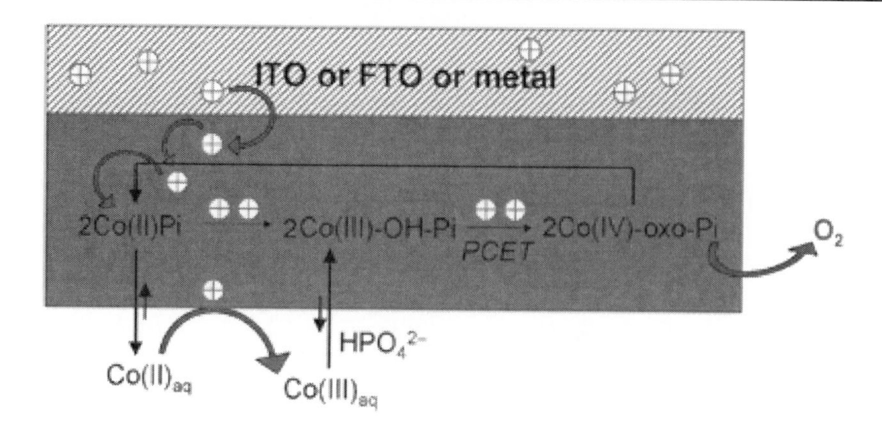

Figure 10. *In situ* formation and mechanism of Co-Pi water oxidation catalyst (Kanan et al., 2009). The Co^{2+} is oxidized to Co^{3+} and deposited on the electrode in the presence of phosphate (Pi). The deposited Co(III)-OH-Pi species may form a cubane-like Co(IV)-oxo complex by a proton-coupled electron transfer step. The Co(IV)-oxo species may rapidly oxidized water to form dioxygen, and cobalt is returned to its 2+ oxidation state.

XAS and EPR spectra of active Co-Pi film point to a mechanism for a molecular cobaltate cluster model as shown in Figure 11 (Kanan et al., 2010; McAlpin et al., 2010). The high catalytic activity of Co-Pi suggests molecular cobaltate cluster structure promot water oxidation and that the Co valency is greater that 3. The "edge" of cobaltate may have terminal waters. By truncating the extended cobaltate lattice, the number of edges is maximized and maximum activity is realized. The extended cobaltate lattices have few terminal oxygens and hence are unable to splitting water.

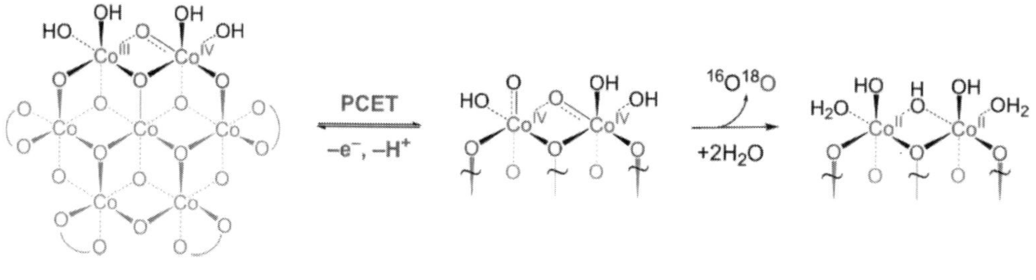

Figure 11. Proposed pathway for oxygen evolution reaction by Co-Pi catalyst (Cook et al., 2010)(Reproduced with permission from the American Chemical Society).

In addition to Co-Pi, a homogeneous catalyst, B-type $[Co_4(H_2O)_2(\alpha\text{-}PW_9O_{34})_2]10-$, which is free of carbon-based ligands, was synthesized and demonstrated high catalytic turnover frequencies for O_2 production at pH 8 (Yin et al., 2010). The key element of the complex is a Co_4O_4 core stabilized by oxidatively resistant polytunstateligands. Although the mechanism of the complex is unclear, the catalytic material provides a basis for further understanding of Co-based water splitting catalysis in general.

CATALYST-COUPLEDNAFION OR SEMICONDUCTORPHOTOANODES IN WATER SPLITTING

The fabrication of efficient catalysts for splitting water into hydrogen and oxygen is one of the most challenges in renewable energy production. The Mn-oxo tetramer cubane-like compound developed by Dismukes and co-workers (Ruettinger and Dismukes, 1997; Ruettinger et al., 1997; Ruettinger et al., 2000) was doped into the Nafion membrane (3-8 μm) to make a Mncubium/Nafionphotoanode, which is able to oxidize water upon activation with visible light (Figure 12) (Brimblecombe et al., 2008; Brimblecombe et al., 2009; Dismukes et al., 2009; Brimblecombe et al., 2010). The key feature of the design is two aspects (1) a photoinduced charge separation system, which is Ru(II)-bipy complex and TiO_2-coated film, and (2) a molecular catalyst, which is Mn-oxo cubic species in a Nafion membrane.

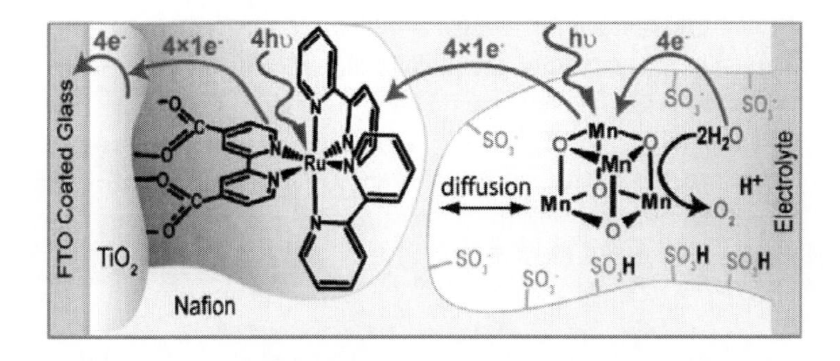

Figure 12. Mn-oxo tetramer/Nafion photo water splitting catalyst (Dismukes et al., 2009; Brimblecombe et al., 2010). On visible light excitation, the TiO_2-supported Ru(II) sensitizer injects an electron into the TiO_2 conduction band and form high oxidized Ru(III)-species with a redox potential of +1.4V (vs SHE). The Ru(III)-species oxidized Mn-oxocubane-like core, and four such steps subsequently oxidize water to form O_2 like PS II in photosynthesis. (Reproduced with permission from the American Chemical Society).

With an energy gap between the conduction and valence bands, semiconductor represents an appealing candidate to effectively absorb photons and transform the optical energy into free charges (electrons and holes). It has been demonstrated that these charges can be readily utilized for water splitting (Fujishima and Honda, 1972). Theoretical calculations have shown that the power conversion efficiency of using semiconductor for water photo-splitting can be as high as that of solid-state solar cells (Bolton, 1996). More recently, significant efforts have been attracted to fabricate nanoscale semiconductor materials as photoelectrodes to further improve the performance of water splitting by adding advantages of high surface area and improved conductivity (Mor et al., 2005; Lin et al., 2009; Yang et al., 2009). Combining semiconductor nanomaterials with the Mn-oxo catalyst overcomes a key challenge in using semiconductor directly – the low catalytic activity of semiconductors. The low reactivity often leads to a high overpotential and results in significant reduction in the overall energy conversion efficiency (Kudo and Miseki, 2009).Using a research scale commercial ALD reactor, various oxides that can be used for photo water splitting were successfully grown, including TiO_2, WO_3, Cu_2O and Fe_2O_3 (Lin et al., 2009). By interfacing these semiconductor materials with a highly conductive nanonet structure, the performance of splitting water is greatly boosted.

A Mn(II)-terpy compound was attached to the surface of TiO_2nanomaterial and showed an efficient electron transfer (Abuabara et al 2007, McNamara et al 2008). A Mn(III/IV)-oxodimer was anchored to TiO_2nanoparticles via direct adsorption or *in situ* synthesis (Li et al 2009). The resulting Mn-oxo dimer/TiO_2 is able to reversibly change mixed valentMn(III/IV) to Mn(IV/IV) state by photoexcitation and interfacial electron injection into the conducting band of TiO_2. This Mn-based TiO_2 material appears to be promising for developing an inexpensive water splitting catalyst in the photocatalytic solar cells.

The Mn-oxooligomer with high catalytic activity, which is the decomposition product of Mn(III/IV)-oxodimer, may be an ideal material for fabricating robust water-splitting catalysts. Our hypothesis is to use n-type semiconductor to generate holes. When irradiated by light, n-type semiconductor will cooperate with Mn-oxo complex to efficiently split water using solar energy (Figure 13). The advantage is the combination of highly active water splitting catalytic ability of Mn-oxooligomer and highly efficient photoconversion of semiconductor. The Mn layer is expected to be within a few nanometer in thickness to ensure the high electric conductivity for photocatalytic water splitting. To test our hypothesis, the nanometer layer of Mn-oxooligomer was coated on the surface of WO_3 material. By controlling the synthetic conditions including reaction temperature and Mn-oxodimer concentration we successfully obtained the Mn layer with thickness of 2, 5, 10, 20, 30, 40 and 50 nm.

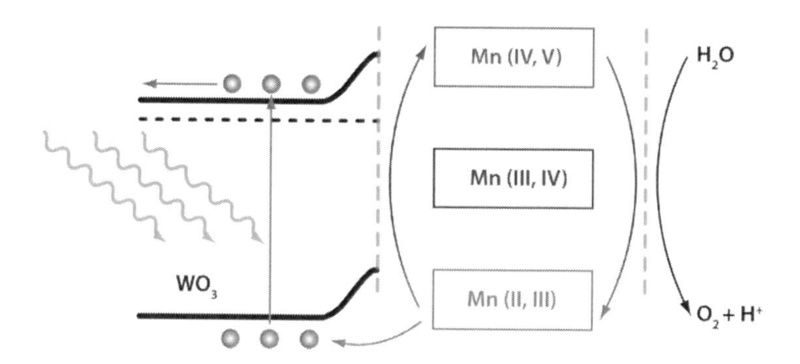

Figure 13. Working model of aMn-oxo oligomer/WO_3 photo water oxidation catalyst. The solar light radiations are absorbed by WO_3 semiconductor and cause the charge separation to produce electrons and holes. The electrons are transferred to the cathode by an electric wire to produce hydrogen gas. The holes receive electrons from Mn-oxooligomer, which is the precipitate of Brudvig catalyst (Mn-oxodimer) under thermal conditions(Liu et al., 2011). (Reproduced with permission from Wiley-VCH).

The Mn-oxo oligomer/WO_3 material is able to directly generate oxygen and hydrogen for solar energy harness(Liu et al., 2011). As shown in Figure 14,various evidence supports that the detected oxygen by capillary GC analysis is the direct product of water splitting (Liu et al., 2011).The amount of hydrogen is approximately twice that of oxygen, consistent with complete decomposition of water. Control experiments with H_218O confirmed that O is the gas phase comes from water. The experimental results also demonstrated that the water splitting reaction requires the cooperation of Mn-oxo catalytic material and WO_3

semiconductor. A possible mechanism of Mn-oxo oligomer/WO_3 system may involve four light photons to oxidize the Mn-oxooligomer accompanying four proton-coupled steps. Each step, the photon causese charge separation to generate a hole and an electron in WO_3. The hole produced in WO_3 receives electron from Mn-oxo complex and oxidzes the Mn ions. Final step of oxidatin in Mn valence produces Mn(V) intermediate species by the fourth photon, and highly active Mn(V) species splits water to form O-O bond and regenerate Mn-oxo catalyst.

The ALD growth ofWO_3without production of corrosive byproducts has not beenreported elsewhere, and the synthetic technique makes it easyto form heteronanostructures. The Mn catalyst derived fromthe oxo-bridged Mndimer is easy to prepare and exhibitsgood stability and catalytic properties. When interfaced withWO_3, it acts as a protecting layer without adverse effect on thewater-splitting properties. To the best of our knowledge, this isthe first time that WO_3photoelectrodes stable in neutralsolution have been prepared. The heteronanostructure designcombines multiple components, each with unique complementaryand critical functions, and offers combinations ofproperties that are not available in single-component materials.The versatility of this method will find applications innumerous areas where the availability of materials is thelimiting factor.

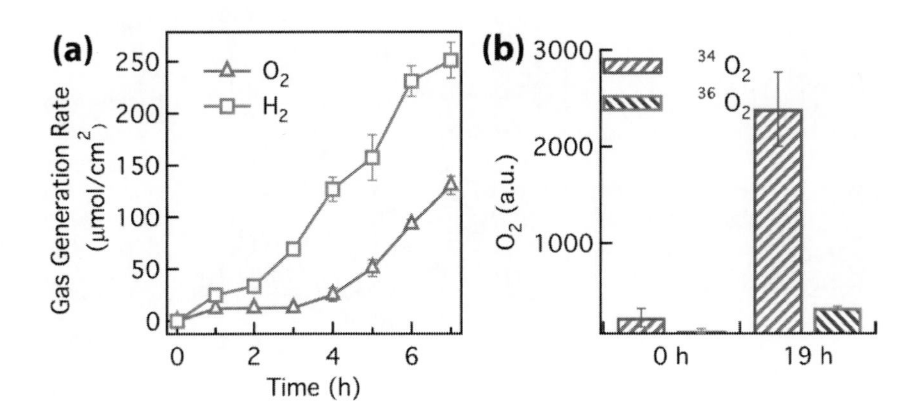

Figure 14. Water splitting reaction by Mn-oxo oligomer/WO_3 catalytic system. (a) the rate of hydrogen production is approximately twice that of oxygen. (b) isotopic labeling experiments verify that oxygen atoms in oxygen come from water (Liu et al., 2011). (Reproduced with permission from Wiley-VCH).

Co-based water splitting catalyst can be electrochemically and photochemically deposited on the surface of semiconductor Fe_2O_3 and ZnO, respectively (Steinmiller and Choi, 2009; Zhong et al., 2009; Zhong and Gamelin, 2010). The resulting Co/Fe_2O_3 and Co/ZnOphotoanodes showed a dramatic improvement in solar water splitting (Figures 15 and 16). These results demonstrate that integration of promising water splitting catalysts with a photo-absorbing substrate can provide a substantial reduction in the external power needed to drive the catalytic water splitting chemistry and can be used as a general route to deposit the molecular catalysts on any semiconductor electrode.

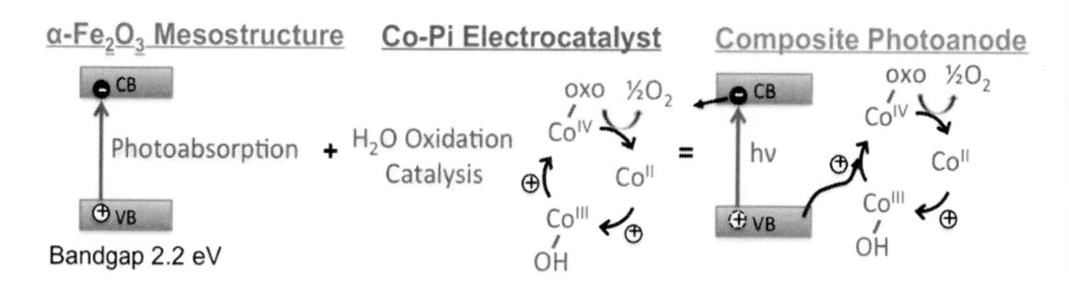

Figure 15. CoPi/Fe_2O_3 photo water splitting catalyst (Zhong et al., 2009; Zhong and Gamelin, 2010). Composite photoanode of the Co-Pi-modified α-Fe_2O_3 semiconductor is able to photooxidize water at low bias (1.0 V vs RHE) and at pH 8 in mild salt water conditions. (Reproduced with permission from the American Chemical Society).

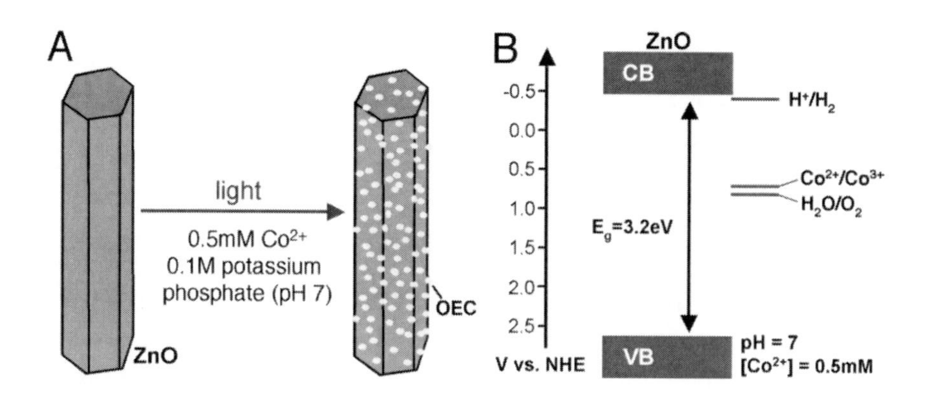

Figure 16. CoPi/ZnO photo water splitting catalyst (Steinmiller and Choi, 2009). Photochemical deposition of Co-based oxygen evolution catalyst on ZnO semiconductor is able to couple oxygen evolution catalysis with semiconductor for solar O_2 evolution. The critical feature is the valence band of semiconductor must locate at a more positive than the oxidatin potential of Co^{2+}. (Reproduced with permission from the National Academy of Sciences).

CONCLUSION

In nature, the water-splitting reaction via photosynthesis driven by sunlight in plants, algae, and cyanobacteria stores the vast solar energy and provides vital oxygen to life on earth. In the recent five years, revolutionary developments in photoelectrochemical water splitting using Mn-oxo complexes and Co-based molecular catalysts (Cady et al., 2008; Dismukes et al., 2009) as well as Ru- and Ir-based compounds (Concepcion et al., 2009; Sala et al., 2009) associated with dye-sensitized semiconductors (Woodhouse and Parkinson, 2008; Youngblood et al., 2009) have been made. In particular, the developed Mn/Nafion, Mn/TiO_2, Mn/WO_3, Co/Fe_2O_3, Co/ZnO systems may be extended to heterostructures of a variety of semiconductors. The protocols are suit for preparing earth-abundant metal/semiconductor catalysts. It is highly likely open a new area of fabricating next generation of highly efficient water splitting catalysts in solar fuel production. In nature, the production of oxygen by oxidation of water is catalyzed by an Mn_4Ca inorganic center in the oxygen-evolving complex of photosystem II. Hence the use of a light harvester, a water splitting catalyst, and an electron acceptor is a promising way for solar energy conversion. Grand challenges remins, including the discovery of inexpensive, robust, and efficient water oxidation catalysts. In particular, the future endeavors will be placed on improvement in efficiency and durability of the catalytic system for its practical application as well as on usage of visible and infrared light. It is highly likely that the progresses in the filed of nanomaterial and photosynthesis will offer novel technology for transforming the solar energy into our future energy systems.

ACKNOWLEDGMENTS

The work was supported by the Alabama State University. The author thanks Professor Gary Brudvig at YaleUniversity for continuous support and Professor Dunwei Wang at Boston College forcollaborative effort.

REFERENCES

Barber J (2009) Photosynthetic energy conversion: natural and artificial. Chem. Soc. Rev. 38:185-196

Bolton JR (1996) Solar photoproduction of hydrogen: A review. Solar Energy 57:37-50

Brimblecombe R, Dismukes GC, Swiegers GF, Spiccia L (2009) Molecular water-oxidation catalysts for photoelectrochemical cells. Dalton Trans. 9374-9384

Brimblecombe R, Koo A, Dismukes GC, Swiegers GF, Spiccia L (2010) Solar-driven water oxidation by a bio-inspired manganese molecular catalyst. J. Am. Chem. Soc. 132:2892-2894

Brimblecombe R, Swiegers GF, Dismukes GC, Spiccia L (2008) Sustained water oxidation photocatalysis by a bioinspired manganese cluster. Angew. Chem. Int. Ed. 47:7335-7338

Brudvig GW (2008) Water oxidation chemistry of photosystem II. Philos. Trans. B 363:1211-1219

Cady CW, Crabtree RH, Brudvig GW (2008) Functional models for the oxygen-evolving complex of photosystem II. Coord. Chem. Rev. 252:444-455

Chen H, Tagore R, Das S, Incarvito C, Faller JW, Crabtree RH, Brudvig GW (2005) General synthesis of di-mu-oxodimanganese complexes as functional models for the oxygen evolving complex of photosystem II. Inorg. Chem. 44:7661-7670

Chen H, Tagore R, Olack G, Vrettos JS, Weng T-C, Penner-Hahn J, Crabtree RH, Brudvig GW (2007) Speciation of the catalytic oxygen evolution system: $[MnIII/IV_2(\mu\text{-}O)_2(terpy)_2(H_2O)_2](NO_3)_3 + HSO_5$. Inorg. Chem. 46:34-43

Concepcion JJ, Jurss JW, Brennaman MK, Hoertz PG, Patrocinio AOT, Murakami Iha NY, Templeton JL, Meyer TJ (2009) Making oxygen with ruthenium complexes. Acc. Chem. Res. 42:1954-1965

Cook TR, Dogutan DK, Reece SY, Surendranath Y, Teets TS, Nocera Daniel G(2010) Solar energy supply and storage for the legacy and nonlegacy worlds.Chem. Rev. 110: 6474-6502

Diner BA, Rappaport F (2002) Structure, dynamics, and energetics of the primary photochemistry of photosystem II of oxygenic photosynthesis.Annu. Rev. Plant Biol. 53:551-580

Dismukes GC, Brimblecombe R, Felton GAN, Pryadun RS, Sheats JE, Spiccia L, Swiegers GF (2009) Development of bioinspired Mn_4O_4-cubane water oxidation catalysts: lessons from photosynthesis. Acc. Chem. Res. 42:1935-1943

Ferreira KN, Iverson TM, Maghlaoui K, Barber J, Iwata S (2004) Architecture of the photosynthetic oxygen-evolving center. Science 303:1831-1838

Fujishima A, Honda K (1972) Electrochemical photolysis of water at a semiconductor electrode. Nature 238:37-38

Geletii YV, Botar B, Kogerler P, Hillesheim DA, Musaev DG, Hill CL (2008) An all-inorganic, stable, and highly active tetraruthenium homogeneous catalyst for water oxidation. Angew. Chem. Int. Ed. 47:3896-3899

Gersten SW, Samuels GJ, Meyer TJ (1982) Catalytic oxidation of water by an oxo-bridged ruthenium dimer. J. Am. Chem. Soc. 104:4029-4030

Hakala M, Tuominen I, Keranen M, Tyystjarvi T, Tyystjarvi E (2005) Evidence for the role of the oxygen-evolving manganese complex in photoinhibition of Photosystem II. Biochim. Biophys. Acta 1706:68-80

Hoganson CW, Babcock GT (1997) A metalloradical mechanism for the generation of oxygen from water in photosynthesis. Science 277:1953-1956

Kanan MW, Surendranath Y, Nocera DG (2009) Cobalt-phosphate oxygen-evolving compound. Chem. Soc. Rev. 38: 109-114

Kanan MW, Yano J, Surendranath Y, Dinca M, Yachandra VK, Nocera DG (2010) Structure and Valency of a Cobalt-Phosphate Water Oxidation Catalyst Determined by in Situ X-ray Spectroscopy. J. Am. Chem. Soc. 132: 13692-13701

Kok B, Forbush B, McGloin M (1970) Cooperation of charges in photosynthetic oxygen evolution. I. A linear four step mechanism. Photochem. Photobiol. 11:457-475

Kudo A, Miseki Y (2009) Heterogeneous photocatalyst materials for water splitting. Chem. Soc. Rev. 38:253-278

Lewis NS, Nocera DG (2006) Powering the planet: chemical challenges in solar energy utilization. Proc. Natl. Acad. Sci. USA 103:15729-15735

Limburg J, Vrettos JS, Chen H, de Paula JC, Crabtree RH, Brudvig GW (2001) Characterization of the O_2-evolving reaction catalyzed by $[(terpy)(H_2O)Mn(III)(O)_2Mn(IV)(OH_2)(terpy)](NO_3)_3$ (terpy = 2,2':6,2"-terpyridine). J. Am. Chem. Soc. 123:423-430

Limburg J, Vrettos JS, Liable-Sands LM, Rheingold AL, Crabtree RH, Brudvig GW (1999) A functional model for O-O bond formation by the O_2-evolving complex in photosystem II. Science 283:1524-1527

Lin Y, Zhou S, Liu X, Sheehan S, Wang D (2009) $TiO_2/TiSi_2$Heterostructures for high-efficiency photoelectrochemical H_2O splitting. J. Am. Chem. Soc. 131:2772-2773

Liu F, Concepcion Javier J, Jurss Jonah W, Cardolaccia T, Templeton Joseph L, Meyer Thomas J (2008) Mechanisms of water oxidation from the blue dimer to photosystem II. Inorg. Chem. 47:1727-1752

Liu R, Lin Y, Chou L-Y, Sheehan SW, He W, Zhang F, Hou HJM, Wang D (2011) Water splitting by tungsten oxide prepared by atomic layer deposition and decoraed with an oxygen-evolving catalyst. Angew. Chem. Int. Ed. 50: 499-502

Loll B, Kern J, Saenger W, Zouni A, Biesiadka J (2005) Towards complete cofactor arrangement in the 3.0Åresolution structure of photosystem II. Nature 438:1040-1044

Lutterman DA, Surendranath Y, Nocera DG (2009) A self-healing oxygen-evolving catalyst. J. Am. Chem. Soc. 131:3838-3839

McAlpin JG, Surendranath Y, Dinca M, Stich TA, Stoian SA, Casey WH, NoceraDG, Britt RD (2010) EPR Evidence for Co(IV) Species Produced During Water Oxidation at Neutral pH. J. Am. Chem. Soc. 132: 6882-6883

McConnell I, Li G, Brudvig GW (2010)Energy Conversion in Natural and Artificial Photosynthesis. Chem. Biol. (Cambridge, MA, U. S.) 17: 434-447

McDaniel ND, Coughlin FJ, Tinker LL, Bernhard S (2008) Cyclometalatediridium(III) aquo complexes: Efficient and tunable catalysts for the homogeneous oxidation of water. J. Am. Chem. Soc. 130:210-217

McEvoy JP, Gascon JA, Batista VS, Brudvig GW (2005) The mechanism of photosynthetic water splitting.Photochem. Photobiol. Sci. 4:940-949

Meyer TJ (2008) The art of splitting water. Nature 451:778-779

Mor GK, Shankar K, Paulose M, Varghese OK, Grimes CA (2005) Enhanced photocleavage of water using titaniananotube arrays. NanoLett. 5:191-195

Nanba O, Satoh K (1987) Isolation of a photosystem II reaction center consisting of D-1 and D-2 polypeptides and cytochrome b-559. Proc. Natl. Acad. Sci. USA 84:109-112

Ruettinger W, Dismukes GC (1997) Synthetic water-oxidation catalysts for artificial photosynthetic water oxidation. Chem. Rev. 97:1-24

Ruettinger W, Yagi M, Wolf K, Bernasek S, Dismukes GC (2000) O_2 Evolution from the Manganese-OxoCubane Core Mn_4O_46+: A molecular mimic of the photosynthetic water oxidation enzyme? J. Am. Chem. Soc. 122:10353-10357

Ruettinger WF, Campana C, Dismukes GC (1997) Synthesis and characterization of $Mn_4O_4L_6$ complexes with cubane-like core structure: A new class of models of the active site of the photosynthetic water oxidase. J. Am. Chem. Soc. 119:6670-6671

Rutherford AW, Boussac A (2004) Water photolysis in biology. Science 303:1782-1784

Sala X, Romero I, Rodriguez M, Escriche L, Llobet A (2009) Molecular catalysts that oxidize water to dioxygen.Angew. Chem. Int. Ed. 48:2842-2852

Sartorel A, Carraro M, Scorrano G, De Zorzi R, Geremia S, McDaniel ND, Bernhard S, Bonchio M (2008) Polyoxometalate embedding of a tetraruthenium(IV)-oxo-core by template-directed metalation of $[\gamma\text{-SiW}_{10}O_{36}]8-$: A totally inorganic oxygen-evolving catalyst. J. Am. Chem. Soc. 130:5006-5007

Sproviero EM, Gascon JA, McEvoy JP, Brudvig GW, Batista VS (2008) Quantum mechanics/molecular mechanics study of the catalytic cycle of water splitting in photosystem II. J. Am. Chem. Soc. 130:3428-3442

Steinmiller EMP, Choi K-S (2009) Photochemical deposition of cobalt-based oxygen evolving catalyst on a semiconductor photoanode for solar oxygen production. Proc. Natl. Acad. Sci. USA 106:20633-20636

Tagore R, Chen H, Crabtree RH, Brudvig GW (2006) Determination of mu-oxo exchange rates in di-mu-oxodimanganese complexes by electrospray ionization mass spectrometry. J. Am. Chem. Soc. 128:9457-9465

Tagore R, Chen H, Zhang H, Crabtree RH, Brudvig GW (2007) Homogeneous water oxidation by a di-μ-oxodimanganese complex in the presence of Ce4+. Inorg. Chim. Acta 360:2983-2989

Tagore R, Crabtree RH, Brudvig GW (2007) Distinct mechanisms of bridging-oxo exchange in di-μ-O dimanganese complexes with and without water-binding sites: Implications for water binding in the O_2-evolving complex of photosystem II. Inorg. Chem. 46:2193-2203

Tagore R, Crabtree RH, Brudvig GW (2008) Oxygen evolution catalysis by a dimanganese complex and its relation to photosynthetic water oxidation. Inorg. Chem. 47:1815-1823

Umena Y, Kawakami K, Shen JR, Kamiya N (2011) Crystal structure of oxygen-evolving photosystem II at a resolution of 1.9 Å. Nature 473: 55-61

Wei Z, Cady C, Brudvig GW, Hou HJM (2011) Photodamage of a Mn(III/IV)-oxo mix valence compound and photosystem II complexes: Evidence that high-valent manganese species is responsible for UV-induced photodamage of oxygen evolving compelx in photosystem II. J. Photochem. Photobiol. B,104:118-125

Woodhouse M, Parkinson BA (2008) Combinatorial approaches for the identification and optimization of oxide semiconductors for efficient solar photoelectrolysis. Chem. Soc. Rev. 38:197-210

Yang X, Wolcott A, Wang G, Sobo A, Fitzmorris RC, Qian F, Zhang JZ, Li Y (2009) Nitrogen-doped ZnOnanowire arrays for photoelectrochemical water splitting. NanoLett. 9:2331-2336

Yano J, Kern J, Sauer K, Latimer MJ, Pushkar Y, Biesiadka J, Loll B, Saenger W, Messinger J, Zouni A, Yachandra VK (2006) Where water is oxidized to dioxygen: structure of the photosynthetic Mn_4Ca cluster. Science 314:821-825

Yin Q, Tan JM, Besson C, Geletii YV, Musaev DG, Kuznetsov AE, Luo Z, Hardcastle KI, Hill CL (2010) A fast soluble carbon-free molecular water oxidation catalyst based on abundant metals. Science 328:342-345

Youngblood WJ, Lee S-HA, Maeda K, Mallouk TE (2009) Visible light water splitting using dye-sensitized oxide semiconductors. Acc. Chem. Res. 42:1966-1973

Zhang F, Cady CW, Brudvig Gary W, Hou HJM (2011) Thermal Stability of $[Mn(III)(O)_2Mn(IV)(H_2O)_2(Terpy)_2](NO_3)_3$ (Terpy = 2,2':6',2"-terpyridine) in aqueous solution. Inorg. Chim. Acta 366: 128-133

Zhong DK, Gamelin DR (2010) Photoelectrochemical water oxidation by cobalt catalyst ("Co-Pi")/α-Fe_2O_3 composite photoanodes: Oxygen evolution and resolution of a kinetic bottleneck. J. Am. Chem. Soc. 132:4202-4207

Zhong DK, Sun J, Inumaru H, Gamelin DR (2009) Solar water oxidation by composite catalyst/α-Fe_2O_3photoanodes. J. Am. Chem. Soc. 131:6086-6087

In: Solar System: Structure, Formation and Exploration
Editor: Matteo de Rossi
ISBN: 978-1-62100-057-0

Chapter 4

A Relativistic Positioning System Exploiting Pulsating Sources for Navigation Across the Solar System And Beyond

Emiliano Capolongo, Matteo Luca Ruggiero** and Angelo Tartaglia
DIFIS Politecnico di Torino, corso Duca degli Abruzzi 24, 10129 Torino, Italy
INFN, Sezione di Torino, Via Pietro Giuria 1, 10125 Torino, Italy

Abstract

We introduce an operational approach to the use of pulsating sources, located at spatial infinity, for defining a relativistic positioning and navigation system, based on the use of null four-vectors in a flat Minkowskian spacetime. We describe our approach and discuss the validity of it and of the other approximations we have considered in actual physical situations. As a prototypical case, we show how pulsars can be used to define such a positioning system: the reception of the pulses for a set of different sources whose positions in the sky and periods are assumed to be known allows the determination of the user's coordinates and spacetime trajectory, in the reference frame where the sources are at rest. In order to confirm the viability of the method, we consider an application example reconstructing the world-line of an idealized Earth in the reference frame of distant pulsars: in particular we have simulated the arrival times of the signals from four pulsars at the location of the Parkes radiotelescope in Australia. After pointing out the simplifications we have made, we discuss the accuracy of the method. Eventually, we suggest that the method could actually be used for navigation across the Solar System and be based on artificial sources, rather than pulsars.

Keywords: relativistic positioning system, astrometry and reference systems, null frames.

1. Introduction

The evocative and romantic image of an ancient carrack with a scanty crew and little else onboard, as it approaches to cross an entire ocean; this can indeed evoke, beyond the greatness of the effort, the sense of confusion and smallness you feel when venturing into the

*E-mail address: matteo.ruggiero@polito.it

exploration of the unknown. All along, man has been eager for knowledge and discovery, and all along he has tried hard to elaborate methods to design devices suited to facilitate his life and direct his steps. Imagine, for a moment, how vital it would be to know, in a hypothetical space travel, where you are and be able to locate yourself in a unique way in space and time. In this case, lost in the deepness of space, even more than on a fragile ship at the mercy of the waves, you would more dramatically feel the need for a positioning system that could guide you and light your way.

Actually, the problem of positioning on the Earth is today solved by systems like GPS and GLONASS (Ashby, 2003; Pascual-Sánchez, 2007); from a physical viewpoint, in these systems positioning takes place in a classical (Euclidean) space and absolute time, over which relativistic corrections are introduced (Ashby, 2003). Even in the space surrounding the Earth up to Low Earth Orbit (LEO) altitude, a complete navigation solution can be provided by the current GPS or, with better performance, by his Differential version DGPS (Parkinson et al., 1996; Kaplan et al., 2006).

For interplanetary and deep space missions, current navigation methods highly rely on Earth ground-based extensive operations for absolute/relative position determination (Emadzadeh et al., 2011; Jordan, 1987; Melbourne, 1976). Different approaches are used to this aim such as radar ranging, optical tracking, and planet imaging (Wertz, 1978; Bate et al., 1971; Battin, 1999).

An important advantage of the radar ground-based system consists in the absence of active hardware on vehicles. Yet,one must also consider the many drawbacks this system may experience: high costs in terms of ground operations; problematic reception of data corrupted by a strong, sometime critical, background noise; need for precise positional information about radar observation stations on the Earth and Solar System objects (Jordan, 1987); position estimation error increasing directly proportional with respect to the distance from the Earth, with a degrading factor given by the angular accuracy. Even if active transmitters were used on the space vehicles to send signals back to the Earth (Jordan, 1987), it would be possible to estimate the radial velocity by measuring the Doppler frequency of this received signal would obtain only some little improvements, provided that most often errors always increase with distance; early experiments using these tracking systems on the Viking spacecraft showed accuracies up to about 50 km in position estimation error for missions to Mars and positional accuracies of the order of 100 km for the outer planets (Emadzadeh et al., 2011; Melbourne, 1976).

The optical tracking system is very similar to the basic principles of the radar tracking, except that it utilizes visible light reflected from a spacecraft (Bate et al., 1971). In some environmental conditions, where optical measurements are favored, this can be an appealing chance; the vehicle's position is calculated by the comparison of a taken photograph with respect to a fixed star background and, therefore, real-time measurements using such methods are typically not easily achieved.

For planetary observation missions in the neighborhood of the examined planet, it is possible to obtain the positioning of the vehicle relative to the planet itself by comparing video images of the planet, taken on board, to the known planetary parameters such as diameter and position relative to the other celestial objects (Battin, 1999).

In order to increase the performance of the navigation system, a combination of these techniques can be employed, still requiring tight human interaction and a not so easy interpretation of the data. Furthermore, as radar-ranging errors increase as the vehicle distance from the Earth increases, accurate navigation becomes more complex because of the required necessity to refine the pointing accuracy of ground antennas. Moreover, the imaging process on vehicles, sufficiently close to the planets, implies complex and expensive on board systems.

Most of actual spacecrafts employ the NASA Deep Space Network (DSN, 2010). This system allows to reach accurate radial position but, as mentioned before, the angular uncertainty increases with distance, leading to low performance in long range navigation. Position accuracies in the order of 10 km per astronomical unit of distance from Earth are achievable using interferometric measurements of the Very Long Baseline Interferometer (VLBI) through the DSN (Emadzadeh et al., 2011).

The requirement of higher accuracy, as well as of reduction of costs associated with the ground system, and also the possibility to augment and improve the current available navigation systems push toward different and alternative methods and more autonomous solutions (Folta et al., 1999; Gounley et al., 1984) without human assistance and communication with the Earth. The ultimate goal would be a complete, accurate, absolute, autonomous navigation solution working throughout the Solar System; and, eventually, at galactic/intergalactic scales.

In this chapter we discuss how it is possible to define a relativistic positioning system, effective for the navigation in the Solar System and beyond, by using electromagnetic signals coming from pulsating sources, located at spatial infinity. Actually, a *fully relativistic* positioning system can be built by exploiting the worldlines of electromagnetic signals to define the so called *emission coordinates* (Coll, 2006; Coll et al., 2006a,b; Rovelli, 2002; Blagojevic et al., 2002; Ruggiero & Tartaglia, 2007; Bini et al., 2008). The simplest way of understanding what emission coordinates are is to consider four emitting clocks, in motion in spacetime, broadcasting their proper times: the intersection of the past lightcone of an event with the worldlines of the emitting clocks corresponds to the proper times of emission along the worldlines of the emitters; these proper times are the emission coordinates of the given event. A positioning system based on the use of emission coordinates can be effective for positioning on the Earth and, in this case, one may think of a set of satellites orbiting around the Earth and equipped with onboard clocks broadcasting their proper time. However, for positioning and navigation in the Solar System, a set of *pulsars* (or, more generally, suitable pulsating sources) could rather be used (Coll & Tarantola, 2004, 2009). In fact, known pulsars emit their signals at a highly regular rate (this is the case, in particular, of the *millisecond-pulsars*, see e.g. Kramer et al. (2004)), which makes them natural beacons for building a relativistic positioning system. What can be measured with great accuracy is the arrival time of the N-th pulse, so that counting these pulses can in principle allow to define something similar to the emission coordinates, even though in this case the actual proper emission time is unknown and inaccessible.

The idea of using pulsars as stellar beacons has been considered since the early years of the discovery of pulsars (Downs, 1974), and also more recently (Sala et al., 2004; Sheikh et al., 2006, 2007): indeed, for these approaches, positioning is not autonomous, but can be referred to a reference frame centered at the Solar System barycenter (SSB), because pulses arrival times or their phases have to be related to their expected values in the SSB.

What we want to show here is that it is possible to operationally define an autonomous positioning system, by building emission coordinates exploiting electromagnetic signals coming from periodic sources (assumed to be at rest at spatial infinity) such as pulsars (Tartaglia, 2009; Tartaglia et al., 2011; Ruggiero et al., 2011). In other words, we will show that, by counting pulses from a set of different (from at least four) sources, whose positions in the sky and periods are assumed to be known; and, then measuring the proper time intervals between successive arrivals of the signals. It is possible to then determine, in the reference frame, where the sources are at rest; the user's coordinates and spacetime trajectory within an accuracy controlled by the precision of the user clock. In doing so, the phases of the received pulses play the role of emission coordinates and we assume that the user worldline is a straight line during a proper time interval corresponding to the reception of a limited number of pulses, which means that the effects of the acceleration are negligibly small. This system can allow the autopositioning with respect to an arbitrary event in spacetime and three directions in space, so that it could be used for space navigation and positioning in the Solar System and even beyond. In practice, the initial event of the self-positioning process is used as the origin of the reference frame, and the axes are oriented according to the positions of the distant sources; all subsequent positions will be given in that precise frame. If one wants to further position the whole section of worldline of the receiver in some other external reference frame the location of the initial event in the external frame has to be known by other means. Our approach is based on the use of null frames in a flat Minkowskian spacetime, but we discuss the validity of this and other approximations we have considered for actual physical situations.

Moreover, we focus on a simple application of our method, in which we imagine that our sources are four millisecond pulsars and, simulating the arrival times of their signals, we show how the worldline of the receiver is reconstructed: in particular, we make use of the TEMPO2 software (Hobbs et al., 2006), a pulsar-timing package that simulates the pulse time arrival at a given location on the Earth, and we determine the trajectory of that location in spacetime, due to the combined motion of the Earth around the Sun and its daily rotation.

The chapter is organized as follows: in Section 2. we show how to build the basic reference frame and describe the localization procedure; in Section 3. we discuss the sources of error; in Section 4. we present an application example to test our method, by using simulated data. Eventually, discussion and conclusion are in Section 5.

2. Definition of the Basic Null Frame Grid and corresponding Localization

In this Section, we first focus on the theoretical framework which enables to define the basic (null) reference frame where positioning takes place; and, then, we show how the localization procedure can be operationally implemented in this frame.

2.1. The Basic Null Frame

Let us consider a number of sources of periodic electromagnetic signals, at rest at spatial infinity, in a four-dimensional Minkowski spacetime. For our purposes, at least four sources are needed. Each of these sources is characterized by the frequency of its periodic signals and by their directions in space; since the sources are supposed to be far away (i.e., at spatial infinity), their signals can be seen as corresponding to plane waves. In the inertial frame where the sources are at rest, once Cartesian coordinates are chosen, we associate to each source a null four-vector[1] $\boldsymbol{f}$ whose Cartesian contravariant components are given by

$$f^{\mu} \doteq \frac{1}{cT}(1, \vec{\mathbf{n}}), \tag{1}$$

T being the (proper) signal period, and $\vec{\mathbf{n}}$ the unit vector describing the direction of propagation in the given frame. If in the same reference frame we consider the position four-vector

$$\boldsymbol{r} \doteq (ct, \vec{\mathbf{x}}), \tag{2}$$

with respect to an arbitrary and yet unspecified origin, then we can define the scalar function X at the spacetime event identified by $\boldsymbol{r}$, the position four-vector

$$X(\boldsymbol{r}) \doteq \boldsymbol{f} \cdot \boldsymbol{r}, \tag{3}$$

where dot stands for Minkowski scalar product. The scalar X might be thought of as the phase difference of the wave described by $\boldsymbol{f}$ with respect to its value at the origin of the coordinates. Four linearly independent four-vectors constitute a basis, or a frame: we may think of choosing four null four-vectors to serve as a basis (see e.g., Blagojevic et al. (2002)), so that the four wave four-vectors $\{\boldsymbol{f}_{(a)}, \boldsymbol{f}_{(b)}, \boldsymbol{f}_{(c)}, \boldsymbol{f}_{(d)}\}$ in the form (1) constitute our *null frame*, or *null tetrad*. Then, according to the general approach to coordinate systems and frames developed by Coll et al. (2009) in connection with positioning systems, the four phase differences

$$X_{(N)} \doteq \boldsymbol{f}_{(N)} \cdot \boldsymbol{r}, \quad N = a, b, c, d \tag{4}$$

obtained at any event $\boldsymbol{r}$ whose coordinates are defined by (2), with a, b, c, d labeling the sources, are *null coordinates*: in other words, they are spacetime functions with null spacetime gradient; and, hence, they define a *null coordinate system*. Furthermore, the $\{X_{(N)}\}$ are *emission coordinates*, since they are physically related to the reception of electromagnetic signals emitted by the sources.

The tetrad formalism (see e.g. Chandrasekhar (1983)) allows to define the symmetric matrix

$$\eta_{(M)(N)} = \boldsymbol{f}_{(M)} \cdot \boldsymbol{f}_{(N)}, \tag{5}$$

which, in this case, has constant components, and whose inverse is determined by the relation

$$\eta_{(M)(P)}\eta^{(P)(N)} = \delta^{(N)}_{(M)}. \tag{6}$$

Tetrad indices $N = a, b, c, d$ are lowered and raised by means of the matrices $\eta_{(M)(P)}$ and $\eta^{(M)(P)}$. We can write the position four-vector $\boldsymbol{r}$ in the form

$$\boldsymbol{r} = X^{(N)}\boldsymbol{f}_{(N)} = X_{(N)}\boldsymbol{f}^{(N)}, \tag{7}$$

[1] Arrowed boldface letters like $\vec{\mathbf{x}}$ refer to spatial vectors, while boldface letters like refer to four-vectors; Greek indices refer to spacetime components, while Latin letters label the sources.

and, as a consequence, we see that the phase differences $X_{(N)}$ are the components of the position four-vector with respect to the vectors

$$\boldsymbol{f}^{(N)} = \eta^{(N)(M)} \boldsymbol{f}_{(M)}, \tag{8}$$

or, differently speaking, the functions

$$X^{(N)} = \eta^{(N)(M)} X_{(M)} \tag{9}$$

are the components of the position four-vector with respect to the null tetrad vectors $\boldsymbol{f}_{(N)}$. It is useful to remark that while the frame $\{\boldsymbol{f}_{(a)}, \boldsymbol{f}_{(b)}, \boldsymbol{f}_{(c)}, \boldsymbol{f}_{(d)}\}$ is constituted by null vectors, the frame $\{\boldsymbol{f}^{(a)}, \boldsymbol{f}^{(b)}, \boldsymbol{f}^{(c)}, \boldsymbol{f}^{(d)}\}$ is constituted by space-like vectors.

In summary, if we consider the hyperplanes conjugated to the null frame $\{\boldsymbol{f}_{(a)}, \boldsymbol{f}_{(b)}, \boldsymbol{f}_{(c)}, \boldsymbol{f}_{(d)}\}$ vectors, we are able to define a spacetime grid (see Tartaglia (2009)), in which each event is identified by the relative phase of the electromagnetic signals with respect to an arbitrary origin; and, in this frame, the coordinates of each event are given by the functions $\{X^{(N)}\}$; equivalently, the phases $\{X_{(N)}\}$ are the coordinates with respect to the space-like frame $\{\boldsymbol{f}^{(a)}, \boldsymbol{f}^{(b)}, \boldsymbol{f}^{(c)}, \boldsymbol{f}^{(d)}\}$.

2.2. Localization within the Grid

After having shown how to build a grid, we want to focus on how localization can be achieved within the grid. In particular, we suppose to deal with periodic signals in the form of electromagnetic pulses, such as those coming from pulsars, and that these signals can be thought of as plane waves locally. Furthermore, we suppose that the user is equipped with a receiver able to recognize and count the pulses coming from the various sources, and a clock, that can be used to measure the proper time span between the arrivals from each source.

Let us start with a toy model, where the emission from the sources is continuous and the phases of any signal can be determined with an arbitrary precision, at any event. We choose a starting event, from which the phases of each signal are measured, and is the origin of our coordinates (in other words, the event with $\boldsymbol{r} = \boldsymbol{0}$, according to what we have described earlier), and three directions in space, defining the Cartesian axes of the inertial frame of the sources. We point out that even though the starting event is arbitrary, in order to correctly define the null frame, the directions of the sources in the sky have to be known: in other words, we have to know the unit vectors $\vec{\mathbf{n}}$ for each source (and their proper frequency ν too), which also enable us to calculate the matrices $\eta_{(N)(M)}, \eta^{(N)(M)}$ of the given frame.

To a subsequent event $\boldsymbol{r}$, we associate the measured phases

$$X_{(N)} = \boldsymbol{f}_{(N)} \cdot \boldsymbol{r}, \tag{10}$$

and, according to eq. (7), it is then possible to obtain the coordinates of the event $\boldsymbol{r}$, in terms of the measured phases:

$$\boldsymbol{r} = X_{(a)} \boldsymbol{f}^{(a)} + X_{(b)} \boldsymbol{f}^{(b)} + X_{(c)} \boldsymbol{f}^{(c)} + X_{(d)} \boldsymbol{f}^{(d)}. \tag{11}$$

and to reconstruct the user's worldline.

Coming to a more realistic situation, such as the one in which the emitters are pulsars, we should consider that the received signals consist in a series of pulses and are not continuous. In this case, we may proceed as follows. First, we call "reception" the event corresponding to the arrival of a pulse from one of the sources. As a consequence, the position in spacetime of an arbitrary reception event can be written in the form

$$\boldsymbol{r} = X_{(a)}\boldsymbol{f}^{(a)} + X_{(b)}\boldsymbol{f}^{(b)} + X_{(c)}\boldsymbol{f}^{(c)} + X_{(d)}\boldsymbol{f}^{(d)}, \tag{12}$$

with

$$X_{(a)} = n_{(a)} + p, \tag{13}$$
$$X_{(b)} = n_{(b)} + q, \tag{14}$$
$$X_{(c)} = n_{(c)} + s, \tag{15}$$
$$X_{(d)} = n_{(d)} + w. \tag{16}$$

The $X_{(N)}$'s in the case of a continuous signal would be phases. Here, they are given by an integer $n_{(N)}$, numbering the order of the successive pulses from a given source, and a fractional value: e.g., p means a fractional value of the cycle in $X^{(a)}$, and the same thing holds for q, s, w, where $0 < p, q, s, w$. In Eqs. (13)-(16), only one of the p, q, s, w will in general be zero: when, for instance, a pulse from source "a"arrives, p will be zero, while q, s, w will not; when a pulse from source "b"comes, q will be zero and the other three fractions will not; and so on. Once we choose an arbitrary origin, we may count the pulses in order to measure the $n_{(N)}$, but we have no direct means to measure the fractional values p, q, s, w. By the way, a procedure to determine these values can be obtained, based on geometric considerations: we suppose that the acceleration of the user is small during a limited series of reception events, so that we may identify the user's worldline with a straight line. Furthermore, we also suppose that by means of his own clock the user can measure the proper time interval τ_{ij} between the i-th and j-th arrivals. By these assumptions, we can proceed as follows to determine the fractional values p, q, s, w. Let us consider two sequences[2] For arrival times from the sources; we have eight events, each of them in the form

$$\boldsymbol{r}_j = X_{(a)j}\boldsymbol{f}^{(a)} + X_{(b)j}\boldsymbol{f}^{(b)} + X_{(c)j}\boldsymbol{f}^{(c)} + X_{(d)j}\boldsymbol{f}^{(d)}, j = 1, .., 8, \tag{17}$$

where $X_{(N)j}$ are expressions like (13–16). The events are arranged in such a way that $\boldsymbol{r}_1$ is the generic arrival of the signal from pulsar "a", $\boldsymbol{r}_2$ is the arrival of the first signal of pulsar "b" after $\boldsymbol{r}_1$, $\boldsymbol{r}_3$ is the arrival of the first signal of pulsar "c" after $\boldsymbol{r}_1$, and $\boldsymbol{r}_4$ is the arrival of the first signal of pulsar "d" after $\boldsymbol{r}_1$ (the pulsars are ordered from largest ("a") to shortest ("d") period); $\boldsymbol{r}_5$ is the arrival of the next signal from pulsar "a", and so on. The flatness hypothesis allows us to write the displacement four-vector between two reception events in the form

$$\boldsymbol{r}_{ij} \doteq \boldsymbol{r}_i - \boldsymbol{r}_j = \left(X_{(N)i} - X_{(N)j}\right)\boldsymbol{f}^{(N)} \doteq \Delta X_{(N)ij}\boldsymbol{f}^{(N)}. \tag{18}$$

Indeed, the assumption that the worldline of the receiver is straight during a limited number of pitches of the signals can be used also to provide further information. In fact, let us

[2]They may be subsequent or not, provided the total time span does not spoil the hypothesis of linearity of the worldline.

consider three successive reception events i, j, k; we have

$$\boldsymbol{r}_{ji} = \Delta X_{(N)ji}\boldsymbol{f}^{(N)}, \qquad \boldsymbol{r}_{kj} = \Delta X_{(N)kj}\boldsymbol{f}^{(N)}. \tag{19}$$

The straight line hypothesis allows us to write

$$\frac{\tau_{ji}}{\tau_{kj}} = \frac{\Delta X_{(a)ji}}{\Delta X_{(a)kj}} = \frac{\Delta X_{(b)ji}}{\Delta X_{(b)kj}} = \frac{\Delta X_{(c)ji}}{\Delta X_{(c)kj}} = \frac{\Delta X_{(d)ji}}{\Delta X_{(d)kj}}, \tag{20}$$

where τ_{ji}, τ_{kj} are the proper times elapsed between the i-th and j-th, and j-th and k-th reception events, respectively. These relations enable us to obtain the values we are interested in: in fact, we may arrange the coefficients of eqs. (17) in an 8×4 matrix (8 events, 4 sources):

$$X_{(N)i} = \begin{pmatrix} n_1^{(a)} & n_1^{(b)} + q_1 & n_1^{(c)} + s_1 & n_1^{(d)} + w_1 \\ n_2^{(a)} + p_2 & n_2^{(b)} & n_2^{(c)} + s_2 & n_2^{(d)} + w_2 \\ n_3^{(a)} + p_3 & n_3^{(b)} + q_3 & n_3^{(c)} & n_3^{(d)} + w_3 \\ n_4^{(a)} + p_4 & n_4^{(b)} + q_4 & n_4^{(c)} + s_4 & n_4^{(d)} \\ n_5^{(a)} & n_5^{(b)} + q_5 & n_5^{(c)} + s_5 & n_5^{(d)} + w_5 \\ n_6^{(a)} + p_6 & n_6^{(b)} & n_6^{(c)} + s_6 & n_6^{(d)} + w_6 \\ n_7^{(a)} + p_7 & n_7^{(b)} + q_7 & n_7^{(c)} & n_7^{(d)} + w_7 \\ n_8^{(a)} + p_8 & n_8^{(b)} + q_8 & n_8^{(c)} + s_8 & n_8^{(d)} \end{pmatrix} \tag{21}$$

As it can be seen, the p, q, s, w are zero along the main diagonals of the upper and lower square half matrices forming the whole matrix. This happens in correspondence to the arrivals of the pulses from the various sources: on the arrival of a pulse from "a" the corresponding p is zero, from "b" q is zero, and so on. Then, on using relations like (20) we obtain the fractional values in terms of observed quantities, i.e., proper time intervals measured by the observer. For instance, we have

$$p_1 = 0, \quad q_1 = n_2^{(b)} - n_1^{(b)} - \left(n_6^{(b)} - n_2^{(b)}\right)\frac{\tau_{21}}{\tau_{62}}, \tag{22}$$

$$s_1 = n_3^{(c)} - n_1^{(c)} - \left(n_7^{(c)} - n_3^{(c)}\right)\frac{\tau_{31}}{\tau_{73}}, \tag{23}$$

$$w_1 = n_4^{(d)} - n_1^{(d)} - \left(n_8^{(d)} - n_4^{(d)}\right)\frac{\tau_{41}}{\tau_{84}}, \tag{24}$$

$$p_2 = n_1^{(a)} - n_2^{(a)} + \left(n_5^{(a)} - n_1^{(a)}\right)\frac{\tau_{21}}{\tau_{51}}, \quad q_2 = 0, \tag{25}$$

$$s_2 = n_3^{(c)} - n_2^{(c)} + \left(n_7^{(c)} - n_3^{(c)}\right)\frac{\tau_{23}}{\tau_{73}}, \tag{26}$$

$$w_2 = n_4^{(d)} - n_2^{(d)} + \left(n_8^{(d)} - n_4^{(d)}\right)\frac{\tau_{24}}{\tau_{84}}, \tag{27}$$

$$p_3 = n_1^{(a)} - n_3^{(a)} + \left(n_5^{(a)} - n_1^{(a)}\right)\frac{\tau_{31}}{\tau_{51}}, \tag{28}$$

$$q_3 = n_2^{(b)} - n_3^{(b)} - \left(n_6^{(b)} - n_2^{(b)}\right)\frac{\tau_{23}}{\tau_{62}}, \tag{29}$$

$$s_3 = 0, \quad w_3 = n_4^{(d)} - n_3^{(d)} + \left(n_8^{(d)} - n_4^{(d)}\right)\frac{\tau_{34}}{\tau_{84}}, \tag{30}$$

$$p_4 = n_1^{(a)} - n_4^{(a)} + \left(n_5^{(a)} - n_1^{(a)}\right) \frac{\tau_{41}}{\tau_{51}}, \tag{31}$$

$$q_4 = n_2^{(b)} - n_4^{(b)} - \left(n_6^{(b)} - n_2^{(b)}\right) \frac{\tau_{24}}{\tau_{62}}, \tag{32}$$

$$s_4 = n_3^{(c)} - n_4^{(c)} - \left(n_7^{(c)} - n_3^{(c)}\right) \frac{\tau_{34}}{\tau_{73}}, \quad w_4 = 0, \tag{33}$$

and so on. Moving the pair of sequences and repeating the elaboration step by step, we are able to reconstruct the whole worldline of the receiver, in terms of measured quantities, i.e., proper times.

3. Biases and Uncertainties

In this Section, the sources of error are discussed and their importance in the positioning process are presented. Before going into further details about our analysis, we would like to preliminarily discuss the nature of errors and inaccuracies that affect our positioning process, in order to clarify how they can be dealt with to improve accuracy.

Roughly speaking we can distinguish between *systematic errors* and *uncertainties* (or *fluctuations*). Systematic errors originate from mis-modeling of the physical processes we are dealing with or from poor knowledge of the system parameters; they globally affect the positioning process and are either constant or time dependent. For instance, errors in the angular positions for the sources in the sky, as well as in the periods of the pulses are systematic errors. Furthermore, if we deal with pulsars, their angular positions may change because of their proper motion and their periods because of energy loss. As for the uncertainties, they are related to the stochastic variations of the quantities involved in the positioning process: for instance in the procedure of measuring the arrival times of the pulses, fluctuations are due to the detection device and to the user's clock, as well as to the emission mechanism at the surface of the star. In principle, the turbulence of the interstellar plasma could also have a role. Systematic errors produce global consequences with an unknown distortion of the re-built spacetime trajectory of the user; the uncertainties transform the worldline in an uncertainty stripe across spacetime. Systematics can be reduced by improving our model and the knowledge of its parameters; a statistical analysis and the best technologies can help to reduce the impact of random disturbances.

3.1. Model Limitations

Let us begin by discussing the validity of our model, which is based on the propagation of electromagnetic signals originating from pulsating sources at rest in a given reference frame, in Minkowski spacetime. The question we would like to address is: how realistic is this model? Can it be used for positioning in actual physical situations? We start with some general observations.

We explicitly work in flat spacetime, thus eliminating the effects of the gravitational field. It is however obvious that, for positioning in the Solar System, the gravitational field of the Sun (and of the other major bodies) influences the propagation of electromagnetic signals[3]. The dimensionless magnitude of the static gravitational field of the Sun is of the order of $\delta_{\odot} \simeq \frac{GM_{\odot}}{c^2 d} \simeq 10^{-8} \left(\frac{1 \text{ A.U.}}{d}\right)$, and reaches its maximum value near the Sun[4], where $\delta_{\odot} \simeq 10^{-6}$. This field produces effects on the times of arrival of the pulses which are relevant for our purposes only if they change in a time comparable with the integration times used for our algorithm; only the radial component (with respect to the Sun) is important. The effects due to the motion of a user who travels with speed $\vec{v}$ in the radial direction $\hat{r}$ (and thus experiences a time varying gravitational field) in a time span δt are expressed in terms of apparent fractional change of the period of the sources[5] as $\delta_{\odot,v} \simeq \frac{GM_{\odot}}{c^2 d^2} \vec{v} \cdot \hat{r}\, \delta t \simeq 6 \times 10^{-8} \left(\frac{1 \text{ A.U.}}{d^2}\right) \left(\frac{v}{30 \text{ km/s}}\right) \text{y}^{-1} \delta t$. This fixes an upper limit to the acceptable δt in order the effect to be compatible with the required tolerance. One more problem is that when the line of sight to a source grazes the Sun the time of flight of the signals depends on the geometric curvature of the rays and on the Shapiro time delay, which is indeed huge depending on the apparent impact parameter (Straumann, 2004). However these disturbances can be dealt with either choosing sources which are located in the sky so that their lines of sight are far away from the gravitating bodies; or having recourse, which is appropriate for many other reasons also, to a redundant number of sources, so that only the best are used each time.

Another systematic error comes from considering the sources as being at rest, at spatial infinity, in a given reference frame, which is a very idealized situation. In fact, actual sources, such as galactic pulsars, have both a proper motion and a finite distance from the observer. Taking into account estimates of the proper motion of real pulsars (Hobbs et al., 2005) it is possible to see that the rate of change of the angular position [6] is of the order $10^{-6} \left(\frac{100 \text{ pc}}{d}\right)$ rad per year. In practice these figures tell us that we are allowed to keep the position nominally fixed for months before correcting the value of the direction cosines, which is possible because the behavior of the sources is known. As for the radial proper motion of the pulsar, which is commonly less known than its transverse motion; it is indeed contained in the arrival of the time pulses, but can reasonably be considered as fixed during very long times, to the extent of not affecting the positioning process. Finally, it is known that the periods of the pulses are generally increasing with time, due to rotation energy loss of the star. However, the decay rate is in general known, so that we may deal with this kind of change in the same way as we do for the position in the sky: keep the period fixed for a time compatible with the required accuracy (which would in any case be rather long), then introduce a corrected new value (which is known once the pulsar is given). As for glitches (sudden and random jumps in the frequency) they can be cured thanks to the redundancy of the number of sources above the minimum of four, corresponding to the dimensions of spacetime.

In the case of pulsars there are many other effects that could affect the arrival times of

---

[3]*En passant*, we notice that since in our procedure only ratios of proper times measured in practice at the same spatial position are involved, the effect of the gravitational field on the user's clock can be safely neglected.

[4]The distance d is here expressed in astronomic units (A.U.) so that a distance equal to 1 corresponds to the position of the Earth.

[5]The velocity is in units of the average speed of the Earth along its orbit (30 km/s) and the rate of change is per year (y^{-1}).

[6]Actually, proper motion affects any observed change in periodicity, see e.g., Straumann (2004), Damour & Taylor (1991), so that an estimate of the relative time variation of the period is of the order $10^{-10} \frac{100 \text{ pc}}{d} \text{y}^{-1} \delta t$.

the signal; a thorough analysis can be found in Kramer et al. (2004). Besides what we have already mentioned, we should, for instance, take into account the emission mechanism at the surface of the star and the propagation across the interstellar plasma: these effects are responsible for the fact that each single incoming pulse is in general different from any other. The method to deal with this is the same adopted at radiotelescopes: the acquisition of the data runs for a high number of pulses, then the signal is analyzed by folding, in order to extract the fiducial sequence with the desired accuracy. In the case of the positioning far shorter integration times are needed, because the pulsar is already known and the aim is simply to recognize it, rather than investigate on it. For completeness, we should finally mention the problems related to the acquisition and elaboration chain in the receivers. However, to discuss the technological aspects of the antenna and acquisition apparatus is beyond the scope of the present work.

Finally, we would like to focus on the approximation that is implicit in our method for the conversion of a sequence of times in the arrival of discrete pulses into the coordinates of the receiver. We exposed the method in Section 2.2., where we limited the extension of the time series of the signals we used at each step of the process to an interval allowing to locally approximate the user's worldline with a straight line: how far is this hypothesis tenable? Given the user's clock accuracy[7] $\delta\tau$, we can define the maximum proper time interval $\Delta\tau_{\max}$ that can be considered in order to be self-consistent with the straight line hypothesis. Developing the worldline of the receiver in powers of its proper time up to the second order we see that, if a is the order of magnitude of the user's acceleration, and v the velocity, the following condition should be satisfied:

$$\Delta\tau_{\max} = \sqrt{2\frac{v}{a}\delta\tau}. \tag{34}$$

For instance, if the user is moving in flat spacetime with $\delta\tau \simeq 10^{-10}$ s, $v = 5 \times 10^5$ m/s and an acceleration $a = 1$ m/s^2, we have $\Delta\tau_{\max} = 10^{-2}$ s, which corresponds to several periods of millisecond pulsars: enough for both averaging away fluctuations of the single pulses, and for the piecewise reconstruction of the worldline. Actually, the deviation from the linearity of the user's worldline can also be due to the curvature of spacetime, i.e., to the presence of the gravitational field. We can give a similar estimate of the corresponding maximum proper time interval $\Delta\tau_{\max}$ by setting $a = |\nabla\Phi|$ in (34) where Φ is the gravitational potential. For instance, for $a = 10^{-3}$ m/s^2, which is the order of magnitude of the gravitational field of the Sun at 1 A.U., we get for $v = 10^3$ m/s, $\Delta\tau_{\max} \simeq 10^{-2}$ s. So, we see that there are actual physical situations where the hypothesis of linearity holds, and the procedure that we have described is meaningful.

3.2. Errors in the procedure of position determination

After having discussed the systematic errors related to the physical model underlying our procedure of position determination, we would like to turn to the analysis of the errors in the

[7] Into this $\delta\tau$ we should actually include also the drifts due to the proper motion and the period decay of the pulsar, which we keep constant during one step of the process; the latter are however expectedly far smaller effects than those due to the acceleration of the receiver.

procedure itself which, as we show are connected both with our knowledge of the system parameters and with the measurement process.

Actually, the coordinates of an event $\boldsymbol{r}$ are determined by solving eq. (10). We notice that eq. (10) can be written in the form

$$A\overline{x} = \overline{y} \tag{35}$$

where[8]

$$A = \begin{pmatrix} f^0_{(a)} & -f^1_{(a)} & -f^2_{(a)} & -f^3_{(a)} \\ f^0_{(b)} & -f^1_{(b)} & -f^2_{(b)} & -f^3_{(b)} \\ f^0_{(c)} & -f^1_{(c)} & -f^2_{(c)} & -f^3_{(c)} \\ f^0_{(d)} & -f^1_{(d)} & -f^2_{(d)} & -f^3_{(d)} \end{pmatrix}, \quad \overline{x} = \begin{pmatrix} x^0 \\ x^1 \\ x^2 \\ x^3 \end{pmatrix}, \quad \overline{y} = \begin{pmatrix} X_{(a)} \\ X_{(b)} \\ X_{(c)} \\ X_{(d)} \end{pmatrix} \tag{36}$$

with $x^0 = ct$, $x^1 = x$, $x^2 = y$, $x^3 = z$.

Eq. (35) is a linear system where the unknown vector $\overline{x}$ is obtained in terms of the matrix A (which has to be non-singular) and the vector $\overline{y}$:

$$\overline{x} = A^{-1}\overline{y}. \tag{37}$$

The entries of matrix A are related to the signal periods $T_{(N)} = (cf^0_{(N)})^{-1}$ and the direction cosines $n^i_{(N)} = cT_{(N)}f^i_{(N)}$ defining the angular positions of the sources; the fact that the $\vec{\mathbf{n}}_{(N)}$'s, $N = a, b, c, d$ are different ensures that the matrix is not singular. As for the vector $\overline{y}$, its entries are the measured phases determined (at each event) by the procedure described in Section 2.2..

The entries of the matrix A are affected by systematic errors, while phase measurements have random errors. It is then possible to evaluate the maximum relative error in the solutions $\overline{x}$ of the linear system (37) by estimating the maximum errors on the entries of A, while a covariance analysis allows to quantify how random errors in phase measurements translate into errors in the determination of the coordinates of the spacetime event $\boldsymbol{r}$, on the basis of the geometric properties of matrix A.

In fact, after defining a suitable norm[9], the following relation holds (see e.g., Demidovic & Maron (1966))

$$\delta\overline{x} \leq k(A)^2 \delta A \tag{38}$$

where $\delta\overline{x} = \frac{\|\Delta\overline{x}\|}{\|\overline{x}\|}$, $\delta A = \frac{\|\Delta A\|}{\|A\|}$ and $k(A) = \|A\|\|A^{-1}\|$ is the condition number of the system (35). If we suppose that the relative errors in periods $\Delta T/T$, and direction cosines Δn^i are roughly the same for all sources in all directions, then Eq. (38) allows to conservatively estimate the relative error in the form

$$\delta\overline{x} \leq k(A)^2 \left[\sqrt{\left(\frac{\Delta T}{T}\right)^2 + \frac{3}{2}(\Delta n)^2} \right] \tag{39}$$

[8]We use $\overline{x}$ to refer to generic vectors belonging to ^{4}, not to be confused with the four-vectors.

[9]For instance: $\|A\| = \overline{\ _{i,j} |a_{ij}|^2}$, $\|\overline{x}\| = \overline{\ _i |x_i|^2}$.

We remember that it is always $k(A) \geq 1$; in our case, we may write

$$k(A) \propto \frac{1}{|\det(A)|^2} \sum_{N=a,b,c,d}^{4} \frac{1}{T_{(N)}^2 c^2} \tag{40}$$

As a consequence, we see that the relative error is minimized when: (i) the determinant is maximized and (ii) periods are minimized; since the spatial components of A are the direction cosines, the determinant is maximized when the volume spanned by these directions is maximized. To fix ideas, for $\Delta T/T \simeq 10^{-4}$, $\Delta n \simeq 10^{-8}$, it is $\delta x \lesssim k(A)^2 10^{-4}$.

Further insight on the accuracy achievable by our positioning procedure can be obtained by discussing the geometric properties of the system (37), which ultimately depends on the actual angular positions of the sources in the sky. We refer to the Geometric Dilution Of Precision (GDOP), which is based on the covariance matrix of the errors in position determination, and provides a measure of how fit the set of sources is: indeed, it is related to the geometric properties of the matrix A and a great accuracy can be obtained if sources are chosen that are sufficiently scattered in the sky (this is pretty much like what happens with GPS satellites, see e.g., Hofmann-Wellenhof et al. (2001)). The covariance matrix of the positioning errors, as determined by the phase measurements, can be expressed as

$$\text{cov}\,\overline{x} = \left[A^T \left(\text{cov}\,\overline{y}\right) A\right]^{-1} \tag{41}$$

If the phase measurements errors from each source are uncorrelated and Gaussian distributed, with zero mean and variances $\sigma_{(N)}^2$, the position covariance matrix turns out to be

$$\text{cov}\,\overline{x} = \left[A^T \text{diag}(\sigma_{(a)}^2, \sigma_{(b)}^2, \sigma_{(c)}^2, \sigma_{(d)}^2) A\right]^{-1} \tag{42}$$

The GDOP is defined by

$$\text{GDOP} = \sqrt{\text{Tr}(\text{cov}\,\overline{x})} \tag{43}$$

To understand the meaning of GDOP, let us consider the case when all phase measurements have the same variance σ^2, then $\text{cov}\,\overline{y} = \text{diag}(\sigma^2, \sigma^2, \sigma^2, \sigma^2)$; furthermore it is $(A^T A)^{-1} = \frac{1}{|\det A|^2} \left(\text{cof} A\right)^T \left(\text{cof} A\right)$ so that we can write

$$\text{GDOP} = \frac{\sigma}{|\det A|} \sqrt{\text{Tr}\left(\text{cofA}\right)^{\text{T}} \left(\text{cofA}\right)} \tag{44}$$

Again, we see the role of $|\det(A)|$ in minimizing the errors: GDOP is minimized when the solid angle spanned by the directions of the sources is maximized.

The error in position determination can be minimized by using a number n of sources greater than four, which will result in an over-determined linear system in the form (35), $B\overline{x} = \overline{y}$, provided that now B is an $n \times 4$ matrix and $\overline{y}$ is a vector of n measured phases. Redundant equations allow to improve the accuracy of the solutions via least squares estimation techniques (see e.g., Stark & Woods (1986)), so that Eq. (41) is generalized to [10]

$$\text{cov}\,\overline{x} = (B^T B)^{-1} B^T \left(\text{cov}\,\overline{y}\right) B (B^T B)^{-1} \tag{45}$$

[10] Notice that if $n = 4$ it reduces to Eq. (41).

If, as before, we consider the phase measurements errors to be identically Gaussian distributed with the same variance and independent, we obtain equation (41) back, also in the case $n > 4$: in summary, we may say that the components of $(B^T B)^{-1}$ determine how phase errors translate into errors of the computed spacetime position.

4. Numerical Simulation

In order to test the procedure that we have previously described, it is necessary to define the null frame, that is to say the basis four-vectors in the form (1) for each source. In other words, we need to know the positions of the sources and their periods. Then, in order to apply the procedure, we need the arrival times of the pulses, as measured by the receiver. Our purpose is to demonstrate how the system works in practice, but we have no actual device at hands so we may follow two strategies: a) simulate the sources, giving them an arbitrary position in the sky and an arbitrary periodicity, then somehow mimicking the uncertainties associated with real sources; b) choose, as an example, four real millisecond pulsars with the data we find in the literature. In practice the difference between the two approaches is not really important, since the second choice is only nominally different from the first, so we decided to use the parameters of four real pulsars as they are listed in Table 1.

Next, we use a software which simulates the arrival times of the pulses received at a given terrestrial location, emitted by our set of pulsars; by this way we try and reconstruct the worldline of the receiver at the chosen position on Earth. The simulator generates sequences of arrival times as they would be obtained at an antenna and we use them, applying our algorithm in order to rebuild the motion of the Earth or, more correctly, the trajectory of a terrestrial location where the pulses would be received, which moves because of the daily rotation and the motion of the Earth around the Sun: this trajectory is then compared to the one obtained by the ephemerides. More precisely we reconstruct the motion of the receiver with respect to the "fixed" stars, assuming as the origin the event where the reception has started: in a sense we produce a self-positioning framework. The position of the initial event with respect to any given reference frame must be known by other means.

Pulsar	T (ms)	Elong (°)	Elat (°)
J1730-2304	8.123	263.19	0.19
J2322+2057	4.808	0.14	22.88
B0021-72N	3.054	311.27	−62.35
B1937+21	1.558	301.97	42.30

Table 1. The parameters of the four pulsars we chose are listed as they were taken from the ATNF Pulsar Catalogue. The basis four-vectors are obtained after computing the direction cosines from the ecliptic coordinates; then use is made of the formula $\boldsymbol{f}_{(N)} = \frac{1}{cT_{(N)}}(1, \vec{\mathbf{n}}_{(N)})$, for $N = a,\ b,\ c,\ d$. Both the periods and the direction cosines are assumed to be known with an accuracy limited by the numerical precision only.

For generating the sequence of arrival times we have used TEMPO2 (Hobbs et al., 2006), a specific software environment, widely used nowadays by the astronomers and astrophysicists studying pulsars. In particular, the TEMPO2 plug-in "*fake*" enables to simulate the time residuals expected from a given pulsar observation session. In fact, this code automatically generates a set of time arrivals for a specific pulsar at a pre-defined location on the Earth surface (corresponding, for instance, to a radio-telescope site), in a time window defined by the user, and starting from the transit time of the given pulsar through the local meridian (superior culmination point). It takes into account the contribution for the timing of the gravitational field in the Solar System due to the Sun and the other bodies, and other kinematical effects (see e.g., (Straumann, 2004)). The possibility to add various types of error, in particular the Gaussian one or the red noise one (a timing noise that is actually negligible for most millisecond pulsars), to the times of arrival is also allowed. Hence, we have simulated the signals coming from the real pulsars described in Table 1 and we have introduced a Gaussian 1 μs uncertainty: this is a conservative estimate of the error in the timing procedure, due both to the detection process and to the fluctuations of the sources. Then TEMPO2 has been for us the equivalent of an antenna where the sequences of pulses from our quartet of sources are received. The arrival times have been simulated during a time interval of about three days, at a given position on the surface of the Earth, that is the one of the Parkes observatory in Australia. In particular, we considered for each pulsar a set of about 28000 pulses, sampled out of the continuous sequence each 10 seconds. The duration of the simulation allows to evidence the actual motion of the observatory, due to the combined motion of the Earth around the Sun and of its daily rotation. The chosen pulsars define the null frame, and they are supposed to be at rest in the International Celestial Reference System (ICRS) (where, in turn, the barycenter of the Solar System is at rest).

By applying this procedure described here, we have rebuilt the trajectory of the observatory. To make a comparison, we consider ζ, that is the trajectory of the observatory, as determined by the ICRS ephemerides having components t, x, y, z, while, as before, the reconstructed trajectory $\bar{\zeta}$ has been obtained according to Eq. (12):

$$\boldsymbol{r}[i] = X[i]_{(a)}\boldsymbol{f}^{(a)} + X[i]_{(b)}\boldsymbol{f}^{(b)} + X[i]_{(c)}\boldsymbol{f}^{(c)} + X[i]_{(d)}\boldsymbol{f}^{(d)}. \tag{46}$$

where $[i]$ is an index labeling the i-th reception event. In particular, from (46) we obtain the Cartesian components $\bar{t}, \bar{x}, \bar{y}, \bar{z}$.

The results are shown in Figure 1 where the reconstructed spatial trajectory is compared with the one determined by the ICRS Ephemerides of the chosen observatory. The scale of the figure does not permit to appreciate the differences between the two trajectories. Actually this application of the method is purely indicative. TEMPO2 has of course not been designed for our purposes, so the sampling of the data each 10 seconds may introduce some additional uncertainty; moreover, as we stressed in Section 3. referring to the Geometric Dilution Of Precision (GDOP), a crucial role in minimizing the uncertainty is played by the geometry of the sources: the uncertainty is minimized when the volume spanned by the sources directions is maximized.

We stress the demonstrative purpose of our work, which has led us to disregard a series of aspects that should be taken into account for a real positioning system (see e.g., Ruggiero et al. (2011))

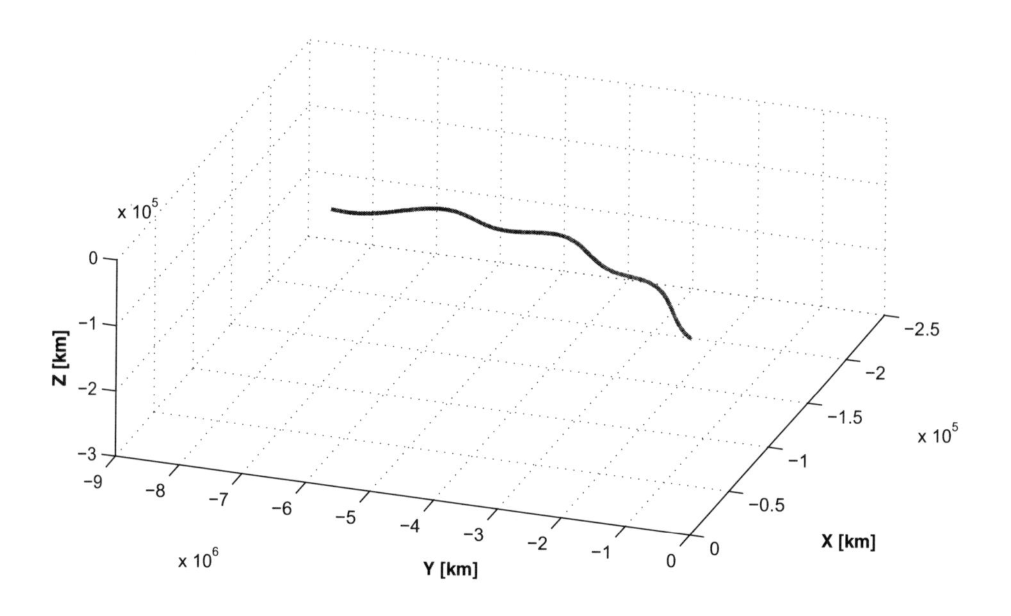

Figure 1. Space trajectory of the Parkes observatory on Earth with respect to the pulsars during three days. At this scale the ideal and the reconstructed curves are indistinguishable.

5. Discussion and Conclusion

We have described an operational approach to the use of pulsating signals for positioning purposes; in particular, pulsars signals can be used. Our procedure is based on the definition of a null frame in flat spacetime, by means of the four-vectors associated to the signals in the inertial reference frame where the sources are at rest (so that the emission directions and the frequencies of the pulsating signals have to be known) and far away (so that their signals can be dealt with as plane waves). The procedure is fully relativistic and allows position determination with respect to an arbitrary event in flat spacetime. Once a null frame has been defined, it turns out that the phases of the electromagnetic signals can be used to label an arbitrary event in spacetime. If the sources emit continuously and the phases can be determined with arbitrary precision at any event, it is straightforward to obtain the coordinates of the user and his worldline. However, actual sources emit signals that consist of a series of pulses: so, we have developed a simple method that can be used to determine the user's worldline by assuming that the worldline is a straight line during a proper time interval corresponding to the reception of a limited number of pulses, which means that the effects of the acceleration are negligibly small. This is indeed true for any solid system when the time span is only a fraction of a second, of the order of, say, one hundredth or less. We have discussed the source of errors that affect the positioning process, taking into account model limitations (including the hypothesis of dealing with a flat spacetime) and uncertainties, due to the stochastic changes of the quantities involved in the procedure.

Then, we have focused on an application of our approach, by which we have reconstructed the trajectory of a given terrestrial location; due to the combined motion of the Earth around the Sun and to its daily rotation. Using a simulating plug-in of the TEMPO2 software, we sampled the arrival times of the signals from a given set of pulsars, expected from an observation session at a specific location on the Earth, which in our case is the Parkes observatory in Australia. By collecting data which simulate an observation of about three days, we determined the trajectory of the observatory due to the combined rotation and revolution motion of the Earth. Then, we compared the reconstructed world-line with the one obtained by the ephemerides. The comparison was only for qualitative purposes since the use of TEMPO2 intermittently and for such a long (from the view point of our method) time may introduce some additional uncertainty; furthermore the choice and the use we made of pulsars corresponds to an idealized situation. However, the results seem promising for the testing of the reliability of our approach.

What we want to discuss, now, is the actual possibility of using our method for a positioning system effective for the navigation in the Solar System.

First, we focus on the use of pulsars, as basic "clocks" for the definition of the reference system: actually, we mostly refer to pulsars because they represent natural sources of regular pulses and they are suggestive of the old navigation at sea using the stars; an advantage of using pulsars is the fact that they can practically be considered as being at infinite distance and occupying fixed positions in the sky. However, a major drawback is the weakness of their signals. Indeed, this is a severe limitation for the design of a positioning system, but we would like to point out that our method can perfectly work with any other artificial source of pulses provided that one knows the law of motion for the source in a given reference frame. As a consequence, for the use in the Solar System, one could for instance think to lay down regular pulse emitters on the surface of some celestial bodies: let us say the Earth, the Moon, Mars, etc. The behavior of the most relevant bodies is indeed pretty well known, so that we have at hand the time dependence of the direction cosines of the pulses: this is enough to apply the method and algorithm we have described and the final issue in this case would be the position within the Solar system. In principle, the same can be done in the terrestrial environment: here the sources of pulses would be onboard satellites, just as it happens for GPS, but without the need of continuous intervention from the ground. Yet, again, the key point is a very good knowledge of the motion of the sources in the reference frame one wants to use.

Another important issue for the practical implementation of our system, is the number of independent sources. In the previous sections, we have many times mentioned the fact that the number of independent sources to be used is at least four. The reason for that number is that the spacetime coordinates needed to localize the user are of course four (three space components plus time). There are however many reasons for using a bigger number of emitters. If we have $n > 4$ pulsars (or equivalent sources) we may apply our method to all possible quadruples contained in n: the position will be determined as an average of the results obtained from all quadruples. This would be one more way to average the effect of random disturbances at the emission of the pulses out. Furthermore, if one of the sources fails or disappears for any reason (e.g., because it is eclipsed) the localization process is not interrupted; if one new source comes in, provided that its position in the sky and proper recursion time are known, the sequence of its pulses is hooked to the main sequence of the arrival times and may be used for further positioning. What matters is that all arrival times from all visible sources are arranged in one single sequence identified by the onboard clock and that the main sequence is not interrupted. To explain this better, the maximum tolerable total blackout of the sequence should not last more than what can be reconstructed by extrapolation from the last portion of the world-line of the observer. With reasonable accelerations at the typical speeds of an interplanetary travel this may be as long as a few seconds.

Eventually, our procedure is based on the measurement of signal arrival times, by means of the user's clock: it is then implicit that everything works if the clock used is a good and reliable clock. We are considering atomic clocks fit for being carried onboard a spacecraft. The problems with the clocks arise from both random and systematic instabilities. The former contribute to the general uncertainty in the positioning, while the latter introduce an apparent shortening of the period of the received pulses. A strategy to minimize the effects of the drift in the frequency of the clock is similar to the one adopted for GPS. It is redundancy: more than one atomic clocks, and of different types, should be carried by the observer (GPS satellites have four clocks onboard): the arrival times would be given by an average of the different readings. The redundancy would partly reduce also the effect of the drift of the frequency of the clocks, but this would emerge over a long period of time. The accuracy of portable present day atomic clocks is in the order of 1×10^{-14} (see e.g., Ashby (2003)) which means that the measured repetition time of a millisecond pulsar would be significantly affected by the frequency drift of the clock over times of the order of months, if not years. It would in any case be advisable to periodically check the accuracy of the measurement from some reference station on the Earth, and the best strategy would be to check the position and compare with the one obtained by self-positioning; rather than to try to directly verify the clock, which would be possible but it is a very delicate task indeed and pre-supposes, any way, a good knowledge of the relative motion between the space observer and the ground station. A periodic check up (after weeks or month intervals) would in any case be cheaper than a continuous guidance from the ground.

In summary, the practical implementation of our method in the real world requires a careful analysis of these issues, together with several technological aspects pertaining to the signals detection and elaboration chain. However, our work suggests that investigation in this direction seems worthwhile, for the definition of a positioning system effective for the navigation across the Solar System and beyond.

References

Ashby, N., Relativity in the Global Positioning System Living Rev. Rel. 6, 1, [http://www.livingreviews.org/lrr-2003-1], 2003.

Bate, R. R., Muller, D. D., White, J. E., *Fundamentals of Astrodynamics*, New York: Dover, 1971.

Battin, R. H., *An Introduction to the Mathematics and Methods of Astrodynamics*, Washington D.C.: American Institute of Aeronautics and Astronautics, Revised Edition, 1999.

Bini, D., Geralico, A., Ruggiero, M.L., & Tartaglia, A., Emission versus Fermi coordinates: Applications to relativistic positioning systems, Class. Quantum Grav. 25, 205011, 2008.

Blagojevic, M., Garecki J. F., Hehl, W., & Obukhov, Yu. N, Real null coframes in general relativity and GPS type coordinates, Phys. Rev. D 65, 044018, 2002.

Chandrasekhar, S., *The Mathematical Theory of Black Holes*, Oxford University Press (Oxford, UK, 1983).

Coll, B., & Tarantola, A., Galactic Positioning System, in "Journées Systèmes de Référence Spatio-Temporels, St. Petersburg", edited by Finkelstein, A., & Capitaine, N., Institute of Applied Astronomy of Russian Academy of Science, 333-334 (St. Petersburg, Russia, 2004).

Coll, B., Relativistic Positioning Systems, in "A century of Relativity Physics ERE 2005; XXVIII Spanish Relativity Meeting", AIP Conf. Proc. 841, 277-284, 2006.

Coll, B., Ferrando, J. J., & Morales, J. A., Two-dimensional approach to relativistic positioning systems, Phys. Rev. D 73, 084017, 2006.

Coll, B., Ferrando, J. J., & Morales J. A., Positioning with stationary emitters in a two-dimensional space-time, Phys. Rev. D 74, 104003, 2006.

Coll B., & Tartantola A., Using pulsars to define space-time coordinates, [arXiv:0905.4121], 2009.

Coll, B., Ferrando, J. J., & Morales J. A., Four Causal Classes of Newtonian Frames, Found. of Phys. 39, 1280-1295, 2009.

Damour, T., & Taylor, J.H., On the orbital period change of the binary pulsar PSR 1913 + 16, Ap. J. 366, 501-511, 1991.

Demidovic, B.P, & Maron, I.A., *Foundations of Numerical Mathematics*, SNTL (Praha, Czech Republic, 1966).

Downs, G.S., NASA Technical Report 32-1594, 1974.

About The Deep Space Network (2010). Jet Propulsion Lab (JPL), California Institute of Technology, Pasadena, CA., see online http://deepspace.jpl.nasa.gov/dsn/

Emadzadeh, A. A., Speyer, J. L., *Navigation in Space by X-ray Pulsars*, Springer Science+Business Media, pp. 3-5, 2011.

Folta, D., Gramling, C., Long, A., Leung, D., Belur, S., Autonomous Navigation Using Celestial Objects, in Americal Astronautical Society (AAS), Girdwood, Alaska, USA, Aug. 1999, pp. 2161-2177.

Gounley, R., White, R., Gai, E., Autonomous Satellite Navigation by Stellar Refraction, Journal of Guidance, Control and Dynamics, vol. 7, no. 2, pp. 129-134, 1984.

Hobbs, G., Lorimer, D. R., Lyne, A. G., & Kramer, M., A statistical study of 233 pulsar proper motions, Mon. Not. Roy. Astron. Soc. 360, 963-973, 2005.

Hobbs, G.B., Edwards, R.T., Manchester, R.N., *Monthly Notices of the Royal Astronomical Society* **369**, 655 (2006), see also the web site `http://www.atnf.csiro.au/research/pulsar/tempo2`

Hofmann-Wellenhof, B., Lichteneeger, H., & Collins, J., *Global Positioning System: Theory and Practice*, Springer (Wien, Austria, 2001)

Jordan, J. F., Navigation of Spacecraft on Deep Space Missions, Journal of Navigation, vol. 40, no. 1, pp. 19-29, Jan. 1987.

Kaplan, E. D., Hegarty, C. J., *Understanding GPS: Principles and Applications* , 2nd ed., Artech House, Inc., 2006.

Lorimer, D., Kramer, M. *Handbook of Pulsar Astronomy*, Cambridge University Press (Cambridge, UK, 2004).

Melbourne, W. G., Navigation Between the Planets, Scientific American, vol. 234, no. 6, pp. 58-74, Mar. 1976.

Parkinson, B. W., Spiker Jr., J. J., *Global Positioning System: Theory and Applications* , Washington D.C.: American Institute of Aeronautics and Astronautics, Inc., 1996, vols. I and II.

Pascual-Sánchez, J. F., Introducing relativity in global navigation satellite systems, Ann. Phys. (Leipzig), 16, 258-273, 2007.

Rovelli, C., GPS observables in general relativity, Phys. Rev. D 65, 044017, 2002.

Ruggiero, M.L., & Tartaglia, A., Mapping Cartesian coordinates into emission coordinates: Some toy models, Int. J. Mod. Phys. D17, 311-326, 2008.

Ruggiero M.L., Capolongo E., Tartaglia A., (2011), *Int. J. of Mod. Phys. D* , at press (2011), [`arXiv:1011.0065`]

Sala, J., Urruela, A., Villares, X., Estalella, R., & Paredes, J. M. Feasibility Study for Spacecraft Navigation System relying on Pulsar Timing Information, ESA Advanced Concept Team. ARIADNA Study 03/4202 (Final Report), June 23, 2004. Available on the ACT website [`http://www.esa.int/gsp/ACT/phy/pp/pulsar-navigation.htm`], 2004.

Sheikh, S.I., Pines, D.J., Ray, P.S., Wood, K.S., Lovellette, M.N., & Wolff M.T., Spacecraft navigation using X-ray pulsars, J. Guid. Control Dynam. 29, 49-63, 2006.

Sheikh, S.I., Hellings, R.W., & Matzner, R.A., High-Order Pulsar Timing For Navigation, in Proceedings of the 63rd Annual Meeting of The Institute of Navigation, Cambridge, MA, April 2007, 432-443, 2007.

Smits, R., Kramer, M., Stappers, B., Lorimer, D. R., Cordes, J., Faulkner, A., *Astronomy and Astrophysics* **493**, 1161 (2009)

Stark, H., & Woods, J.W., *Probability, Random Processes and Estimation Theory for Engineers*, Prentice Hall (New Jersey, USA, 1986)

Straumann N., *General Relativity*, Springer (Berlin, Germany, 2004).

Tartaglia, A., Emission coordinates for the navigation in space, arXiv:0910.2758, 2009; Acta Astronaut. 67, 539-545, 2010.

Tartaglia, A., Ruggiero, M.L., Capolongo, E., (2010), *Advances in Space Research* **47**, 645 (2011), [arXiv:1001.1068]

Wertz, J. R., *Spacecraft Attitude Determination and Control* , Dordrecht: Kluwer, 1978.

In: Solar System: Structure, Formation and Exploration ISBN: 978-1-62100-057-0
Editor: Matteo de Rossi

Chapter 5

AN ESTIMATE OF THE LONG-TERM TENDENCY OF TIDAL EVOLUTION OF EARTH-LUNAR SYSTEM

Gao Yi and Xiao Nai-Yuan*

Abstract

According to the conservation principle of angular momentum, we calculate in this paper the revolution period and the distance between the Earth and the Moon in the equilibrium state of the tidal evolution in the Earth-Moon system. The difference of energy between the current state and the equilibrium state is used to compute the time needed to fulfil the equilibrium state. Then the long-term variations of the Earth-Moon distance and of the Earth rotation rate are further estimated.

Keywords: solar system: Earth—solar system: Moon— celestial mechanics: tidal theory

1. Introduction

Since the "secular acceleration" of the Moon was found by Halley in 1695, many methods and data have been dedicated to the study of the long-term evolution of the Earth-Moon system. According to the time scales , these data can be divided into three categories:

(1) Geological data, including mainly the fossils of ancient corals and bivalves, which can be used to study the variation of Earth's rotation over times ranging from hundreds of millions to billions years [1,2];

(2) Historical data, consisting mainly of records of lunar and solar eclipses, star eclipses by the Moon and planetary motions, which are used to analyze the variations during the recent hundreds to thousands of years [3,4];

(3) Modern data from new technologies (e.g., VLBI, LLR, SLR and GPS) that can measure directly the rotation of the Earth so that the UT1 and daily duration can be derived [5].

Many achievements on the Earth-Moon system have been obtained from the analysis of all these data[6]. However, so far these achievements are on the past history of this system. In this paper, we will give an estimate of the tendency of the tidal evolution in this system in future time.

*E-mail address: nyxiao@nju.edu.cn

Due to the tidal evolution, the Moon has now reached a state in which its rotation period equals the period of revolution around the mass of center of the Earth-Moon system. As a result, the Moon always faces to the Earth with the same side. Thanks to the friction during this ongoing tidal evolution, the Earth rotation rate decreases, resulting in a lengthening of the day. The dissipative friction also consumes the rotating energies of both the Earth and the Moon. The potential energy in this system increases as the Earth-Moon distance increases, while the total kinematic energy buried in this system reduces. If we regard the Earth and Moon as an isolated system, then the evolution will arrive finally at an equilibrium where both the Moon and the Earth will rotate at the same rate as they revolve around their mass centers. However, this final equilibrium will definitely be destroyed by perturbations from other bodies in the solar system (especially the Sun), and the system will continue the dynamical evolution forever. To simplify the model, in this paper we treat the Earth-Moon system as an isolated system. From the current state of this isolated system, we will discuss below when the system can reach the equilibrium state, what are the rotation period and mutual distance at equilibrium, and the changing rate of the Earth-Moon distance.

2. Conservation of Angular Momentum in the Earth-Moon System

According to Reference [1], the secular deceleration of Earth rotation is mainly due to tidal friction. On the other hand, since the decelerations resulting from the variation of the Earth's moment of inertia and from the solar tidal effects have the same magnitude but opposite directions, so it is reasonable to ignore these two effects. Perturbations from the other planets are much smaller (e.g., the tidal effect from Jupiter is only 0.7 millionth that from the Sun) and can reasonably be neglected. Thus, we can approximately treat the Earth-Moon system as an isolated system.

We denote the mass and radius of the Earth and Moon by M, R and m, r respectively. The current rotating rate of the Earth, the Earth-Moon distance, the revolution angular velocity of the Moon (around the mass center) and the revolution period are denoted by ω, D_0, Ω_0 and P_0 respectively, while at the equilibrium, the distance, the Earth rotation rate (the same as the Moon rotation rate and as the revolution rate at equilibrium) and the revolution period , by D, Ω and P. Obviously,

$$\Omega_0 = \frac{2\pi}{P_0}, \quad \Omega = \frac{2\pi}{P}, \tag{1}$$

and Kepler's third law says

$$\frac{D^3}{P^2} = \frac{G(M+m)}{4\pi^2} = GM\frac{1+\mu}{4\pi^2}, \tag{2}$$

where $\mu = 1/81.30$ is the Moon-Earth mass ratio, G is the gravitational constant. From Eq. (2), we have

$$D = \sqrt[3]{\frac{GM}{4\pi^2}(1+\mu)P^2}\,. \tag{3}$$

The total angular momentum of the Earth-Moon system consists of the angular momenta from the Earth rotation, H_1, the Earth revolution (around the mass center), H_2, the Moon's revolution, H_3 and the Moon's rotation, H_4. It is easy to calculate the these angular momentum components as follow:

$$H_1 = \frac{2}{5}MR^2\omega\,,$$

$$H_2 = \left(\frac{2}{5}MR^2 + \frac{\mu^2}{(1+\mu)^2}MD_0^2\right)\Omega_0\,,$$

$$H_3 = \left(\frac{2}{5}\mu Mr^2 + \frac{\mu}{(1+\mu)^2}MD_0^2\right)\Omega_0\,,$$

$$H_4 = \frac{2}{5}\mu Mr^2\Omega_0\,.$$

Set the orbital plane of the Earth-Moon revolution as the reference plane. Although the angle between the Earth equator and this reference plane is not a constant, the fast precession of the Moon's orbital plane with respect to the ecliptic (with a period of 18.6 years) gives us a good reason to regard it coplanar with the ecliptic on average (in fact the angle between them is only $5°9'$). Therefore, the angular momentum conservation in this system can be expressed as:

$$H_1\cos\epsilon + H_2 + H_3 + H_4\cos I = H(\text{constant})\,,$$

where $\epsilon = 23°26'$ is the inclination between the Earth equator and the ecliptic, and $I = 1°32'$ is the angle between the Moon's equator and the reference plane (the revolution orbital plane). Finally, we get

$$\frac{2}{5}MR^2\omega\cos\epsilon + \frac{2}{5}MR^2\Omega_0 + \frac{\mu}{1+\mu}MD_0^2\Omega_0 + \frac{2}{5}(1+\cos I)\mu Mr^2\Omega_0 = H\,. \quad (4)$$

Similarly, we get the corresponding equation at the equilibrium:

$$\frac{2}{5}(1+\cos\epsilon)MR^2\Omega + \frac{\mu}{1+\mu}MD^2\Omega + \frac{2}{5}(1+\cos I)\mu Mr^2\Omega = H\,. \quad (5)$$

The value of the total angular momentum can be obtained from Eq. (4). Substituting Eq. (1) and Eq. (3) into Eq. (5), we get

$$\frac{4\pi}{5}(1+\cos\epsilon)MR^2P^{-1} + (2\pi)^{-1/3}\frac{\mu}{1+\mu}[GM(1+\mu)]^{2/3}MP^{-1/3} + \frac{4\pi}{5}(1+\cos I)\mu Mr^2P^{-1} = H\,. \quad (6)$$

If we let

$$a = \frac{4\pi}{5}(1+\cos\epsilon)MR^2\,,$$

$$b = (2\pi)^{-1/3}(GM)^{2/3}(1+\mu)^{-1/3}\mu M\,,$$

$$c = \frac{4\pi}{5}(1+\cos I)\mu Mr^2\,,$$

then Eq. (6) can be simplified to a 4th order algebraic equation:

$$b^3P^4 - H^3P^3 + 3H^2(a+c)P^2 - 3H(a+c)^2P + (a+c)^3 = 0\,.$$

Substituting all the quantities we know into this equation, we can further simplify it to (with period P in days):

$$P^4 - 51.3582P^3 + 59.0948P^2 - 22.6656P + 2.8978 = 0\,.$$

After some simple calculation we get the root of this equation:

$$P = 50.19\,\mathrm{d}\,. \tag{7}$$

This gives the common period of the Earth's rotation, the Earth-Moon revolution and the Moon's rotation at the equilibrium state. After introducing this period in Eq. (3), the Earth-Moon distance at equilibrium can be obtained:

$$D = 5.771 \times 10^5\,\mathrm{km}\,. \tag{8}$$

3. Calculation of the Dissipative Energy

As a result of dissipative tidal friction, the total energy (kinetic energy of rotation and the gravitational potential energy) of the Earth-Moon system at equilibrium will be smaller than the current value. In order to simplify the calculation, we set the origin of the reference coordinate at the center of mass of the Earth. Considering the angular momentum conservation with respect to the Earth spin axis, we have

$$I_E\omega + \mu M D^2\Omega = C\,, \tag{9}$$

where I_E is the momentum of inertia of the Earth and Ω is the angular speed of the Moon's revolution around the Earth. Calculating the time derivatives of both sides of Eq. (9) leads to

$$I_E\dot{\omega} + \mu M\frac{\mathrm{d}}{\mathrm{d}t}(D^2\Omega) = 0\,. \tag{10}$$

The Moon revolves around the Earth in a circular orbit at equilibrium, therefore the centrifugal force that it suffers is balanced by the gravity from the Earth, that is:

$$G\frac{\mu M^2}{D^2} = \mu M D\Omega^2\,. \tag{11}$$

Thus the revolving rate is

$$\Omega = \left[\frac{GM(1+\mu)}{D^3}\right]^{1/2}\,. \tag{12}$$

Substituting Eq. (12) into Eq. (10), we have

$$I_E\dot{\omega} + \mu M\sqrt{GM}\frac{\mathrm{d}}{\mathrm{d}t}(D^{1/2}) = 0\,,$$

that is,

$$\frac{1}{2}D^{-1/2}\dot{D} = -\frac{I_E\dot{\omega}}{\mu M\sqrt{GM}}\,.$$

Finally, we get

$$\dot{D} = -\frac{2\sqrt{D}I_E\dot{\omega}}{\mu M\sqrt{GM}}\,. \tag{13}$$

This shows that the mutual distance D will increase as $\dot{\omega}$ decreases.

The total energy E consists of the rotating kinetic energies of the Earth and the Moon, the kinetic energy of the Moon rotating around the center of mass of the Earth-Moon system, and the gravitational potential energy, which can be written as:

$$E = \frac{1}{2}I_E\omega^2 + \frac{1}{2}I_m\Omega^2 + \frac{1}{2}\mu MD^2\Omega^2 - G\frac{\mu M^2}{D}\,.$$

Since both the mass and spinning rate of the Moon are much smaller than those of the Earth, the rotating kinetic energy of the Moon is smaller than the Earth by several orders of magnitude. We neglect the Moon's rotation energy in our calculation. The energy finally is:

$$E = \frac{1}{2}I_E\omega^2 + \frac{1}{2}\mu MD^2\Omega^2 - G\frac{\mu M^2}{D}\,. \tag{14}$$

Substituting Eq. (12) into Eq. (14), we have

$$E = \frac{1}{2}I_E\omega^2 - G\frac{\mu M^2}{2D}\,. \tag{15}$$

The variation of the total energy with respect to time is

$$\frac{\mathrm{d}E}{\mathrm{d}t} = I_E\omega\dot{\omega} + G\frac{\mu M^2}{2D^2}\dot{D}\,,$$

Introducing $\dot{\omega}$ into this equation, we have

$$\frac{\mathrm{d}E}{\mathrm{d}t} = \frac{M}{2}\left[G\frac{\mu M}{D^2} - \omega\sqrt{G\frac{M}{D}}\right]\dot{D}\,. \tag{16}$$

We calculate from Eq. (1) $\Omega = 1.4489\times10^{-6}\mathrm{s}^{-1} = 1.2518\times10^{-1}\mathrm{d}^{-1}$. Denote the current kinetic energy of the Earth-Moon system by E_0 and the energy at equilibrium by E, then the difference between them, ΔE, can be computed as follow:

$$\begin{aligned}\Delta E &= E - E_0\\ &= \left(\frac{1}{2}I_E\Omega^2 - \frac{1}{2}I_E\omega^2\right) - \left(G\frac{\mu M^2}{2D^2} - G\frac{\mu M^2}{2D_0^2}\right) + \left(\frac{1}{2}\mu MV^2 - \frac{1}{2}\mu MV_0^2\right)\\ &= -2.5801\times10^{29}\,\mathrm{J}\,,\end{aligned} \tag{17}$$

where V_0 and V are the Moon's speed on its revolving orbit around the Earth at present and at the equilibrium, respectively. Eq. (17) means that an energy with value -2.5801×10^{29} J shall be dissipated before the system reaches the final equilibrium state.

4. Time Needed to Achieve the Equilibrium

Based of the above analysis on the energy difference and the energy dissipating rate, we will give a rough estimate of the time from present to the equilibrium.

A simple mathematical model can be constructed as follows. As a first order approximation, the $\dot{\omega}$ in Eq. (13) is a constant. We adopt the value given in Reference [1], $\dot{\omega} = -6.145 \times 10^{-22} \mathrm{rad\,s^{-2}}$. If we set

$$-\frac{2I_E\dot{\omega}}{\mu M\sqrt{GM}} = A = 2.193 \times 10^{-10}\,,$$

we can obtain from Eq. (13),

$$\dot{D} = AD^{1/2} \ \ (t = 0, D = D_0)\,. \tag{18}$$

Integrating this equation, we have

$$D^{\frac{1}{2}} = \frac{1}{2}At + C$$

with $C = \sqrt{D_0} = 1.961 \times 10^4$. And finally the variation of distance between the Earth and the Moon is

$$\sqrt{D} = 2.193 \times 10^{-10}t + 1.961 \times 10^4\,. \tag{19}$$

Substituting this into Eq. (16) and integrating it, we obtain

$$\int_0^T \frac{\mathrm{d}E}{\mathrm{d}t}\mathrm{d}t = \Delta E\,, \tag{20}$$

where the energy difference $\Delta E = -2.5801 \times 10^{29}$ J, and T is the time from present to the equilibrium. The value of T can be numerically computed, and it is $T \approx 2.1760 \times 10^{17}\mathrm{s} = 6.8956 \times 10^9\mathrm{yr}$.

5. Estimation of the Variation of Earth Rotation Rate

The Earth-Moon distance increases with time. We will give in this section an estimate of this variation as well as the variation of the Earth spin rate.

According to the analysis in Reference [1] and the laser lunar ranging data in the last 30 years, the velocity of the Moon leaving the Earth is ~ 4 cm/yr. At the present moment, the Earth-Moon distance is $D = 384400$ km, thus the difference between this value and the distance at equilibrium state is

$$\Delta D = 5.7710 \times 10^5 - 3.844 \times 10^5 = 1.927 \times 10^5\,\mathrm{km}\,. \tag{21}$$

We make estimates of the time needed for attaining the equilibrium under two extreme situations:

(1) The distance D increases linearly. Then ΔD will be accomplished in $T_1 = 4.8175 \times 10^9$ yr.

(2) The velocity of distance increasing decelerates uniformly from the current value (4 cm/yr) to 0 at equilibrium. Then the time for ΔD is $T_2 = 9.6350 \times 10^9$ yr.

The former assumption is obviously unreasonable, because we can not imagine an abrupt stop after a constant variation. The latter assumption can not happen either, because the necessary time calculated from the view of energy consumption is between these two extremes. We may simulate the velocity variation of the Earth-Moon divergence by a parabolic curve as illustrated in Fig. 1. Let the velocity of the divergence be v. Then the parabolic approximation is

$$v = v_0 - \frac{v_0}{T^2}t^2, \tag{22}$$

where t is the time measured from now on and $v_0 = 4$ cm/yr is the current divergence velocity.

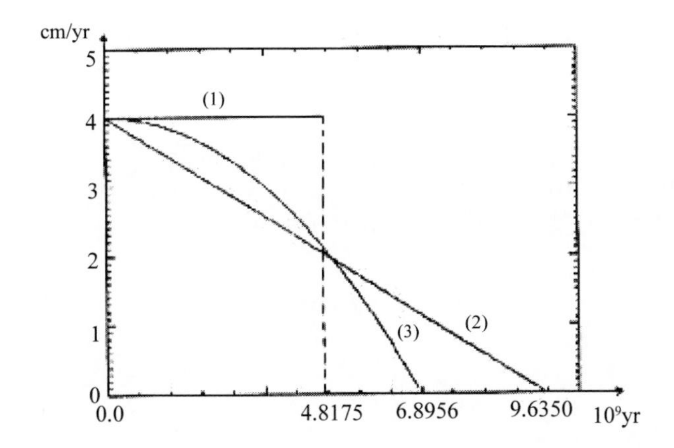

Figure 1. Variation of the rate of divergence of the Earth-Moon distance. The abscissa is time in a unit of 10^9 yr and the ordinate is the velocity v in 10^{-5} km/yr. The line labeled (1) is for a constant, increasing distance; the line (2) is for a constant deceleration of the velocity; and the curve (3) is part of a parabola.

The rationality and reliability of the parabolic approximation can be checked as follows. The area S under the parabola in Fig. 1 is the distance shift so that it should be equal to the ΔD in Eq. (21). Integrating Eq. (22), we get

$$S = \frac{2}{3}v_0 T = 1.838 \times 10^5 \text{ km}. \tag{23}$$

The relative error between this value and ΔD is only 4.6%. It confirms that our conclusion, that is, Eq. (22), approximates very well the variation of the velocity of Earth-Moon divergence.

Generally, the $\dot{\omega}$ is regarded as a constant, but we can see from Eq. (13) that $\dot{\omega}$ is related to $\dot{D}$ and v in Eq. (22). Our results above show that the velocity v varies very slowly. In fact Eq. (22) implies that it takes 2.18×10^9 yr for v to lower the current value by 10%. As a result, it is reasonable to regard $\dot{\omega}$ as a constant.

Acknowledgments

Supported by National Natural Science Foundation.

A translation of *Acta Astron. Sin.* Vol. 48, No. 4, pp. 456–462, 2007.

References

[1] Gao Bu-xi. ChA&A, 2006, 30, 111

[2] GAO Bu-xi. AcASn, 2005, 46, 331

[3] Munk W. H., MacDonald G.J.F., The Rotation of the Earth. New York: Cambridge University Press, 1960

[4] Lambeck K., The Earth's Variable Rotation: Geophysical Causes and Consequences. New York: Cambridge University Press, 1980

[5] Gao Bu-xi, Principles of Astro-Terrestrial Dynamics, Beijing: Science Press, 1997

[6] Morrison L.V., Stephenson F.R., Observations of sucular and decade changes in the Earths rotation. In Earth Rotation: Solved and Unsolved Problems, edited by Anny Cazenave, D Reidel Publishing Company, 1986

In: Solar System: Structure, Formation and Exploration ISBN: 978-1-62100-057-0
Editor: Matteo de Rossi

Chapter 6

A ROLE OF THE SOLAR WIND IN DYNAMICS OF INTERSTELLAR DUST IN THE SOLAR SYSTEM*

M. Kocifaj[1†] ***and J. Klačka***[2‡]
[1]Departmnt of Interplanetary Matter, Astronomical Institute
Slovak Academy of Sciences, Dúbravská cesta 9, 845 04 Bratislava,
Slovak Republic
[2]Department of Astronomy, Physics of the Earth, and Meteorology
Faculty of Mathematics, Physics and Informatics
Comenius University, Mlynská dolina, 842 48 Bratislava,
Slovak Republic

Abstract

Interstellar dust grains have been detected by the dust detectors onboard the *Ulysses* and *Galileo* spacecrafts. Motion of the interstellar dust particles in the Solar System is driven by gravitational and nongravitational forces. As for gravity, the action of the Sun is the dominant gravitational effect. Nongravitational forces are represented by solar electromagnetic radiation force, similar effect of the solar wind, and, Lorentz force for submicrometer-sized dust grains. Lorentz force originates from the action of interplanetary magnetic field on electrically charged grains and solar wind velocity plays a crucial role in this nongravitational force.

Keywords: interstellar dust particles, electromagnetic radiation, Poynting-Robertson effect, solar wind, interplanetary magnetic field, Lorentz force

1. Introduction

Interstellar dust grains have been detected by the dust detector onboard the *Ulysses* spacecraft (Grün *et al.* 1993, Baguhl *et al.* 1995). The interstellar dust particles (IsDPs) enter

*A version of this chapter was also published in Handbook on Solar Wind: Effects, Dynamics and Interactions, edited by Hans E. Johannson published by Nova Science Publishers, Inc. It was submitted for appropriate modifications in an effort to encourage wider dissemination of research.

†E-mail address: kocifaj@savba.sk
‡E-mail address: klacka@fmph.uniba.sk

the Solar System with a speed of about $v_\infty = 26$ km/s (Landgraf *et al.* 1999) and they are arriving from direction of $\lambda_{ecl} = 259°$ (heliocentric ecliptic longitude) and $\beta_{ecl} = +\,8°$ (heliocentric ecliptic latitude) (Landgraf *et al.* 2000). Masses of IsDPs detected by *Ulysses* and *Galileo* spacecrafts range from 10^{-18} kg to 10^{-13} kg (Sykes *et al.* 2004). During their motion in the Solar System, IsDPs are moving under the action of gravitational and nongravitational forces. The action of the Sun is the most important gravitational force. As for nongravitational forces, the effect of solar electromagnetic radiation influences the motion of IsDPs: the papers by Jackson (2001) and Kocifaj and Klačka (2003) discuss the condition under which the IsDPs can be captured in the Solar System. While the paper by Jackson considers only spherical dust particles, the newer paper takes into account that interstellar dust particles are arbitrarily shaped, they are not spherical but rather have an elongated structure (Wurm and Schnaiter 2002). Bearing in mind that IsDPs are mainly smaller than 1 micron, the electrical charge of the particles cannot be ignored and the effect of interplanetary magnetic field plays an important role (e. g., Dermott *et al.* 2001). The orbital evolution of IsDPs in the Solar System, under the action of the gravity of the Sun, solar electromagnetic radiation and Lorentz force due to the effect of interplanetary magnetic field, was investigated by Kocifaj and Klačka (2004).

The effect of solar wind enters in two ways into the equation of motion of dust particle evolving in the Solar System. The first case is the effect similar to the Poynting-Robertson effect (Poynting 1903, Robertson 1937, Klačka 2004, 2008a, 2008b), when the solar wind partially decelarates the motion of the particle orbiting the Sun (the pressure generated by the solar wind is in three orders of magnitude smaller than the solar electromagnetic radiation pressure; see, e. g., Dohnanyi 1978). The second case is the effect of Lorentz force, since the relevant velocity vector in the Lorentz force is given by the relative velocity of the particle with respect to the velocity of the solar wind. The effect of the solar wind is physically justified in heliosphere. We will consider its effect on evolution of IsDPs if their heliocentric distances are smaller than 150 AU (Opher *et al.* 2004).

2. Electrically Neutral Spherical Particles

The most simple case in treating the motion of IsDPs takes into account spherical dust grains entering the Solar System with the speed v_∞. The equation of motion of the spherical grain with respect to the Sun is

$$\begin{aligned} \frac{d\vec{v}}{dt} &= -\frac{GM_\odot}{r^3}\,\vec{r} + \beta\,\frac{GM_\odot}{r^2}\left\{\left(1-\frac{\vec{v}\cdot\vec{e}_R}{c}\right)\vec{e}_R - \frac{\vec{v}}{c}\right\}, \\ \beta &= 5.76\times10^{-4}\,\frac{\bar{Q}'_{pr}}{R[m]\;\varrho[kg/m^3]}, \end{aligned} \tag{1}$$

where $\vec{v}$ is the heliocentric velocity of the grain, $\vec{r}$ is its heliocentric position, $\vec{r} = r\,\vec{e}_R$, G is the gravitational constant, $M_\odot$ is mass of the Sun, c is the speed of light, R is the radius of the particle, ϱ is mass density and $\bar{Q}'_{pr}$ is dimensionless efficiency factor of electromagnetic radiation pressure (Mie 1908, van de Hulst 1981, Bohren and Huffman 1983) averaged over the solar spectrum as follows

$$\bar{Q'}_{pr} = \frac{\int_0^\infty I(\lambda)\, Q'_{pr}(\lambda)\, d\lambda}{\int_0^\infty I(\lambda)\, d\lambda} \quad , \tag{2}$$

where $I(\lambda)$ is the flux of monochromatic (λ is the wavelength) radiative energy. The first part of the acceleration on the right-hand side of Eq. (1) is the gravitational acceleration of the Sun, the second term is the P-R effect. If the gravity of the Sun would exist alone, then the interstellar dust particle would move on hyperbolic orbit and no capture in the Solar System could exist. However, there is an important part in Eq. (1): the term proportional to $-\vec{v}/c$ enables capture of interstellar dust particles in the Solar System (Jackson 2001, Kocifaj and Klačka 2003).

It is easy to include also the effect of the solar wind, into Eq. (1). At first, the solar wind causes also deceleration of dust particle, in the form similar to the P-R effect. This term can be easily added to the electromagnetic radiation force. The simultaneous action of the solar electromagnetic and corpuscular (solar wind) radiation can be obtained by a gentle substitution $\beta \rightarrow \beta\,(1 + \eta/\bar{Q'}_{pr})$, where β is the ratio between the radial component of the electromagnetic radiation pressure force and the gravitational force of the Sun, $\eta \doteq$ 1/3 (Dohnanyi 1978, Gustafson 1994). The equation of motion, including the effect of the solar wind, for an electrically neutral spherical dust particle, is

$$\begin{aligned} \frac{d\vec{v}}{dt} &= -\frac{GM_{\odot}\,(1-\beta)}{r^3}\,\vec{r} \\ &\quad - \beta\left(1 + \frac{\eta}{\bar{Q'}_{pr}}\right)\frac{GM_{\odot}}{r^2}\left(\frac{\vec{v}\cdot\vec{e}_R}{c}\,\vec{e}_R + \frac{\vec{v}}{c}\right) , \\ \beta &= 5.76 \times 10^{-4}\,\frac{\bar{Q'}_{pr}}{R[m]\;\varrho[kg/m^3]} , \\ \eta &= \frac{1}{3} . \end{aligned} \tag{3}$$

Eq. (3) yields acceleration of the spherical dust particle under the action of solar gravity, the P-R effect and the solar corpuscular effect, i.e., the effect of the solar wind in the conventional approach when the radial component of solar wind is considered.

2.1. Numerical Results

This section summarizes the most relevant results on orbital evolution of IsDPs. The equation of motion of an interstellar dust particle is given by Eq. (3). The initial conditions for it are given for the heliocentric position vector $\vec{r}_0$ and heliocentric velocity vector $\vec{v}_\infty$, $|\vec{v}_\infty| = 26$ km/s. The heliocentric radius vector $\vec{r}_0$ is of effectively infinite radial heliocentric distance and the particle is moving at velocity $\vec{v}_\infty$ along a line that misses the center of the Sun by a perpendicular distance b. The quantity b is the impact parameter. As for the material properties of the IsDPs, our calculations are limited to magnesium-rich silicate particles as these could be representative for interstellar dust grains (Dorschner *et al.* 1995) and the mass density is $\varrho = 2.5$ g/cm^3. Particles of various radii are taken into account. The important values of dimensionless efficiency factor for radiation pressure, averaged over solar spectrum (see Eq. 2), are calculated from Mie's solution of Maxwell's equations (Mie 2008, van de Hulst 1981, Bohren and Huffman 1983) and the results are presented in Fig. 1.

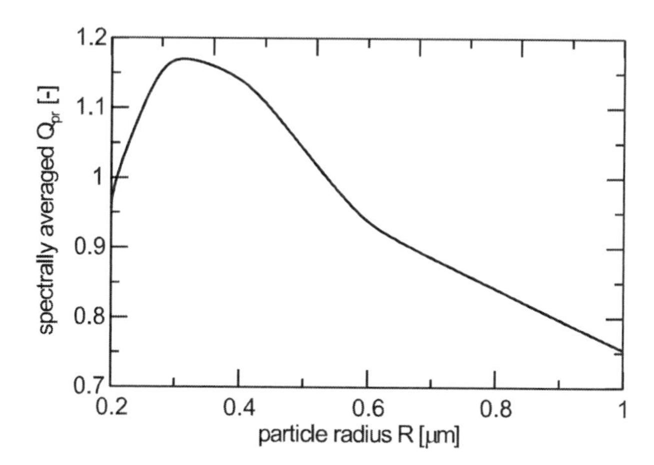

Figure 1. Values of dimensionless efficiency factor for radiation pressure (averaged over solar spectrum) as a function of radius of an interplanetary dust particle. The composition of the particle corresponds to magnesium-rich silicate.

As it follows from Eq. (3), the motion of an interplanetary dust particle is planar. The plane is defined by its normal vector and this can be represented by angular momentum vector $\vec{H} = \vec{r}_0 \times \vec{v}_\infty$, $|\vec{H}| = b\,|\vec{v}_\infty|$. If the impact parameter b is less than a minimum value b_{min}, then the particle hits the Sun. If the value of b is greater than a value b_{max}, then only hyperbolic motion of the particle exists. The value $b \in (b_{min}, b_{max})$ corresponds to the case when the particle is orbiting the Sun, i. e., the particle is captured in the Solar System. This case can really occur, as it is shown on Fig. 2 (moreover, such capture exists also for the particle of radius $R = 0.1$ μm: $b \in (8.1, 8.9)$ solar radii for the pure electromagnetic case and $b \in (8.1, 11.0)$ solar radii for the case when also solar wind is taken into account).

Fig. 2 shows that the IsDPs of radii between 0.3 μm and 1.0 μm can be captured in the Solar System and it also documents the corresponding interval of impact parameter for which the capture exists: $b_{min} < b < b_{max}$. The dotted area denotes the pure electromagnetic case, while the larger area corresponds to the more realistic case, when also the effect of solar wind is taken into account. The pure electromagnetic case yields that only IsDPs of radii 0.3 μm to 0.6 μm can be captured by the Sun, but consideration of the solar wind yields that the IsDPs of radii 0.3 μm to 1.0 μm can be captured. As for the particles of radii 0.4 μm (the peak in size distribution of the detected particles on Ulysses and Galileo), the probability of the capture is 1.6-times larger for the solar wind case than for the pure electromagnetic case. This follows from the ratio of ring areas: $([b_{max}(s.w.)]^2 - [b_{min}(s.w.)]^2)/([b_{max}(elmg)]^2 - [b_{min}(elmg)]^2) \doteq (18.5^2 - 14.1^2)/(17.0^2 - 14.1^2) \doteq 1.6$. Fig. 2 generalizes the results of Jackson (2001) and Kocifaj and Klačka (2003).

The capture of an interstellar dust particle by the Sun, the P-R effect and the effect of radial solar wind is depicted in Fig. 3. Fig. 3 shows orbital evolution of the particle including the first orbit around the Sun both for the pure electromagnetic case and the case with solar wind. The velocity term $-\vec{v}/c$ present in Eq. (3) causes gradual spiralling of the

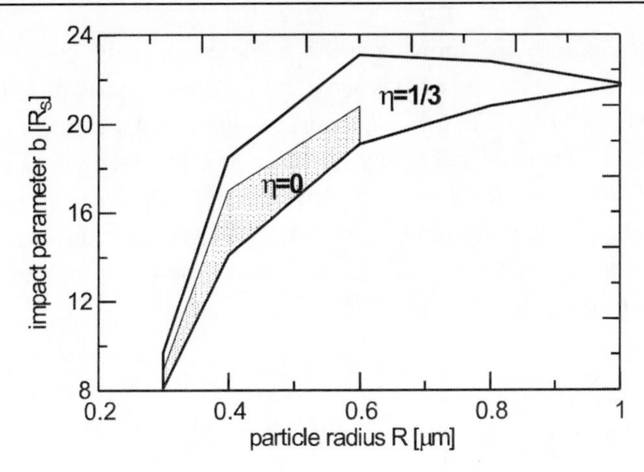

Figure 2. Values of impact parameters (measured in solar radii) when an interstellar dust particle is captured in the Solar System and the particle revolves the Sun. The dotted area holds for the pure electromagnetic case (plus the gravity of the Sun), while the larger area holds for the case when also solar wind effect is considered. Inclusion of the solar wind yields not only higher probability of the capture for particles of a given radius, but also larger particles can be captured and move around the Sun than it is in pure electromagnetic case.

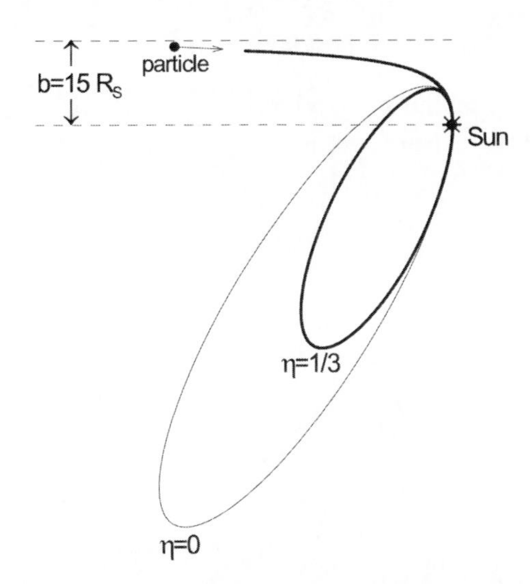

Figure 3. Orbital evolution of an interplanetary dust particle (radius 0.4 μm), captured in the Solar System and orbiting the Sun. The pure electromagnetic case (plus gravity of the Sun) and the case when also solar wind is taken into account, are depicted. The stay of the particle in the evaporation zone requires 5.07 hours (0.0343 % of the total orbital period) for the electromagnetic case ($\eta = 0$), while it is 5.08 hours (0.0919 % of the total orbital period) for the case when solar wind is also considered ($\eta = 1/3$). Real scale is depicted and given by the value of impact parameter $b = 15\ R_S$ (solar radii).

particle toward the Sun. Fig. 3 shows that the particle can move near the Sun for a short time interval. The length of this interval when particle's heliocentric distance is smaller than 3 solar radii (the area where the particle evaporates due to solar electromagnetic radiation) is presented in Fig. 4. There is practically no difference between the cases $\eta = 0$ and $\eta = 1/3$ for particle radii 0.3 μm - 0.6 μm. The particles with radii $R > 0.6$ μm will revolve the Sun only when $\eta = 1/3$. The minimal heliocentric distances for the case of the capture in the Solar System are depicted in Fig. 5: the difference between the pure electromagnetic case and inclusion of the solar wind is evident.

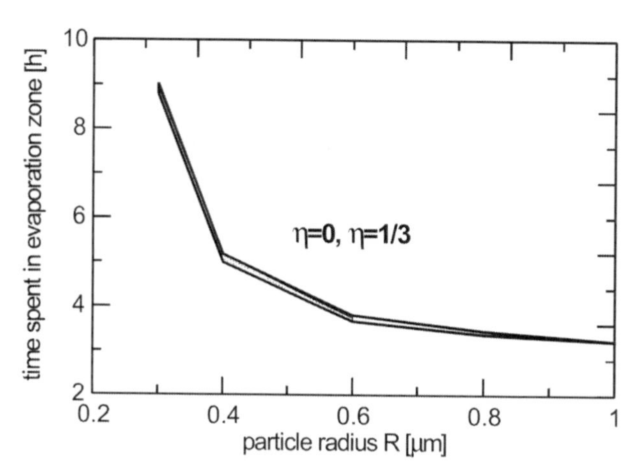

Figure 4. Time interval which an interstellar dust particle captured in the Solar System orbits the Sun in its vicinity, in the evaporation zone. The time interval is given in hours.

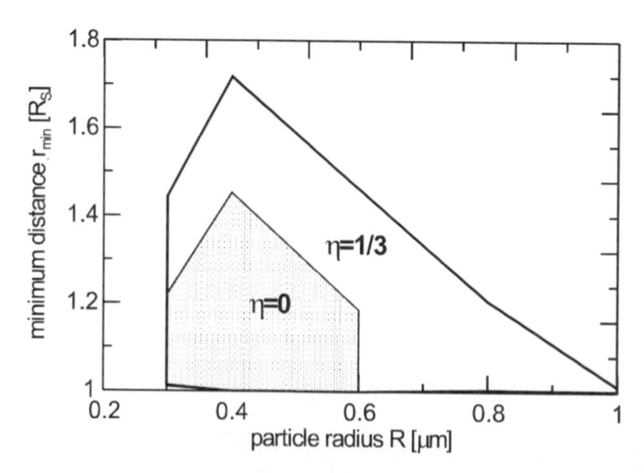

Figure 5. Minimal heliocentric distance of an interstellar dust particle captured in the Solar System, corresponding to the first circulation around the Sun. The distances are smaller than the evaporation heliocentric distance 3 solar radii.

2.2. Comparison of Numbers of IsDPs Orbiting the Sun

We have already found that the probability of the capture of IsDPs of radii 0.4 μm is 1.6-times larger for the more realistic solar wind (SW) case than for the pure electromagnetic (ELMG) case (see also Fig. 2). Let us ignore the motion in the evaporation zone. The SW case is characterized by smaller orbital periods T and semimajor axes a than the ELMG case: $a(SW)/a(ELMG) = [T(SW)/T(ELMG)]^{\kappa} \doteq (3.43/9.19)^{\kappa}$, according to Fig. 3. The value of κ equals to 2/3 for the Keplerian motion. However, Fig. 3 shows that the Sun is not situated in foci of the ellipses (approximation of the trajectory) and the ratio of semimajor axes is, approximately, $a(SW)/a(ELMG) \doteq 1.78$. Thus, Fig. 3 suggests that $\kappa = 0.59$. Moreover, according to Eqs. (1) and (3), the rate of inspiralling toward the Sun is higher for the SW case than for the ELMG case. The ratio of the times of inspirallling is $\tau(SW)/\tau(ELMG) \doteq [a(SW)/a(ELMG)]^2 / (1 + \eta/\bar{Q}'_{pr})$. Substituting from the ratio $a(SW)/a(ELMG)$, we obtain $\tau(SW)/\tau(ELMG) \doteq [T(SW)/T(ELMG)]^{2\kappa} / (1 + \eta/\bar{Q}'_{pr})$. The ratio of the number of IsDPs orbiting the Sun is $1.6 \times \tau(SW)/\tau(ELMG) \doteq 0.3$ for the Keplerian motion and 0.4 for the more realistic case corresponding to Fig. 3. In these calculations the values $T(SW)/T(ELMG) = 3.43/9.19$, $\eta = 1/3$ and $\bar{Q}'_{pr} = 1.14$ for the particle of radius 0.4 μm were used.

When solar wind is considered, the probability of the capture of IsDPs is greater than in the pure electromagnetic case (1.6-times for particles of radii 0.4 μm). But the number of particles orbiting the Sun is 3-times smaller for the solar wind case than for the pure electromagnetic case. These results correspond to the situation when the motion in the evaporation zone is ignored.

3. Electrically Neutral Arbitrarily Shaped Particles

If the particle is not spherical, then one has to use a more general form of the interaction between dust particle and the incident electromagnetic radiation. The acceleration of the particle with respect to the Sun, including solar gravity and solar wind, reads

$$
\begin{aligned}
\frac{d\vec{v}}{dt} &= -\frac{GM_{\odot}(1-\beta)}{r^3}\vec{r} - \beta\left(1 + \frac{\eta}{\bar{Q}'_{pr}}\right)\frac{GM_{\odot}}{r^2}\left(\frac{\vec{v}\cdot\vec{e}_R}{c}\vec{e}_R + \frac{\vec{v}}{c}\right) \\
&\quad + \frac{GM_{\odot}}{r^2}\sum_{j=2}^{3}\beta_j\left[\left(1 - 2\frac{\vec{v}\cdot\vec{e}_R}{c} + \frac{\vec{v}\cdot\vec{e}_j}{c}\right)\vec{e}_j - \frac{\vec{v}}{c}\right], \\
\beta &= 7.61\times 10^{-4}\,\frac{\bar{C}'_{pr}[m^2]}{m[kg]}, \\
\beta_j &= 7.61\times 10^{-4}\,\frac{\bar{C}'_{pr,j}[m^2]}{m[kg]},\quad j = 2, 3, \\
\eta &= \frac{1}{3},
\end{aligned} \tag{4}
$$

where also nonradial components of radiation pressure are considered and thermal emission force is neglected (Klačka 2004, 2008b). In agreement with Eq. (2) the spectrally averaged cross sections of radiation pressure $\bar{C}'_{pr}$, $\bar{C}'_{pr,1}$, $\bar{C}'_{pr,2}$ are obtained as weighted integral quantities

$$\bar{C'}_{pr} = \frac{\int_0^\infty I(\lambda)\, C'_{pr}(\lambda)\, d\lambda}{\int_0^\infty I(\lambda)\, d\lambda} ,$$

$$\bar{C'}_{pr,j} = \frac{\int_0^\infty I(\lambda)\, C'_{pr,j}(\lambda)\, d\lambda}{\int_0^\infty I(\lambda)\, d\lambda} , \quad j = 1,\, 2 , \tag{5}$$

where $I(\lambda)$ is the flux of monochromatic radiation energy. Neckel's and Labs's (1981) data can be taken into account when adequately accurate values of $I(\lambda)$ are required, especially in the spectral region 0.2 μm $\leq \lambda \leq$ 0.6 μm. The analytical function $A\lambda^{-\varrho} \exp\left\{-s\lambda^{-t}\right\}$ is applicable to model the rest of the $I(\lambda)$ curve in near-IR. Here the coefficients A, ϱ, s, and t are published in Kastrov (1979). The symbol m used in Eq. (5) denotes the mass of the particle. If $\beta_2 \equiv \beta_3 \equiv 0$, then Eq. (4) reduces to Eq. (3), i.e., the effect of electromagnetic radiation reduces to the P-R effect.

In practically important cases it is necessary to evaluate cross sections for rapidly rotating particles. One has to take into account that orientation of the particle's rotational axis is practically constant in space, during the particle's motion (Krauss and Wurm 2004, Krauss 2005) and any wavelength-dependent cross sections for radiation pressure need to be obtained from integral

$$\langle\, C'_{pr,i}(\lambda)\, \rangle = \frac{\int_{\Phi_{min}}^{\Phi_{max}} d\Phi \int_{\theta_{min}}^{\theta_{min}} \sin\theta\, d\theta \int_{\Psi_{min}}^{\Psi_{min}} C_{pr,i}(\lambda, \Phi, \theta, \Psi)\, d\Psi}{\int_{\Phi_{min}}^{\Phi_{max}} d\Phi \int_{\theta_{min}}^{\theta_{min}} \sin\theta\, d\theta \int_{\Psi_{min}}^{\Psi_{min}} d\Psi} , \tag{6}$$

where Euler angles (Φ, θ, and Ψ) characterize the orientation of the particle in respect to the laboratory reference frame (Draine and Flatau 2004).

In respect to a nature of physical processes participating in formation of dust particles, the spherical or spheroidal obstacles almost never occur in the space. Thus, an approximate formula for $C_{pr,i}(\lambda, \Phi, \theta, \Psi)$ (e.g. Mie- or Rayleigh-based model) can scarcely be employed. The optical properties of homogeneous and compact particles may be efficiently determined by T-matrix method (Mishchenko et al. 2002) which follows the Huygens principle. T-matrix approach decomposes the electromagnetic field on a basis of functions adapted to the more-complex geometry of the scatterer. Once the T-matrix is determined, the numerical solution of electromagnetic scattering problem is straightforward and very rapid because a change of direction of incidence and scattering and/or change of polarization state of the incident field do not require the repetition of the entire set of calculations. However, T-matrix is limited in arbitrariness of the particle shape and composition. There are really great problems with coding the T-matrix for porous, inhomogeneous or very complex aggregates. To overcome these problems some kind of volume integral equation method can be adopted. One of the world-wide known techniques is so called Discrete dipole approximation (DDA; Purcell and Pennypacker 1973, Draine 1998) which is applicable to various geometries and material configurations (Wriedt and Comberg 1998). The principles of DDA are quite easy: any continuum target is replaced by an array of point dipoles (Purcell and Pennypacker 1973) and the scattering is then solved for this dipole array (Draine and Goodman 1993). A profit derived from automatic satisfaction of radiation condition is that arbitrarily shaped and, also, inhomogeneous particle can be used for calculations. But the method is counterbalanced by a set of disadvantages. In particular, extremely large number of dipoles is needed to guarantee the satisfactory accuracy of scattering matrix elements. Thus, the DDA is time-consuming and very MEM and CPU intensive tool. In addition, the entire set of calculations needs to be repeated for each new geometrical position or polarization state of the incident beam. In spite of these facts the attractiveness, publicity and simplicity of the physical idea of the DDA makes this tool widely applied in solving many light scattering problems.

Whereas the interstellar grains are typically irregularly shaped and composite aggregates, there is no many alternatives for modeling the light scattering processes for such obstacles. The DDA method appears to be one of the best choices. To accelerate the solution of Eq. (4), it is convenient to precalculate the optical properties of a particle and keep obtained optical data in local database. Specifying the initial position and velocity vectors of the particle, the differential equation Eq (4) can be solved (e. g., by the Runge-Kutta's method of the fourth order with adaptive step-size), i.e., new position and velocity vectors of the particle can be determined. Dealing with the probability of the capture of the IsDPs, the impact parameter b, position angle α and orientation of the particle rotation axis need to be specified (see Fig. 6). The capture probability for the particle depends on these parameters. Thus, the situation for irregularly shaped dust particle is more complicated than for the spherical interstellar dust grains treated in the previous section.

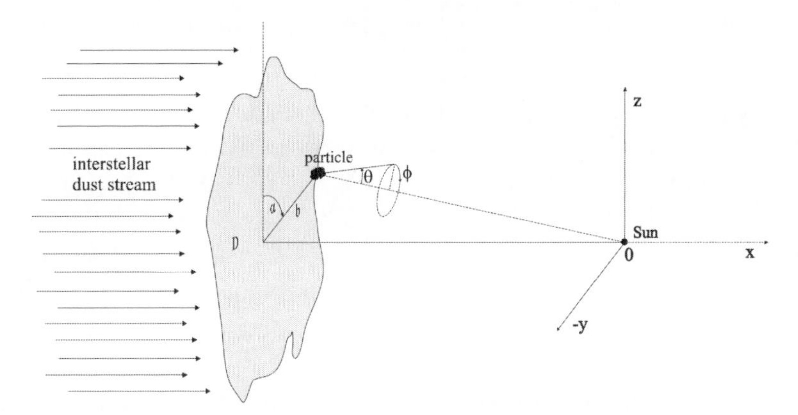

Figure 6. The scheme documents an interaction of interstellar dust particle with the Solar System (according to Kocifaj and Klačka, 2003). The initial position of the particle in dust stream is deffined by the values of impact parameter b and position angle α. The orientation of the particle rotation axis with respect to the incident solar radiation corresponds to the angles θ and ϕ.

The total mass $M(R)$ of the captured IsDPs with radius R is proportional to the capture area $D(R, \theta, \phi)$

$$M(R) \sim N(R)\, m(R) \int_{\theta=0}^{\pi} \int_{\phi=0}^{2\pi} D(R, \theta, \phi)\, \sin\theta \, d\theta \, d\phi \tag{7}$$

where $m(R)$ is a mass of single particle with radius R, and $N(R)$ is the number of interstellar particles crossing the unit area in a unit time. The angles θ and ϕ characterize the

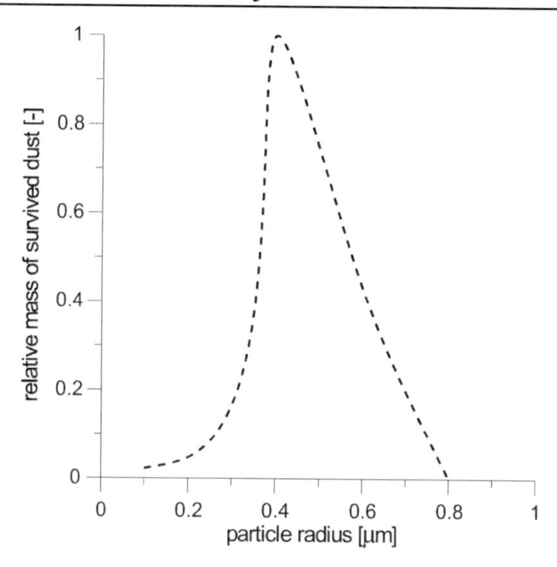

Figure 7. Density function for the mass of non-spherical electrically neutral magnesium-rich silicate particles captured in the Solar System ($\eta = 0$).

orientation of the particle's rotation axis in respect to the direction of incidence of solar electromagnetic radiation (see Fig. 6). The capture area $D(R, \theta, \phi)$ is defined unambiguously as follows: all particles with radius R intersecting this area are captured in the Solar System. Using Eq. (7) we calculated the normalized mass of captured magnesium-rich particles of the shape identical to cosmic dust grain U2015 B10 (Kocifaj et al. 1999). The aspect ratio of the particle U2015 B10 (archived in NASA catalogue; Clanton et al. 1984) coincides very well with typical aspect ratios obtained from mid-infrared spectropolarimetry (Hildebrand and Dragovan 1995). The results of numerical runs presented in Fig. 7 document that non-spherical particles with effective radius 0.4 μm are the most preferrably captured IsDPs in the Solar System if the solar wind is not taken into account (i.e., $\eta = 0$). These results are, however, relevant only for electrically neutral particles. The incorporation of the electric charge is important for generalization of the obtained results. This kind of investigation is presented in the following section.

4. Electrically Charged Arbitrarily Shaped Particles

If dust particle is smaller than 1 micron, and arbitrarily shaped, then one has to take into account also the effect of Lorentz force, since the interplanetary magnetic field plays an important role in motion of the particle. One has to add Lorentz force to Eq. (4). The acceleration of the particle with respect to the Sun, including solar gravity, solar wind and

Lorentz force, is

$$\begin{aligned}
\frac{d\vec{v}}{dt} &= -\frac{GM_{\odot}(1-\beta)}{r^3}\,\vec{r} - \beta\left(1 + \frac{\eta}{\bar{Q}'_{pr}}\right)\frac{GM_{\odot}}{r^2}\left(\frac{\vec{v}\cdot\vec{e}_R}{c}\,\vec{e}_R + \frac{\vec{v}}{c}\right) \\
&+ \frac{GM_{\odot}}{r^2}\sum_{j=2}^{3}\beta_j\left[\left(1 - 2\,\frac{\vec{v}\cdot\vec{e}_R}{c} + \frac{\vec{v}\cdot\vec{e}_j}{c}\right)\vec{e}_j - \frac{\vec{v}}{c}\right] + \\
&\frac{q}{m}\,(\vec{v} - \vec{u}) \times \vec{B}\,, \\
\beta &= 7.61 \times 10^{-4}\,\frac{\bar{C}'_{pr}[m^2]}{m[kg]}\,, \\
\beta_j &= 7.61 \times 10^{-4}\,\frac{\bar{C}'_{pr,j}[m^2]}{m[kg]}\,,\quad j = 2, 3\,, \\
\eta &= \frac{1}{3}\,,
\end{aligned} \tag{8}$$

$\vec{u} = u\,\vec{e}_R$ is velocity of the solar wind (conventionally, the value $u = 400$ km/s is used when dealing with motion of dust particles in the Solar System, e. g., Dohnanyi 1978, Gustafson 1994) and the electric charge of the particle is given by relation $q = 4\,\pi\,\varepsilon_0\,U\,R_{eff}$, where ε_0 is the permittivity of the vacuum, U is surface potential of the particle (Kimura and Mann 1998) with effective radius R_{eff} (volume of the particle equals volume of the sphere of the effective radius). The magnetic field $\vec{B}$ is (Grün *et al.* 1994):

$$\begin{aligned}
\vec{B} &= B_R\,\vec{e}_R + B_T\,\vec{e}_T\,, \\
B_R &= B_{R0}\left(\frac{r_0}{r}\right)^2 \cos\left(\pi \times t[years]/11 + \varphi_0\right)\,, \\
B_T &= B_{T0}\,\frac{r_0}{r}\,\cos\vartheta\,\cos\left(\pi \times t[years]/11 + \varphi_0\right)\,, \\
B_{R0} &= B_{T0} = 3nT\,,\quad r_0 = 1AU\,,
\end{aligned} \tag{9}$$

$\vec{e}_T = \vec{e}_A \times \vec{e}_R$, $\vec{e}_A$ defines magnetic axis of the Sun, ϑ is the heliographic latitude and different polarity of the magnetic field for northern and southern hemispheres is also considered. The phase angle of magnetic field $\varphi_0 = 0$ can be considered.

To recognize the influence of solar wind on capture probability of electrically charged dust particles we performed the calculations equivalent to those discussed in Fig. 7. The actual results are presented in Fig. 8 for two main solar wind components (Zirker, 1981; Hanslmeier, 2002), i.e. for slow ($u = 400$ km/s) and fast ($u = 800$ km/s) solar winds (but $\eta = 0$). Detail analysis shows that capture area $D(R, \theta, \phi)$ found in the case of fast solar wind is approximately two times larger than the corresponding capture area corresponding to the slow solar wind component. In contrary to this fact the mass of the surviving charged dust particles for the fast solar wind is two times smaller than that for the slow solar wind – as it is documented in Fig. 8. One of extremely interesting findings of our numerical simulations is that particles with radii 0.3 μm – 0.6 μm are unambiguosly captured in the Solar System, independent of whether they are charged or electrically neutral (compare Figs. 7 and 8). Such result shows sizes of particles contributing to the dust cloud in the Solar System. These theoretical resuls are consistent with experimental data which indicate that characteristic radius of interstellar dust particles moving in Solar System is about 0.3 - 0.4 μm (Grün et al., 1997).

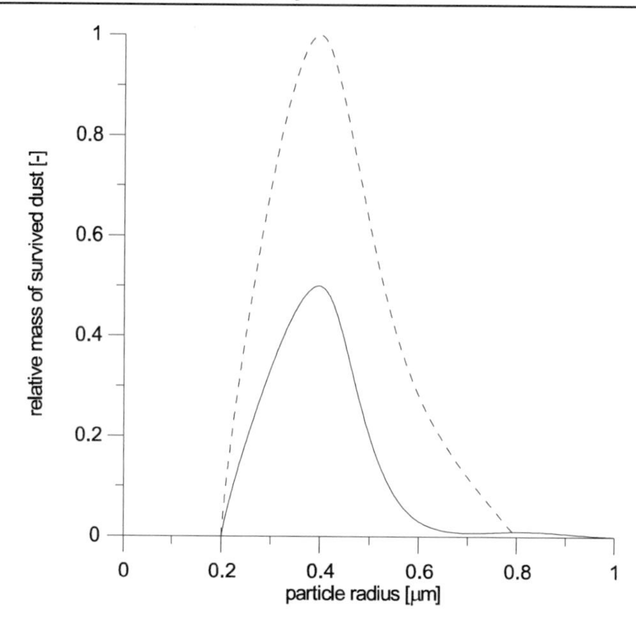

Figure 8. Density function for the mass of non-spherical charged magnesium-rich silicate particles captured in the Solar System (according to Kocifaj and Klačka, 2004). The dashed curve corresponds to slow solar wind component (400 $km\ s^{-1}$), the solid curve to fast solar wind component (800 $km\ s^{-1}$).

5. Conclusion

The effect of solar wind causes decrease (in 60 % – 70 %) of concentration of spherical IsDPs orbiting the Sun in comparison with the action of the Poynting-Robertson effect. This holds for particles of radii about 0.4 μm, which represent the most frequent IsDPs entering the Solar System. However, the solar wind also enriches the amount of spherical dust grains orbiting the Sun: IsDPs of radii 0.6 μm – 1.0 μm can revolve around the Sun thanks to the effect of the solar wind, the pure P-R effect cannot generate such kind of behavior. In any case, spherical IsDPs can revolve around the Sun not far from the evaporation zone (not more than several tens of solar radii from the Sun).

As for non-spherical IsDPs rapidly rotating around fixed orientations of their rotational axes, the particles can be captured in the Solar System and orbit the Sun also in regions far from the evaporation zone of the Sun. The effect of solar wind corresponding to η = 1/3 in Eqs. (4) and (8) plays the less significant role in comparison with the effect of solar electromagnetic radiation, the greater is the aspect ratio of the non-spherical IsDPs (aspect ratio of a particle is defined as the ratio between the maximal and minimal size of the particle). If the aspect ratio is 1.4 or greater, then the effect of the solar wind is at least in one order of magnitude less important that the effect of the electromagnetic radiation: the solar wind effect defined by the value $\eta = 1/3$ in Eqs. (4) and (8) is negligible in secular evolution of orbital elements. However, the solar wind effect is important in equation of motion for IsDPs in the form of the Lorentz force. The capture of the IsDPs orbiting the Sun is most effective for dust particles of effective radii about 0.4 μm, which also corresponds to the peak of size distribution of IsDPs entering the Solar System.

Acknowledgment

This work was supported by the Scientific Grant Agency VEGA, Slovakia, grant No. 1/3074/06.

References

[1] Baguhl M., Grün E., Hamilton D., Linkert G., Riemann R., Staubach P., 1995. The flux of interstellar dust observed by Ulysses and Galileo. *Space Sci. Rev.* **72**, 471-476.

[2] Bohren C. F., Huffman D. R., 1983. Absorption and Scattering of Light by Small Particles, John Wiley & Sons, Inc., New York.

[3] Clanton V. S., Gooding J. L., McKay D. S., Robinson G. A., Warren J. L., Watts L. A. (Eds.), 1984. Picture In: Cosmic Dust Catalog (Particles from collection ¿ag U2015), Vol. 5/1 10. NASA Johnson Space Center, Houston, TX.

[4] Dermott S. F., Grogan K., Durda D. D., Jayaraman S., Kehoe T. J. J., Kortenkamp S. J., Wyatt M. C., 2001. Orbital evolution of interplanetary dust. In: Interplanetary Dust. E. Grün, B. A. S. Gustafson, S. F. Dermott and H. Fechtig (eds.), Springer-Verlag, Berlin, 569-639.

[5] Dohnanyi J. S., 1978. Particle dynamics. In: Cosmic Dust, J. A. M. McDonnell (Ed.), Wiley-Interscience, Chichester, 527-605.

[6] Dorschner J., Begemann B., Henning Th., Jäger C., Mutschke H., 1995. Steps toward interstellar silicate mineralogy. II. Study of Mg-Fe-silicate glasses of variable composition. *Astron. Astrophys.* **300**, 503-520.

[7] Draine B. T., Flatau P. J., 2004. User Guide for the Discrete Dipole Approximation Code DDSCAT.6.1. http://arxiv.org/abs/astro-ph/0409262

[8] Draine B. T., 1988. The discrete-dipole approximation and its application to interstellar graphite grains. *Astrophys. J.* **333**, 848-872.

[9] Draine B. T., Goodman J., 1993. Beyond Clausius-Mossotti: wave propagation on a polarizable point lattice and the discrete dipole approximation. *Astrophys. J.* **405**, 685-697.

[10] Grün E., Zook H. A., Baguhl M., Balogh A., Bame S. J., Fechtig H., Forsyth R., Nanner M. S., Horanyi M., Kissel J., Lindblad B.-A., Linkert D., Linkert G., Mann I., McDonnell J. A. M., Morfill G. E., Phillips J. L., Polanskey C., Schwehm G., Siddique N., Staubach P., Svestka J., Taylor A., 1993. Discovery of jovian dust streams and interstellar grains by the Ulysses spacecraft. *Nature* **362**, 428-430.

[11] Grün E., Gustafson B., Mann I., Baguhl M., Morfill G. E., Staubach P., Taylor A., Zook H. A., 1994. Interstellar dust in the heliosphere. *Astron. Astrophys.* **286**, 915-925.

[12] Grün E., Staubach P., Baguhl M., Hamilton D. P., Zook H. A., Dermott S., Gustafson B. A., Fechtig H., Kissel J., Linkert D., Linkert G., Srama R., Hanner M. S., Polanskey C., Horanyi M., Lindblad B. A., Mann I., McDonnell J. A. M., Morfill G. E., Schwehm G., 1997. Southnorth and radial traverses through the interplanetary dust cloud. *Icarus* **129**, 270-288.

[13] Gustafson, B. A. S., 1994. Physics of Zodiacal Dust. *Ann. Rev. Earth and Planet. Sci.* **22**, 553-595.

[14] Jackson A. A. 2001. The capture of interstellar dust: the pure Poynting-Robertson case. *Planet. Space Sci.* **49**, 417-424.

[15] Hanslmeier A., 2002. The Sun and Space Weather. Kluwer Academic Publishers, Dordrecht, 243pp.

[16] Hildebrand R. H., Dragovan M., 1995. The shapes and alignment properties of interstellar dust grains. *Astrophys. J.* **450**, 663-666.

[17] Kastrov V. G., 1979. Selected papers on atmospheric physics, Gidrometeoizdat, Leningrad (in Russian).

[18] Klačka J., 2004. Electromagnetic radiation and motion of a particle, *Celestial Mech. and Dynam. Astron.* **89**, 1-61.

[19] Klačka J., 2008a. Mie, Einstein and the Poynting-Robertson effect. arxiv:08072795

[20] Klačka J., 2008b. Electromagnetic radiation, motion of a particle and energy-mass relation. arxiv:0807.2795

[21] Kocifaj M. Klačka J., 2003. The capture of interstellar dust: the pure electromagnetic radiation case. *Planet. Space Sci.* **51**, 617-626.

[22] Kocifaj M. Klačka J., 2004. The capture of interstellar dust: the Lorentz force case. *Planet. Space Sci.* **52**, 839-847.

[23] Kocifaj M., Kapišsinský I., Kundracík F., 1999. Optical effects on irregular cosmic dust particle U2015 B10. *J. Quant. Spectrosc. Radiat. Transfer* **63**, 1-14.

[24] Kimura H., Mann I., 1998. The electric charging of interstellar dust in the solar system and consequences for its dynamics. *Astrophys. J.* **499**, 454-462.

[25] Krauss O., 2005. private communication

[26] Krauss O., Wurm G., 2004. Radiation pressure forces on individual micron-size dust particles: a new experimental approach *J. Quant. Spectrosc. Radiat. Transfer* **89**, 179-189.

[27] Landgraf M., 2000. Modeling the motion and distribution of interstellar dust inside the heliopshere. *J. Geophys. Res.* **105**, 10302-10316.

[28] Landgraf M., Augustsson K., Grün E., Gustafson B. S., 1999. Deflection of the local interstellar dust flow by solar radiation pressure. *Science* **286**, 2319-2322.

[29] Mie G., 1908. Beitrâge zur Optik trűber Medien speziell kolloidaler Metalősungen. *Ann. Phys.* **25**, 377-445.

[30] Mishchenko M. I., Travis L. D., Lacis A. A., 2002. Scattering, Absorption, and Emission of Light by Small Particles, Cambridge University Press, Cambridge.

[31] Neckel H., Labs D., 1981. Improved data of solar spectral irradiance from 0.33 to 1.25 microns. *Solar Phys.* **74**, 231-249.

[32] Opher M., Liewer P. C., Velli M., Bettarini L., Gombosi T. I., Manchester W., DeZeeuw D. L., Toth G., Sokolov I., 2004. Magnetic effects at the edge of the Solar System: MHD instabilities, the de Laval Nozzle effect, and an extended jet. *Astrophys. J.* **611**, 575-586.

[33] Purcell E. M., Pennypacker C. R., 1973. Scattering and absorption of light by non-spherical dielectric grains. *Astrophys. J.* **186**, 705-714.

[34] Poynting J. M., 1903. Radiation in the Solar System: its Effect on Temperature and its Pressure on Small Bodies. *Philosophical Transactions of the Royal Society of London* **Series A 202**, 525-552.

[35] Robertson H. P., 1937. Dynamical effects of radiation in the Solar System. *Mon. Not. R. Astron. Soc.* **97**, 423-438.

[36] Sykes M. V., 2007. Infrared Views of the Solar System from Space. In: Encyclopedia of the Solar System, L.-A. McFadden, P. R. Weissmann and T. V. Johnson (eds.), Academic Press (Elsevier), San Diego, 2nd ed., 681-694.

[37] Sykes M. V., Grün E., Reach W. T., Jenniskens P., 2004. The interplanetary dust complex and Comets. In: Comets II, M. C. Festou, H. U. Keller and H. A. Weaver (eds.), The University of Arizona Press, Tucson, in collaboration with Lunar and Planetary Institute, Houston, 677-693.

[38] van de Hulst H. C., 1981. Light Scattering by Small Particles. Dover Publications, Inc. New York, 470 pp. (originally published in 1957 by John Wiley & Sons, Inc., New York)

[39] Wriedt T., Comberg U., 1998. Comparison of computational scattering methods. *J. Quant. Spectrosc. Radiat. Transfer* **60**, 411-423.

[40] Wurm G., Schnaiter M., 2002. Fractal aggregates in space. In: Optics of Cosmic Dust, G. Videen and M. Kocifaj (eds.), Kluwer Academic Publishers, Dordrecht, 89-102.

[41] Zirker J. B., 1981. The solar corona and the solar wind. In: Jordan, S. (Ed.), The Sun as a Star. NASA, Washington, DC, pp. 135-162.

In: Solar System: Structure, Formation and Exploration
Editor: Matteo de Rossi
ISBN: 978-1-62100-057-0

Chapter 7

A PROBABILISTIC MODEL FOR THE SOLAR SYSTEM PLANETARY DISTRIBUTION DRIVEN BY THE CENTRAL STAR FEATURES

Andrea Bolle [*] ***and Christian Circi*** [†]
"Sapienza", University of Rome, Italy

ABSTRACT

The discovery of several extrasolar systems, each one characterized by its own planetary distribution around a central star, made the scientific interest addressed to the analysis of models permitting to predict, or at least estimate, the orbital features of the extrasolar planets. The main purpose of this work is to describe a mathematical model, inspired by quantum mechanics, able to provide a probability distribution of planets placing in a star system, mainly driven by the central star mass. More in detail, for any given eigenvalue of the model discrete spectrum, a distinct probability distribution with respect to the central star distance can be buildt. As per the Solar System, it has been possible to prove that both inner and outer planets belongs to two different spectral sequences, each one originated by the minimum angular momentum owned by silicate/carbonate and icy planetesimals respectively. In both sequences, the peak of the probability distributions almost precisely coincided with the average planets distance from Sun; further more, the eigenvalue spectrum of the inner planets thickens in an accumulation point corresponding to the asteroids belt, thus showing a striking similarity to the real matter distribution in the Solar System. From this point of view, the Titius-Bode law for the Solar System planets distribution is nothing but an exponential interpolation of the eigenvalues of both inner and outer sequences.

* "Sapienza" University of Rome, Aerospace Engineering, PhD, Via Salaria 851, 00138 Rome, Fax 39-06-49919757, Tel 39-06-49919766

† "Sapienza" University of Rome, Assistant Professor, Via Salaria 851, 00138 Rome, Fax 39-06-49919757, Tel 39-06-49919766

OUTLINES

By taking the cue from the well known Titius-Bode law, the present work aims to assess whether the present Solar System planetary distribution is the result of a purely-stochastic phenomenon, or rather it fulfils a deterministic criterion. In this sense, the Titius-Bode law would be a mathematical rule catching the meaning of a deeper order driving the Solar System formation towards a "preferred" configuration.

By assuming that the present Solar System planetary distribution answers for a probabilistic criterion, it is possible to build a probabilistic model, analogous to that designed for the quantum mechanics, whose eigenpairs fully matches the observed planetary orbital distribution. It is important to notice that the present work offers a reading of the Titius-Bode law, without claiming to be considered as its mathematical proof.

To clearly describe the evolution from the Titius-Bode law to the final probabilistic model, this work is outlined as follows. Section I provides the historical context during which the Titius-Bode law has been edited, tested, discussed and eventually cast asides. In Section II, a possible alternative expression together with the rational behind it is proposed. This Section paves the way to the assumption of the probabilistic criterion driving the planets formation and that finds its mathematical expression in Section III. Indeed, in this Section the complete development of the wave equation applied to the Solar System dynamics is shown and deeply explained, through a step-by-step approach. Once the model has been introduced, the corresponding solutions will be derived in Section IV. This Section deserves a detailed analysis, as it contains the description of a key parameter for the planetary distribution: the equivalent Planck's constant for planets motion. According to its value, the sequence of eigensolutions of the wave equation change completely. Instead of being extrapolated from numerical analysis, the Authors provide in this Section a criterion to determine the equivalent Planck's constant based on the analysis of the chemical composition of the original solar nebula and on its temperature radial profile. Section V offers a discussion of the results obtained through the probabilistic model, together with the future developments necessary to gain a rigorous proof of the assumed deterministic criterion running the Solar System formation.

I. HISTORY OF THE TITIUS-BODE LAW

The study of the Solar System origin, especially its evolution up to present configuration, has been a widely discussed topic, debated by physicists and before them by philosophers, belonging to any historical period.

Although ancient astronomers lacked an advanced physical model, the obserations collected throughout the centuries allowed a complete characterization of the Solar System geometry: the three Kepler laws, written so to fit the observation collected by Thyco Brahe, provide the evidence of a deep knowledge of the cinemathics of the planets motions. Once the Newtonian model of gravity had been introduced[1], planets motion was given a "dynamic" characterization: the ellipses described by Kepler ceased to apper as a purely geometric

[1] Sir Isaac Newton released his laws of motion on Jult 5th, 1687, in the booklet *Philosophiae Naturalis Principia Matematica.*

results, to become a solution of the differential equations of motion. The Newtonian approach to physics, and the origin of the *classical physics*, paved the way to a century characterized by a deep inquiry in physics and in mathematics.

In this mathematical framework, under the novel, deterministic approach to the physical and natural problems, the Titius-Bode law has been proposed, based on the observations collected by analyzing the motion of the Solar System's planets. According to this laws, edited by the astronomers Johann Daniel Titius and Johann Elbert Bode, the orbital spacing of the Solar System planets fulfils a simple, mathematical sequence, given by the following expression:

$$r_m[A.U.] = 0.4 + 0.3 \cdot 2^m, \; m = -\infty, 0, 1, 2, ... \quad (1)$$

being r_m the Solar System planets semi-major axis, expressed in Astronomical Units. A summary of the planets mean radii as predicted by the laws, in comparison with the real ones, is given in Table 1.

Table 1. A comparison of the mean Solar System planets radii, evaluated as per the Titius-Bode law, and the real ones

m	Planetary body	Radius[A.U.] fromTitius-Bode law	Observed Mean Radius[A.U.]
-∞	Mercury	0.4	0.39
0	Venus	0.7	0.72
1	Earth	1	1.00
2	Mars	1.6	1.52
3	Caeres	2.8	2.77
4	Jupiter	5.2	5.20
5	Saturn	10	9.54
6	Uranus	19.6	19.18
-	Neptune	-	30.06
7	Pluto	38.8	39.44

When originally published, the law seemed to predict with sufficient precision the semi-major axis of all the known planets, i.e. Mercury, Venus, Earth, Mars, Jupiter and Saturn, but it left a gap between the fourth and fifth planets. This mismatch did not weaken the Titius-Bode law; moreover, the discovery of Uranus in 1781, fitting neatly into the series, provided the ultimate confirmation of the law. Based on this discovery, Bode urged a search for a fifth planet. Ceres, the largest object in the asteroid belt, was found at Bode's predicted position in 1801. Bode's law has been then widely accepted, until *Neptune was discovered* in 1846: this planet was found not to fulfill to the Titius-Bode sequence, and this was the first blow to the law. Simultaneously, the large number of known asteroids in the belt resulted in Ceres no longer being considered a planet. Bode's law was discussed as an example of fallacious reasoning by the astronomer and logician Charles Sanders Peirce in 1898. The discovery of Pluto in 1930 confounded the issue still further. While nowhere near its position as predicted by Bode's law, it was roughly at the position the law had predicted for Neptune. However, the subsequent discovery of the *Kuiper belt*, and in particular of the object *Eris*, which is larger than Pluto yet does not fit Bode's law, have further discredited the formula.

The actual knwoledge of the Solar System dynamics and formation allows to regard the prediction sequence described by Titius and Bode as an "empiric rule", more than a "law": in other terms, the sequence provides an indirect clue of a not purely randomic formation of the Solar System planets, and it shed a light on the hypothesis that the present Solar System configuration is someway "preferred" in comparison to all the available ones.

It is now understood that the actually observed Solar System planets distribution matches the hypothesis of long-period resonances that eventually led to the mot stable configuration. To test this assumption, several simulations reproducing the aggregation process that led to the planets formation and based on a randomic distribution of the initial conditions have been carried put. Interestingly, in a significant number of cases such simulations led to a planet distribution analogous to that observed in the Solar System. Therefore, this result suggest that the present planetary distribution might fulfill a probabilistic criterion: in other terms, the actual planetary distribution may be regarded as the most likely, rather than the most stable. These two distinct assumptions do not mismatch completely one with the other; but notably, the power rule by Titius and Bode gets the meaning of a not purely stochastic phenomenon.

II. Other Power Rules Expressing the Planetary Distribution

II.1. The Murray and Dermott's Law

An indirect approach to take a step forward towards a deeper comprehension of the Titius-Bode law might be the research for some alternative expressions to the usual Titius-Bode rule. Indeed, a different formulation might lead to a deeper comprehension of the supposed criterion driving the Solar System formation.

Murray and Dermott (1999), provided the following expression

$$R_n = R_0 \cdot A^n, \qquad n = 1,2,3,\dots. \tag{2}$$

Such a formulation has been obtained through a numeric interpolation of planets orbit period, by assuming their values change according to a geometric progression. In the Murray and Dermott formulation, the integer parameter n ranges between 1 and 10, and not between $-\infty$ and 10, as it happens in the Titius-Bode rule. This is not a casualty, but an authors' choice, as in their opinion the position $n = -\infty$ seemed too unperceptual.

By minimizing the norm of the deviations χ^2 with respect to R_0 and A:

$$\chi^2 = \frac{1}{k}\sum_{n=1}^{k}\left[\log R^{observed} - \left(\log R_0 + n\log A\right)\right]^2, \tag{3}$$

Murray and Dermott obtained $R_0 = 0.2139$ and $A = 1.706$. Their novel expression for the Titiuus-Bode rule became therefore:

$$R_n = 0.2139 \cdot (1.706)^n, \qquad n = 1,2,3,... \qquad (4)$$

that gives results analogous to that by Titius and Bode, as summarized in **Table 2**

Table 2. Mean planets radii from Murray and Dermott's rule in comparison with those from the Titius-Bode one

n	Planetary body	Radius[A.U.] from Dermott law	Observed Mean Radius[A.U.]
1	Mercury	0.37	0.39
2	Venus	0.63	0.72
3	Earth	1.07	1.00
4	Mars	1.83	1.52
5	Caeres	3.13	2.77
6	Jupiter	5.36	5.20
7	Saturn	9.17	9.54
8	Uranus	15.68	19.18
9	Neptune	26.82	30.06
10	Pluto	45.88	39.44

The planetary distribution according to Murray and Dermott resembles that of Titius-Bode in that it encompasses also a major celestial body between Mars and Jupiter; at the same time, it fits better as per the outer planets, Neptune and Pluto. Indeed, the Murray and Dermott rule provides a better scheme able to predict the sequence of the plates orbits in the Solar System.

These results were so encouraging that made worthy even the research for analogous rules in satellites system, as the Jovian, Saturn and the Uranus ones. As per these systems, Murray and Dermott provided an interpolation formula in terms of orbital period T instead of the mean orbits radii:

$$T_n = T_0 A^n, \qquad n = 1,2,3,.. \qquad (5)$$

being T_0, A summarized below for each system:

- *Jovian system*: T(0) = 0.444; A = 2.03
- *Saturn system*: T(0) = 0.462; A = 1.59
- *Uranus system*: T(0) = 0.488; A = 2.24

The presence of alternative formulations to the original Titius-Bode rule seems to strengthen the hypothesis of an underlying criterion driving the Solar System formation towards a not-randomic configuration. In that, the Murray and Dermott's rule is noteworthy. Nevertheless, being numerically extrapolated from observable data, even this formulation cannot provide a straight connection to such a criterion.

II.2. A Third Fitting Expression for Planetary Distribution

Whereas both the Titius-Bode and the Murray-Dermott rules had been derived from observations, there still exists another formulation deriving from the dynamic model of a moving mass in motion under the solely Sun gravity. Such a rule, obtained by the Author and explained below, not only seems to fit the effective planetary distribution as well as the Murray and Dermott rule, but it derives from the analysis of the orbital periods and some consideration about the dynamics, listed below:

1) As planets orbits are nearly circular, the Titius-bode rule still holds for the planets semi-major axes;
2) By noting that the semi-major axis is related to the energy per unit mass owned by every planet in that it is placed in a central gravity field, the Titius-Bode rule allows interpolating the progressively increase of the energy of the celestial bodies in the Solar System;
3) The numerical sequence interpolating both the planets radii and energy per unit mass depends explicitly on the first planet, Mercury, chosen as a pivotal element;
4) The interpolating sequence for the planets radii, and therefore for the Energy per unit mass, depends on a discrete variable.

Altogether, these considerations could lead one to think that the Titius-Bode rule describes a sort of "discrete spectrum" of the energy per unit mass owned by a planet moving inside a central gravity field. Such energy values, being aside from the initial conditions of the protoplanets that eventually formed the Solar System as is, and depending only on the mass responsible for the central gravity field, may be thought as eigenvalues of the motion equations of a protoplanet moving inside the solar gravity field.

Let $Oxyz$ be a 3-dimensional, inertial reference frame, with the origin O coincident with the Sun. The motion equation of a moving mass under the solely gravity of the Sun, according to the vector notation, is given by:

$$\frac{d^2\boldsymbol{r}}{dt^2} = -\frac{\mu}{\boldsymbol{r}^3}\boldsymbol{r}, \tag{6}$$

being μ the Sun gravity constant and r the mass distance from the origin O[2].

Without any loss of generality, let us assume that the orbits are circular. This last hypothesis allow us to replace r^3 with a^3, being a the semi-major axis of the protoplanet. Under these hypotheses, the new equation system becomes in vector notation becomes:

$$\frac{d^2\boldsymbol{r}}{dt^2} = -\frac{\mu}{\boldsymbol{a}^3}\boldsymbol{r}. \tag{7}$$

[2] Such an equation of motion cannot be applied as is to describe the behaviour of a protoplanet moving within the moulding Solar System, as it does not take into account the "viscous" interaction with other masses in the neighbourhood. Nevertheless, by considering only the presence of Sun, such a dynamic allows obtaining an eigenvalues spectrum for the energy per unit mass that close resembles the Titius-Bode rule.

Now, let us look for a complete set of solutions to this system: to do so, we choose to neglect the set of initial conditions and look only for eigenpairs. From the theory of linear differential equations, it is known that the linearized oscillator equation has the following frequence ω_0:

$$\omega_0 = \sqrt{\frac{\mu}{a^3}},$$

Let T_0 be the period of the planet of the sequence, and let us suppose that the n-th planet in the Solar System has a period multiple of T_0 by a factor n:

$$n \cdot T_0 = 2\pi \cdot \sqrt{\frac{a_n^3}{\mu}}, \tag{8}$$

while Eq. 7 leads to

$$a_n = \left(\frac{T_0^2 \mu}{4\pi^2} \right)^{1/3} \cdot n^{2/3} \tag{9}$$

or simply:

$$a_n = a_{\min} \cdot n^{2/3} \tag{10}$$

being $a_{\min}$ the first semi-major axis of the sequence.

Eq. 8 provides a sequence of consecutive semi-major axes as a function of an integer parameter n. In analogia con la legge di Titius-Bode. Interestingly, if one choose the integer n as a power of 2, as per Titius-Bode law, the corresponding semi-major axis sequence match very closely the true planets mean radii:

$$a_n = a_{\min} \cdot \left(2^p\right)^{2/3} \tag{11}$$

This means, in terms of orbital period, that they increase according to the power of 2. Results are summarized in Table 3.

These results deserve an in-depth examination.

a) The sequence chosen according to Eq. 9 seems to fit the position of Uranus, Neptune and Pluto too, even if with a not negligible error. Whereas the Murray and Dermott rule is a numeric inteporlation of the observed data, and the Titius-Bode one depends on an integer varying in a too-wide range, the present rule descend from simple analytic considerations and depends on an integer p whose starting value is $p = 0$;

b) As per the Asteroid belt, the sequence (11) provides the semi-major axes of 2 major asteroids: Caeres and Sylvia, as it occurs for both the Titius-Bode rule and the Murray and Dermott one. This is more than a coincidence: nowadays it is well-known that the mass of Jupiter prevented the Asteroids from forming a planet between Mars and Jupiter itself;

c) The dependence on the powers of 2 still holds, although it has not been justified mathematically.

Table 3. Comparison between semi-major axes evaluated through Eq. (11) and the true plantets ones

n	2^p	Planetary body	Radius[A.U.] from formula (8)	Observed Mean Radius[A.U.]
1	2^0	Mercury	0.39	0.39
2	2^1	Venus	0.62	0.72
4	2^2	Earth	0.98	1.00
8	2^3	Mars	1.56	1.52
16	2^4	Caeres	2.48	2.77
32	2^5	Sylvia	3.93	3.50
64	2^6	Jupiter	6.24	5.20
128	2^7	Saturn	9.91	9.54
256	2^8	Uranus	15.72	19.18
512	2^9	Neptune	24.96	30.06
1024	2^{10}	Pluto	39.62	39.44

As per the rule (11), the most notable feature – beside the dependence on the powers of 2 - is the dependence on the period of Mercury, the first planet of the sequence.

Hayes and Tremaine (1998), through the testing of several interpolating functions, concluded that the regularity in the present planetary distribution depends on the long term resonances responsible for the decay of unstable orbits: in other terms, the present configuration is stable, other than the most likely.

These considerations, as well as the presence of multiple alternative formulations to the original Titius-Bode rule, suggest the hypothesis that the driving criterion leading to the actual planet spacing may be a not only a likelihood criterion, but an asymptotic stability criterion. The present orbits of the Solar System planets are the asymptotic result of an evolution occurred during the formation of the Solar System and heading towards stability, and at the same time they fulfill a likelihood principle whose main agent is the Solar mass.

With these assumptions, it is straightforward to search for a prediction model based probability density and an asymptotic evolution in time. The probabilistic model employed in the quantum physics, whose solution fulfill the well known as Schrödinger's equation, seems to meet the requirements described up to now. Furthermore, the results of such a model would be eigensolutions of the system, depending only on the central force field applied and not on the particular boundary/initial conditions. Nevertheless, such an approach encounters two major objections.

The first critic to the application of such a model to the physics of the Solar System is that whereas the state of electrons in an isolated atom changes with discrete variations, multiple of the Planck's constant, the physical state of planet do not. However, if one assume that only the asymptotically stable orbits of planets increase with discrete variations, the model holds. With this assumption, the traditional idea of quantized energy and momentum, well known for the electrons of an isolated atom, changes significantly. As per the Solar System dynamics, the quantization of the planets energy and momentum holds only for the asymptotically stable orbits, as the long-term resonances caused the decay of unstable ones. Stable states, whose components are the energy and the momentum - and therefore the semi-major axis and the eccentricity – are regularly spaced, and vary as multiple of an elementary quantity, i.e. a sort of Planck's constant for the Solar System. And this is the second critical issue. Whereas the Planck's constant is *de facto* a physical evidence, there is no corresponding evidence in the Solar System dynamics and formation. To build a "quantistic" model able to justify the planetary distribution, it is necessary to extrapolate compensation from the analysis of the Solar System formation. In the following Sections, a detailed description of the quantistic model for gravity and a criterion for the determination of the corresponding Planck's constant will be provided.

III. Transposition of the Quantum Model to the Planetary Spacing Motion

III.1. Wave Functions for Gravity

The quantistic model is based on the assumption that a system state can be completely described through a generally complex function of its coordinates q, from now on named with $\boldsymbol{\Psi}(q)$. Such a function is referred as the system wave function, and it is characterized by the following properties:

1) $|\boldsymbol{\Psi}(q)|^2 dq$ is the probability density that a measure executed on the system provides values falling within a range of *dq* within the configuration space.
2) The sum of $|\boldsymbol{\Psi}(q)|^2 dq$ over the whole configuration space must be equal to unity:

$$\int |\boldsymbol{\Psi}(q)|^2 dq = 1, \tag{12}$$

Eq. (12) is also known as the normalization condition;

1) The wave functions depend on time;
2) The superimposition of states must hold. According to this properties, any intermediate state $\boldsymbol{\Psi}_n(q)$ between the state 1 and 2 can be expressed as the linear

combination of $\boldsymbol{\Psi}_1, \boldsymbol{\Psi}_2$. Furthermore, this property implies that the wave functions fulfill linear differential equations;

3) As a consequence of 4), a wave function depending on both the two coordinates q_1, q_2 and named with $\boldsymbol{\Psi}_{12}(t, q_1, q_2)$ can be expressed through the product of two simpler functions $\boldsymbol{\Psi}_1(q,t)$ and $\boldsymbol{\Psi}_2(q,t)$:

$$\boldsymbol{\Psi}_{12}(q_1, q_2, t) = \boldsymbol{\Psi}_1(q_1, t) \cdot \boldsymbol{\Psi}_2(q_2, t), \ \forall t > 0. \tag{13}$$

This last property is notably important when trying to approach the solution to the Schrödinger's equation.

III.2. The Hamiltonian Operator

Let f be a given component of a quantitstic state, i.e. the energy per unit mass owned by a planitesimal orbit around the Sun, and let us assume that the spectrum of the eigenvalues of f changes with discretely. Let f_n, $n = 0,1,2,3,...$ be such eigenvalues, and $\boldsymbol{\Psi}_n$ the wave function corresponding to f_n, namely its eigensolution. According to the properties 1) to 5), any arbitrary state $\boldsymbol{\Psi}$, in any instant of time, can be expressed as the linear combination of the eigensolutions $\boldsymbol{\Psi}_n$ as follows:

$$\boldsymbol{\Psi} = \sum_n a_n \boldsymbol{\Psi}_n \tag{14}$$

being a_n a set of parameters attainable through the dot product of $\boldsymbol{\Psi}$ and $\boldsymbol{\Psi}_n$ in the configuration space:

$$a_n = \int \boldsymbol{\Psi} \boldsymbol{\Psi}_n^* dq, \tag{15}$$

being $\boldsymbol{\Psi}_n^*$ the complex conjugate of $\boldsymbol{\Psi}_n$. Such parameters a_n have also an important physical meaning, ashey express the likelihood that the corresponding eigenpair $f_n, \boldsymbol{\Psi}_n$ will occur. As the sum of all the probabilities must be equal to unity, it is easily to understand that:

$$\sum_n |a_n|^2 = 1. \tag{16}$$

Let $\bar{f}$ be the average value of f. Such a value can be obtained through the linear combination of the eigenvalues spectrum f_n, weighted according to the corresponding probability functions a_n:

$$\bar{f} = \sum_n f_n \cdot |a_n|^2 . \tag{17}$$

Let $\hat{f}$ be a linear operator, that applied to a generic wave function $\boldsymbol{\Psi}$ permits to obtain the average value $\bar{f}$ through the scalar product:

$$\bar{f} = \int \boldsymbol{\Psi}^* \cdot \left(\hat{f}\boldsymbol{\Psi}\right) dq \tag{18}$$

By noting that:

$$\bar{f} = \sum_n f_n \cdot |a_n|^2 = \sum_n f_n \cdot a_n a_n^* = \int \boldsymbol{\Psi}^* \cdot \left(\sum_n f_n a_n \boldsymbol{\Psi}_n\right) dq , \tag{19}$$

one gets:

$$\left(\hat{f}\boldsymbol{\Psi}\right) = \sum_n f_n a_n \boldsymbol{\Psi}_n . \tag{20}$$

It is straightforward that:

$$\left(\hat{f}\boldsymbol{\Psi}_n\right) = \hat{f}\boldsymbol{\Psi}_n = f_n \boldsymbol{\Psi}_n , \tag{21}$$

i.e., by applying the operator $\hat{f}$ on a given eigensolution $\boldsymbol{\Psi}_n$ one gets eigensolution itself times the corresponding eigenvalue f_n This means that $\boldsymbol{\Psi}_n$ are the solution to a typical eigenvalue problem, that can be written in the form:

$$\hat{f}\boldsymbol{\Psi} = f\boldsymbol{\Psi} , \tag{22}$$

which gives exactly the steady part of the Schroedinger'equation, when we replace the generic linear operator $\hat{f}$ with the Hamiltonian operator $\hat{H}$.

To introduce the Hamiltonian operator, it is necessary to provide some additional considerations. In quantum mechanics, the wave function fully characterizes the system state. This means that the wave function $\boldsymbol{\Psi}(q, \bar{t})$, known at a given instant $\bar{t}$, allows describing the properties of the system not only at the instant $\bar{t}$ but also its evolution in time. Mathematically, this means that $\frac{\partial \boldsymbol{\Psi}}{\partial t}$ must depend on the wave function $\boldsymbol{\Psi}$, and given the model linearity, the link between $\frac{\partial \boldsymbol{\Psi}}{\partial t}$ and $\boldsymbol{\Psi}$ must be linear.

In other terms:

$$i\kappa \cdot \frac{\partial \boldsymbol{\Psi}}{\partial t} = \hat{H}\boldsymbol{\Psi}, \tag{23}$$

being $\hat{H}$ a linear, hermitian operator and i is the complex unit. The meaning of the additional parameter κ will be given later.

As the integral $\int \boldsymbol{\Psi}^* \boldsymbol{\Psi} dq$ is constant in time, one can write:

$$\frac{\partial}{\partial t}\int \boldsymbol{\Psi}^* \boldsymbol{\Psi} dq = \int \frac{\partial \boldsymbol{\Psi}^*}{\partial t} \boldsymbol{\Psi} dq + \int \boldsymbol{\Psi}^* \frac{\partial \boldsymbol{\Psi}}{\partial t} dq = 0. \tag{24}$$

By combining Eq. with , and introducing the complex conjugate operator $\tilde{\hat{H}}^*$ one gets:

$$-\frac{i}{\kappa}\left[\int\left(\hat{H}\boldsymbol{\Psi}^*\right)\boldsymbol{\Psi} dq + \int \boldsymbol{\Psi}^*\left(\hat{H}\boldsymbol{\Psi}\right)dq\right] = \frac{i}{\kappa}\left[\int\left(\hat{H}^*\boldsymbol{\Psi}^*\right)\boldsymbol{\Psi} dq - \int \boldsymbol{\Psi}^*\left(\hat{H}\boldsymbol{\Psi}\right)dq\right] =$$
$$= \frac{i}{\kappa}\left[\int \boldsymbol{\Psi}^*\left(\tilde{\hat{H}}^*\boldsymbol{\Psi}\right)dq - \int \boldsymbol{\Psi}^*\left(\hat{H}\boldsymbol{\Psi}\right)dq\right] = \frac{i}{\kappa}\int \boldsymbol{\Psi}^*\left(\tilde{\hat{H}}^* - \hat{H}\right)\boldsymbol{\Psi} dq = 0$$

that is an identity if:

$$\tilde{\hat{H}}^* = \hat{H} \tag{25}$$

i.e., if the $\hat{H}$ operator is hermitian.

To fully understand the meaning of the operator $\hat{H}$, it is necessary to recall the meaning of the quasiclassical wave function formulation. Accordingly, the electron was thought vibrating as a wave with a constant energy level. Its corresponding wave function should therefore be given by:

$$\boldsymbol{\Psi} = a \cdot \exp\left[\frac{i}{\kappa} S\right], \tag{26}$$

where a is by far a function of time, κ a dimensional factor and S the system mechanical action. By deriving with respect to time, one gets:

$$\frac{\partial \boldsymbol{\Psi}}{\partial t} = \frac{i}{\kappa} \cdot \frac{\partial S}{\partial t} \cdot \boldsymbol{\Psi}, \tag{27}$$

that combined with (26), gives the remarkably result:

$$\hat{H}\boldsymbol{\Psi} = -\frac{\partial S}{\partial t}\boldsymbol{\Psi}. \tag{28}$$

At this point, according to the Least Action Principle, the term $-\frac{\partial S}{\partial t}$ coincides with the Hamiltonian operator.

If we assume this approach holds for the electron, it should be valid, all the more so, for a classical problem like the planet motion. Indeed, the position vector of a planet can be written in the form:

$$\boldsymbol{P}(t) = R \cdot \exp\left[\frac{i}{\kappa} S\right], \tag{29}$$

being R is the mean radius and S, the mechanical action, is related to the planet anomaly during its motion around the Sun. This show how the quantum model by itself is not strictly related only to electrons state around an atom nucleus, but it can be applied without twisting its original sense to classic problems, when one is interested to examine the most likely states of a given system in time.

As a conclusion, one is allowed to guess that the most likely states in an isolated system ruled by the gravity of a central body, like the Sun, can be successfully described by the sequence of the solution of a Schrödinger-like differential equation. Indeed, such solutions are eigensolutions of the Schrödinger-like equation, thus revealing the inner features of the system itself. Furthermore, as per the Solar System, the most likely solutions are also steady in time, as altogether they define an asymptotic stable state.

One final consideration: as the mechanical action is dimensionally equal to an angular momentum, also the parameter κ must be an angular momentum. We are now introducing what we can define as the equivalent Planck's constant for gravity.

III. 3. The Meaning of Steady States

The Hamiltonian function written for an isolated system, as well as per a system ruled by a a constant force field, cannot depend explicitly on time; the corresponding hold quantity is the overall mechanical energy. The corresponding states, each one characterized by its own level of energy, are defined as steady states, and they are described b the eigenfunctions $\boldsymbol{\Psi}_n$ fulfilling the equation:

$$\hat{H}\boldsymbol{\Psi}_n = E_n \boldsymbol{\Psi}_n, \tag{30}$$

being E_n the corresponding eigenvalues. Thus, the wave equation becomes:

$$i\kappa \frac{\partial \boldsymbol{\Psi}_n}{\partial t} = \hat{H}\boldsymbol{\Psi}_n = E_n \boldsymbol{\Psi}_n, \tag{31}$$

This equation can be straightforwardly solved, thus giving:

$$\boldsymbol{\Psi}_n = \exp\left[-\frac{i}{\kappa}E_n t\right]\cdot \psi_n(q), \tag{32}$$

where $\psi_n(q)$ depends only on the spatial coordinates of the object. Such functions, as well the corresponding eigen-energy, can be obtained by:

$$\hat{H}\psi = E\psi\,; \tag{33}$$

This means that the least energy level among all the available ones is a normal state or also main states of the system. Energy eigenvalues are also referred as energetic levels.

As it has been proved in the previous Section, any wave function can be expressed as a linear combination of all the steady states, with coefficients given by a_n. This allows rewriting Eq. .. as follows:

$$\boldsymbol{\Psi} = \sum_n a_n \cdot \exp\left[-\frac{i}{\kappa}E_n t\right]\cdot \psi_n(q). \tag{34}$$

Up to now, only the general structure of the solution to the quantum model has been described, but we still have not entered deeply into the solution shape: in order to do so, one has to write explicitly the Hamiltonian operator, as it will be shown in the next Sections.

III.4. The Momentum Operator

The Hamiltonian operatori is related to the system overall mechanical energy: if the system is ruled only by inner force fields, not depending explicitly on time, the mechanical energy is constant in time.

Let us now looking for an operator $\hat{p}$, whose corresponding eigenvalue is the momentum. It is possible to show that such an operator would commute with the Hamiltonian one. This means that also the momentum operator $\hat{p}$ is linear, hermitian, and subject to the conservation of its related quantity.

Such an operator is related to the space homogeneity. This means that if the space were not constant, the momentum of an isolated particle would change even if no forces are acting on it. For a single object moving in the force field, the momentum operator $\hat{p}$ shall assume the form:

$$\hat{\boldsymbol{p}} = -i\kappa\nabla\,, \tag{35}$$

or, in cartesian components:

$$\hat{p}_x = -i\kappa\frac{\partial}{\partial x}\hat{\boldsymbol{i}}, \qquad \hat{p}_y = -i\kappa\frac{\partial}{\partial y}\hat{\boldsymbol{j}}, \qquad \hat{p}_z = -i\kappa\frac{\partial}{\partial z}\hat{\boldsymbol{k}}. \quad (36)$$

being κ the angular momentum. As showed for the Hamiltonian opertaor $\hat{H}$, the momentum eigenpairs can be found by solving the equation:

$$-i\kappa\nabla\psi = \boldsymbol{p}\psi. \quad (37)$$

One can easy argue that the corresponding solutions will assume the form:

$$\psi = c \cdot \exp\left[\frac{i}{\kappa}\boldsymbol{p}\cdot\boldsymbol{r}\right], \quad (38)$$

where c is a constant factor and $\boldsymbol{p}$ the momentum.

III.5. The Angular Momentum Operator

Beside the space homogeneity, if the space is also isotrope, any possible direction for motion is equivalent to any other. This means that the Hamiltonian function cannot change after an arbitrary rotation of the system around any axis. But this means also that the operator $\hat{\boldsymbol{l}}$, associated to the angular momentum, can commute with the Hamiltonian operator, thus it shall be linear, hermitian and subject to the conservation of the corresponding quantity.

Such an operator $\hat{\boldsymbol{l}}$, being related to arbitrary rotations in space, shall be written as follows:

$$\kappa\hat{\boldsymbol{l}} = -i\kappa(\boldsymbol{r}\times\nabla) \quad (39)$$

It must be noted $\left|\hat{l}\right|$ is dimensionless. By recalling the definition of the momentum operator $\hat{p}$, one gets:

$$\kappa\hat{\boldsymbol{l}} = -i\kappa(\boldsymbol{r}\times\nabla) = \boldsymbol{r}\times\hat{\boldsymbol{p}}, \quad (40)$$

or in cartesian components:

$$\kappa\hat{l}_x = y\hat{p}_z - z\hat{p}_y, \qquad \kappa\hat{l}_x = z\hat{p}_x - x\hat{p}_z, \qquad \kappa\hat{l}_x = x\hat{p}_y - y\hat{p}_x \quad (41)$$

By far, the angular momentum of a particle subject to an arbitrary force field changes in time; however, as per the motion of a planet in the Solar System, as the gravity field is spherical, any radial direction will be equivalent to the other, and the angular momentum shall hold with respect to the sphere centre.

Named with ϑ the colatitude, and with φ the longitude, one gets, in polar coordinates:

$$\hat{l}^2 = -\left[\frac{1}{\sin^2\vartheta}\frac{\partial^2}{\partial\varphi^2} + \frac{1}{\sin\vartheta}\frac{\partial}{\partial\vartheta}\left(\sin\vartheta\frac{\partial}{\partial\vartheta}\right)\right]. \tag{42}$$

In other terms, the nglar momentum operator coincides with the angular part of the Laplace operator: this means the Hamiltonian operator $\hat{H}$ will encompass also the Laplace one.

The corresponding eigenpairs can be found by solving the equation:

$$\kappa^2\hat{l}^2 Y(\vartheta,\phi) = \kappa^2\left|\hat{l}\right|^2 Y(\vartheta,\varphi), \tag{43}$$

being $\left|\hat{l}\right|^2$ the squared modlus of the angular momentum operator. It is possible to prove that:

$$\left|\hat{l}\right|^2 = l(l+1) \tag{44}$$

This last result, by combining Eq. (, with () and (), allows one to write:

$$\left[\frac{1}{\sin^2\vartheta}\frac{\partial^2 Y(\vartheta,\varphi)}{\partial\varphi^2} + \frac{1}{\sin\vartheta}\frac{\partial}{\partial\vartheta}\left(\sin\vartheta\frac{\partial Y(\vartheta,\varphi)}{\partial\vartheta}\right)\right] + l(l+1)Y(\vartheta,\varphi) = 0. \tag{45}$$

whose solution is given by the well known spherical harmonics $Y_{l,m}(\vartheta,\varphi)$:

$$Y_{l,m}(\vartheta,\varphi) = (-1)^{(m+|m|)/2} \cdot \sqrt{\frac{2l+1}{4\pi}\frac{(l-|m|)!}{(l+|m|)!}} \cdot P_l^{|m|}(\cos\vartheta)\cdot e^{im\varphi}, \tag{46}$$
$$given\, l, \quad m = -l, -l+1, -l+2, \ldots, 0, 1, 2, \ldots, l$$

$P_l^m(\cos\vartheta)$ are the associated Legendre functions, defined as follows:

$$P_l^m(\cos\vartheta) = \frac{(-1)^{(|m|-m)/2}}{2^l \cdot l!}\frac{(l+m)!}{(l+|m|)!}\left(1-\cos^2\vartheta\right)^{|m|/2} \cdot \frac{d^{l+|m|}}{d(\cos\vartheta)^{l+|m|}}\left(\cos^2\vartheta - 1\right)^l \tag{47}$$
$$m = -l, -l+1, \ldots, -1, 0, 1, \ldots, l-1, l$$

Spherical harmonics have been already normalized, so that:

$$\int Y_{l,m}^2(\vartheta,\varphi)d\Omega = 1. \tag{48}$$

III.6. The Schrödinger Wave Equation for Gravity Field

In the previous Sections it has been shown that the squared wave functions $\boldsymbol{\Psi}\cdot\boldsymbol{\Psi}^*$ express the probability density of finding an object subject to a force field in a given point in space; if we consider an eigensolution $\boldsymbol{\Psi}_n$, instead of a generic wave function $\boldsymbol{\Psi}$, one can assess the magnitude of a given energetic level E_n, being also a main, steady state for the energy owned by the object. Such eigenfunctions $\boldsymbol{\Psi}_n$ are defined in an homogenous and iosotrope space, so the Hamiltonian operator $\hat{H}$ shall warranty the momentum and the angular momentum conservation, under the action of a spherical force field.

The differential equation whose solutions show all the above mentioned criteria is named wave equation:

$$i\kappa\cdot\frac{\partial\boldsymbol{\Psi}}{\partial t} = \hat{H}\boldsymbol{\Psi}, \tag{31}$$

Through the knowledge of both the momentum and the angular momentum operators, we can write explicitly the Hamiltonian operator. Let us write the mechanical energy for a planet under the action of the solely gravity potential:

$$E = \frac{p^2}{2m} - U(r). \tag{49}$$

where, of course, $U(r)$ is the Sun gravity potential as a function of distance r from the Sun itself:

$$U(r) = -m\frac{\mu}{r}.$$

Thus, it is straightforward to switch the conserved quantities with the corresponding operators, to obtain:

$$\hat{H} = \frac{\hat{\boldsymbol{p}}^2}{2m} + U(r). \tag{50}$$

At this point, one can write the full wave equation explicitly:

$$i\kappa \cdot \frac{\partial \Psi(q,t)}{\partial t} = -\frac{\kappa^2}{2m}\Delta\Psi(q,t) + U(r)\Psi(q,t), \tag{51}$$

And if only the steady states are considered:

$$\frac{\kappa^2}{2m}\Delta\psi(q) + [E - U(r)]\psi(q) = 0. \tag{52}$$

Eq. 51, 52 are the wave equations describing an object motion under the action of a central gravity field, produced by the Sun, in the unsteady and steady form respectively. The eigensolutions to these equations shall provide a set of most likely energy and angular momentum values that the object can own as a consequence of being under the action of the Sun gravity. It must be noted that the whole procedure explained up to now do not imply the exclusive application of the present model to quantum mechanics; on the contrary, this model offers an approach having a wide application range. As per quantum mechanics, electrons belonging to an isolated atom are subject to have a discrete state described in terms of energy and angular momentum. As per the Solar System dynamics, an object moving under the action of the Sun gravity can *de facto* changes continuously its energy and angular momentum; nevertheless, the most likely states, as well as the most stable configurations, changes discretely and form a discrete eigenvalues spectrum.

IV. Solution to the Wave Equation Applied to the Solar System Dynamics

IV.1. The Equivalent Plank's Constant for the Solar System

Up to now, a description of the angular momentum κ has not been given. In quantum mechanics, such a parameter is the reduced Planck constant, and it is usually denoted as $\hbar$: it defines the minimum angular momentum variation related to an electron motion, and its value has been established by observing the energy emitted or absorbed during "jumps" from a starting energy level to another.

As per the planet motion, to the best of authors knowledge, there is not a criterion to determine the value of κ : nevertheless, it is clear that such a value must depend on the process leading to the Solar System formation, as well as on the Sun mass.

A compliant criterion is to assume that such a constant express the minimum angular momentum that a solid matter can own during the formation of planets, i.e. according to the temperature radial profile in the Solar System. In other terms, κ depends on the minimal radius where solid grain can aggregate to form the actual planets.

At this point, two major issues must be noted. First of all, κ does not express the angular momentum of the planet closet to the Sun, but the least angular momentum owned by the solid grain closest to Sun: in other terms, this angular momentum is smaller than that of

Mercury. Secondly, if one assumes that κ depends on the solid grain formation, the chemical composition of the grain itself cannot be neglected. As per inner planets, solid grains are composed by oxides of aluminum and silicates, whereas outer planets are mostly composed by hydrogen, helium, methane, ammonia, in both gaseous and liquid state. These considerations suggest that perhaps two different values of κ, from now on denoted with χ_1, χ_2 should be taken into account: one for inner, and the other for outer planets.

Let us examine χ_1 first. Aluminum, Nickel and Magnesium oxides were the first element to achieve the solid state, with an external temperature of nearly 1300 K. The corresponding distance from the Sun, can be found by imposing the thermal equilibrium between the solar flux and the thermal flux emitted by the element of discoid nebula, proportional to its temperature:

$$4\sigma T_e^4 = (1-a)\sigma T_s^4 \left(\frac{R_s}{r}\right)^2, \tag{53}$$

where r is the unknown distance from Sun, T_s is the Sun surface temperature, R_s the Sun radius, α is the albedo and finally σ is the Stefan-Boltzmann constant. By imposing $T_s = 1300$ K, and solving for r, it is possible to obtain:

$$r \approx 0.04 \text{ A.U.} \tag{54}$$

Such a distance is a sort of threshold, below which the solid grain formation was forbidden. The corresponding angular momentum per unit mass, denoted as χ_1 ,owned by a grain moving along a circular orbit of radius equal to $r \approx 0.04$ A.U. is therefore given by:

$$\chi_1 = 9.1e+14 \quad J \cdot \sec \cdot kg^{-1}. \tag{55}$$

Analogous considerations can be expressed as per the χ_2 value.

As per the Hydrogen, it could condensate in the regions where the temperature was below 220 K. According to Eq. (53), the estimated distance r where the equilibrium temperature was equal to 220 K will therefore be equal to:

$$r \approx 1.3 \text{ A.U.} \tag{56}$$

To note that this distance is smaller than the frozen line, placed at 2.7 A.U. from the Sun. The corresponding angular momentum χ_2 owned by a frozen hydrogenous grain orbiting around the Sun along a circular orbit twill therefore be

$$\chi_2 = 4.9e+15 \quad J \cdot \sec \cdot kg^{-1}. \tag{57}$$

Such parameters χ_1, χ_2 will allow generating two distinct sequences, one for inner planets and the other for outer ones, as it will be shown in the next Section.

IV.2. The Solution to the Wave Equation for Gravity Field

The linearity of the wave equation makes it possible to separate the dependence between the radial and the angular variables, and search for a solution in the form:

$$\psi = R(r) \cdot Y(\vartheta, \varphi). \tag{58}$$

The solution to the angular part of the wave equation is given by the spherical harmonics, and provides useful information about the probability density related to the orbit inclination and eccentricity. The radial part of the wave equation, on the contrary, provides the radial probability density of finding a given planet at a certain distance from Sun, being this the central purpose of the present chapter. This is the main reason why particular attention will be given only to this part of the solution of the wave equation.

Let us write the radial part of the wave equation in polar coordinates, as the central force field is spherical:

$$\frac{1}{r^2}\frac{d}{dr}\left(r^2\frac{dR(r)}{dr}\right) - \frac{l(l+1)}{r^2}R(r) + \frac{2}{\chi^2}\left[e + \frac{\mu}{r}\right]R(r) = 0, \tag{59}$$

Here we considered the energy per mass unit, denoted by e. As the modulus of the angular momentum operator, namely $\left|\hat{l}\right|^2 = l(l+1)$, is still present in the radial part of the wave equation, the corresponding eigensolutions of the angular momentum will depend on that of the momentum.

Before entering the detail of the resolution, the introduction of non-dimensional variables is advisable. By imposing the following measure units:

$$[length] = \frac{\chi^2}{\mu}, \qquad [en.\ per\ unit\ mass] = \frac{\mu^2}{\chi^2} \tag{60}$$

the corresponding, non-dimensional variables will be

$$\bar{r} = r\frac{\mu}{\chi^2}, \qquad \bar{e} = e\frac{\chi^2}{\mu^2}. \tag{61}$$

Eq. (59) then becomes:

$$\frac{1}{\bar{r}^2}\frac{d}{d\bar{r}}\left(\bar{r}^2\frac{dR(\bar{r})}{d\bar{r}}\right)-\frac{l(l+1)}{\bar{r}^2}R(\bar{r})+2\left[\bar{e}+\frac{1}{\bar{r}}\right]R(\bar{r})=0. \tag{62}$$

By assuming negative values for Energy, as we are interested in eigensolutions bounded within the Sun gravity field, one can choose another dimensionless radial variable:

$$\rho=2\bar{r}\sqrt{-2\bar{e}}\,. \tag{63}$$

At this point, the solution to the radial part of the wave equation will be searched in the form:

$$R(\rho)=\rho^l e^{\frac{\rho}{2}}w(\rho), \tag{64}$$

being ρ^l the asymptotic behavior when $\rho\rightarrow 0$, $e^{\frac{\rho}{2}}$ that when $\rho\rightarrow\infty$, and $w(\rho)$ the patching function between these two distinct trends.

It is easy to prove that $w(\rho)$ must fulfill the equation:

$$\rho w''+(2l+2-\rho)w'-\left(-\frac{1}{\sqrt{-2\bar{e}}}+l+1\right)w=0. \tag{65}$$

This last equation provides the evidence that $\frac{1}{\sqrt{-2\bar{e}}}$ must be an integer value, in order to be summed with the others inside the brackets.

Therefore, one can define:

$$n=\frac{1}{\sqrt{-2\bar{e}}}, \tag{66}$$

So that Eq. becomes:

$$\rho w''+[(2l+2)-\rho]w'-(-n+l+1)w=0. \tag{67}$$

Named with $\alpha=-n+l+1$ and $\gamma=2l+2$, the solution is given by the confluent hypergeometric function, written below:

$$F(\alpha,\gamma,\rho)=1+\frac{\alpha}{\gamma}\frac{\rho}{1!}+\frac{\alpha(\alpha+1)}{\gamma(\gamma+1)}\frac{\rho^2}{2!}+\frac{\alpha(\alpha+1)(\alpha+2)}{\gamma(\gamma+1)(\gamma+2)}\frac{\rho^3}{3!}+\ldots+\frac{\alpha(\alpha+1)(\alpha+2)\ldots(1)}{\gamma(\gamma+1)(\gamma+2)\ldots(\gamma+|\alpha+1|)}\frac{\rho^\alpha}{\alpha!} \tag{68}$$

which holds if ad only if $\alpha = -n + l + 1$ is a non positive integer number. Such a constraint limits the possible values of l, given n, as follows:

$$l = 0,1,2,...,n-1, \tag{69}$$

This equation shows how the angular momentum eigenvalues are bounded by the Energy eigenvalues.

The complete solution is finally given by:

$$R_{n,l}(\rho) = C \cdot \rho^l e^{-\frac{\rho}{2}} F(-n + l + 1, 2l + 2, \rho), \tag{70}$$

where C is a parameter for normalization, obtained by imposing

$$\int_0^{+\infty} R_{n,l}^2(\rho) \cdot \rho^2 d\rho = 1. \tag{71}$$

By combining the singular eigensolutions, the complete eigenfunctions defined by a set of three integers n, l, m is given by:

$$\begin{cases} \psi_{n,l,m}(\rho, \vartheta, \varphi) = R_{n,l}(\rho) Y_{l,m}(\vartheta, \varphi), \\ R_{n,l}(\rho) = \dfrac{2}{n^2(2l+1)!} \sqrt{\dfrac{(n+l)!}{(n-l-1)!}} (\rho)^l e^{-\frac{\rho}{2}} F(-n+l+1, 2l+2, \rho) \\ Y_{l,m}(\vartheta, \varphi) = (-1)^{(m+|m|)/2} \cdot \sqrt{\dfrac{2l+1}{4\pi} \dfrac{(l-|m|)!}{(l+|m|)!}} \cdot P_l^{|m|}(\cos\vartheta) \cdot e^{im\varphi} \\ n = 0,1,2,...; \qquad l = 0,1,2,...,n-1; \qquad m = -l, -l+1, ..., 0, 1, 2, ..., +l \end{cases} \tag{72}$$

IV.3. Quantum Numbers in the Solar System

The three indexes used for series expansion of the complete solution are referred in literature as the quantum numbers; as per quantum mechanics, a set of three quantum numbers fully defines the state of a certain orbital in the isolated atom. *Mutatis mutandis*, such quantum numbers maintains their physical meanings when the wave equation is written to find the asymptotic stable motion within the gravity action of Sun.

n is the main quantum number, and it is related to the energy per unit mass through the equation:

$$e = -\frac{\mu^2}{2 \cdot \chi^2 \cdot n^2} \tag{73}$$

It must be noted that the first allowed value for this quantum number is $n = 1$, defining the least energy corresponding to a steady state that can be owned by a planet at the end of its formation:

$$e_{\min} = -\frac{\mu^2}{2 \cdot \chi^2}, \tag{74}$$

This eigenvalue takes the name of main or basic status. Such a status depends on the Sun gravity constant and on the basic angular momentum χ: by changing this parameter, the basic status and the following sequence will change accordingly. When $n \to \infty$, finally, the energy per unit mass becomes null, and the object is free to escape from the Sun attraction.

l is the second quantum number, and is related to the angular momentum per unit mass h owned by the body orbiting around the Sun. It is easy to show that:

$$\sqrt{l(l+1)}\chi = h, \tag{75}$$

The second quantum number l can be null, but it cannot exceed $n-1$. This mathematic constraint has a deep physical meaning. From orbit mechanics it known that:

$$\begin{cases} a = -\dfrac{\mu}{2e} \\ h = \sqrt{\mu \cdot a \cdot \left(1 - ecc^2\right)} \end{cases} \tag{76}$$

being ecc the orbit eccentricity. By combining the system (76) with Eq. (75), one obtains:

$$n^2\left(1 - ecc^2\right) = l(l+1),$$

that gives:

$$ecc = \sqrt{1 - \frac{l(l+1)}{n^2}} \tag{77}$$

In other terms, if $l > n-1$, the orbit eccentricity becomes complex. This in an interesting result, applied to celestial mechanics, even if it has been derived from a model typically used only in quantum mechanics. Besides, Eq. (77) has also another important meaning: it shows how to build an eigen-orbit by knowing only the corresponding first and second quantum number.

Another important result is that both inner and outer planetary sequence, generated according to the different values of χ_1, χ_2, cannot start by imposing $n=1$. Indeed, in this case $l=0$, so the object won't have in this case any angular momentum, and it will be doomed to be swallowed by the Sun. This means that each sequence shall start at least by imposing $n=2$, or even $n=3$

Finally, the azimuth quantum number m. It provides the vertical component of the angular momentum. As planets orbits are almost fully co-planar, their angular momentum can be considered nearly coincident with a vertical axis orthogonal to the Ecliptic plane. This allow us searching for solutions characterized by $m=+l$.

IV.4. Planets Orbits as a Sequence of Eigensolutions

In this Section, the radial density profile as described by $R_{n,l}^2(\rho)$ will be provided. Surprisingly, when considering nearly circular orbit, the two sequence of eigensolutions generated by imposing $\chi=\chi_1$ and $\chi=\chi_2$ respectively, fit with notably coherence the true planetary distribution. The peak of the probability density coincides almost exactly with the observed mean radius.

When $\chi=\chi_1$, the sequence of eigensolutions for the inner planets is obtained by imposing $n=3,4,5,6$; as per outer planets, i.e. when $\chi=\chi_2$, the sequence is given by $n=2,3,4,5,6$. Results are summarized in **Table 4a, b**. the corresponding graphs are shown in **Fig 1** to 10. To note that even if Pluto is no longer considered a planet, it has been considered in the analysis in order to make a full comparison with the original Titius-Bode formulation.

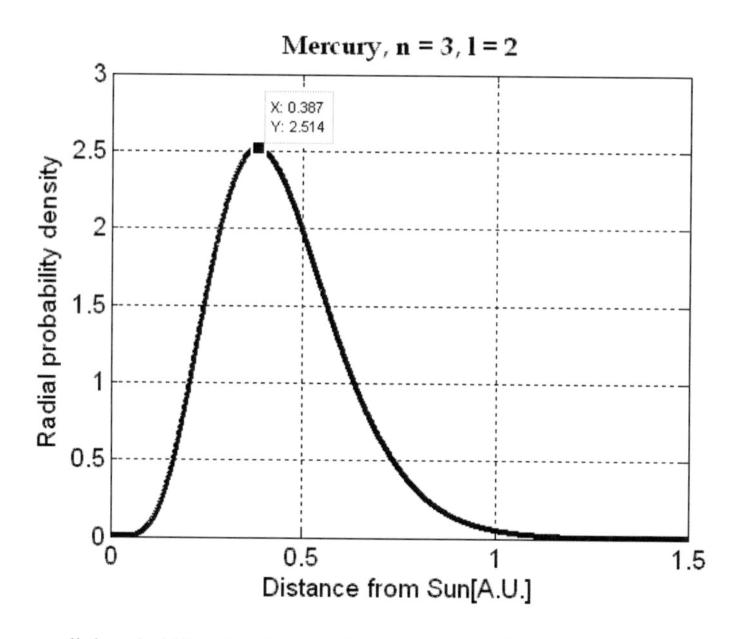

Figure 1. Mercury: radial probability density as a function of the distance from Sun r.

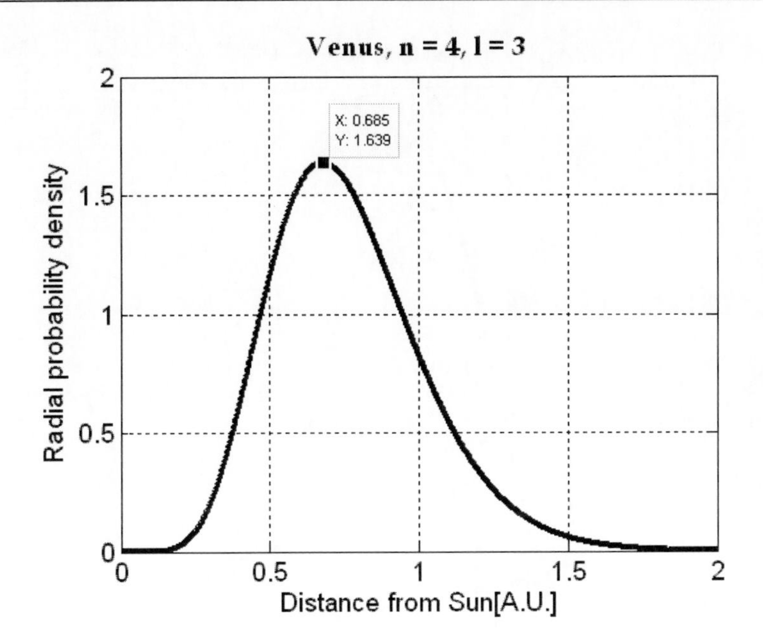

Figure 2. Venus: radial probability density as a function of the distance from Sun r.

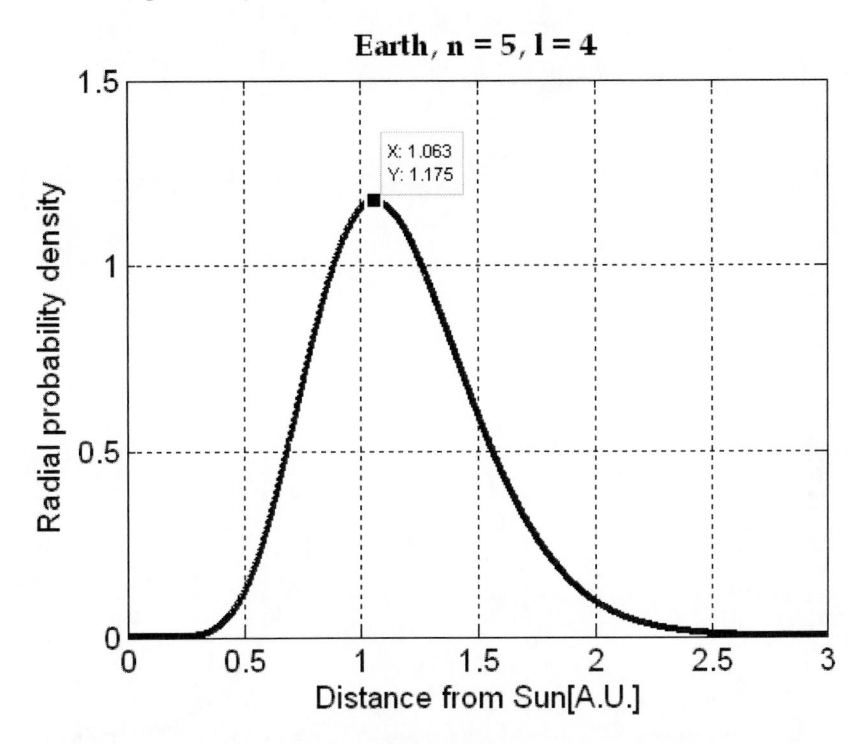

Figure 3. Earth: radial probability density as a function of the distance from Sun r.

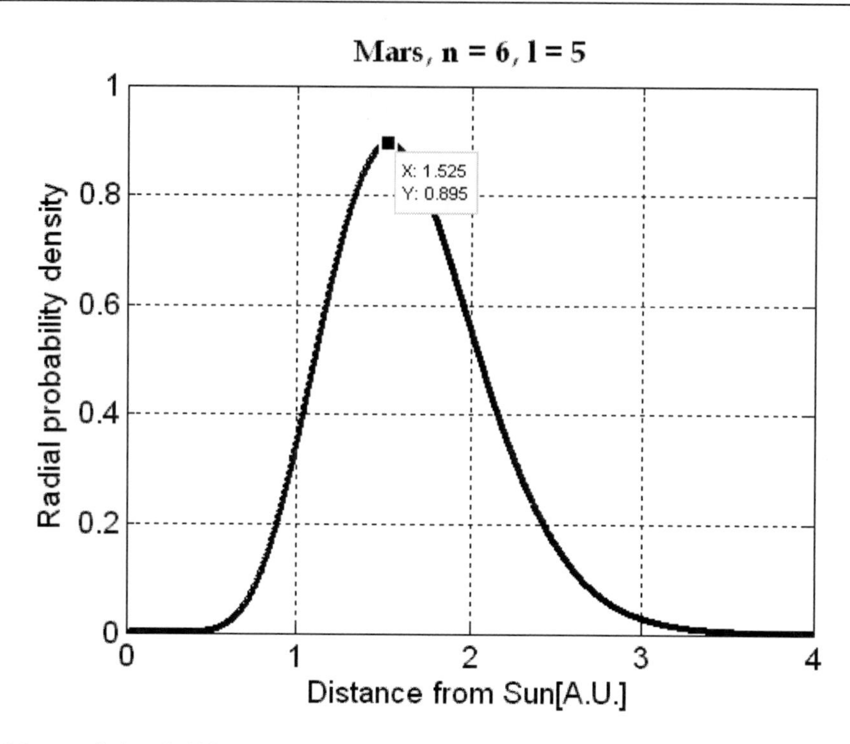

Figure 4. Mars: radial probability density as a function of the distance from Sun r.

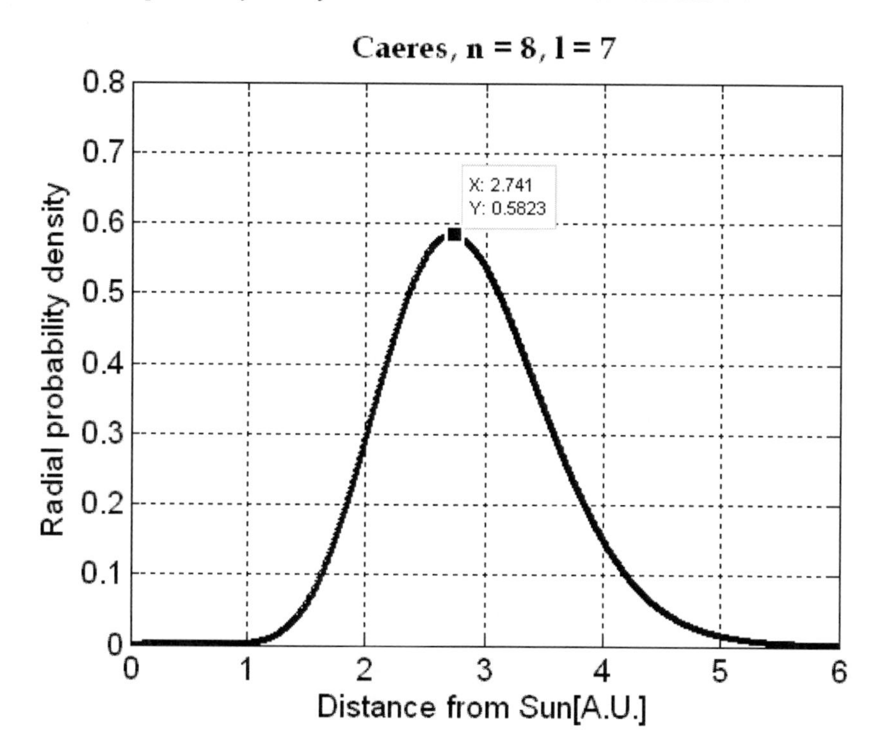

Figure 5. Caeres: radial probability density as a function of the distance from Sun r.

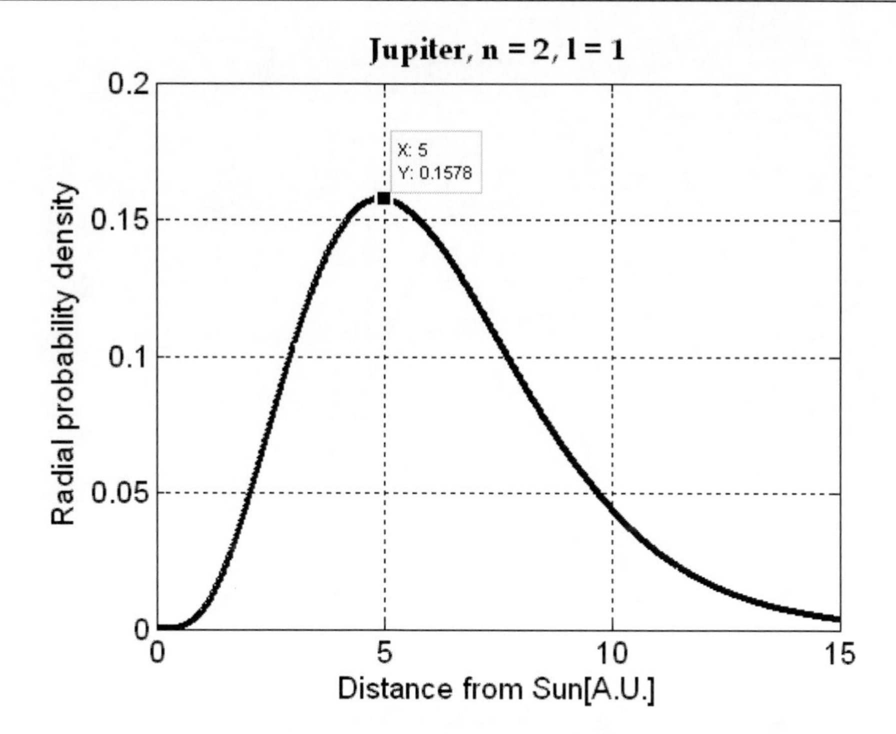

Figure 6. Jupiter: radial probability density as a function of the distance from Sun r.

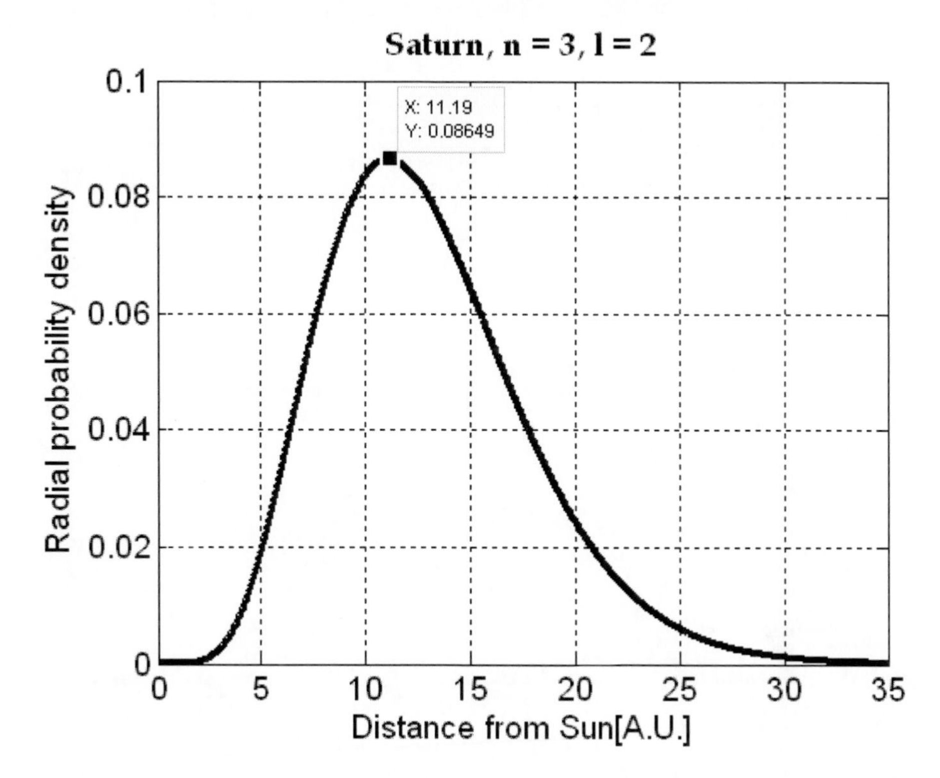

Figure 7. Saturn: radial probability density as a function of the distance from Sun r.

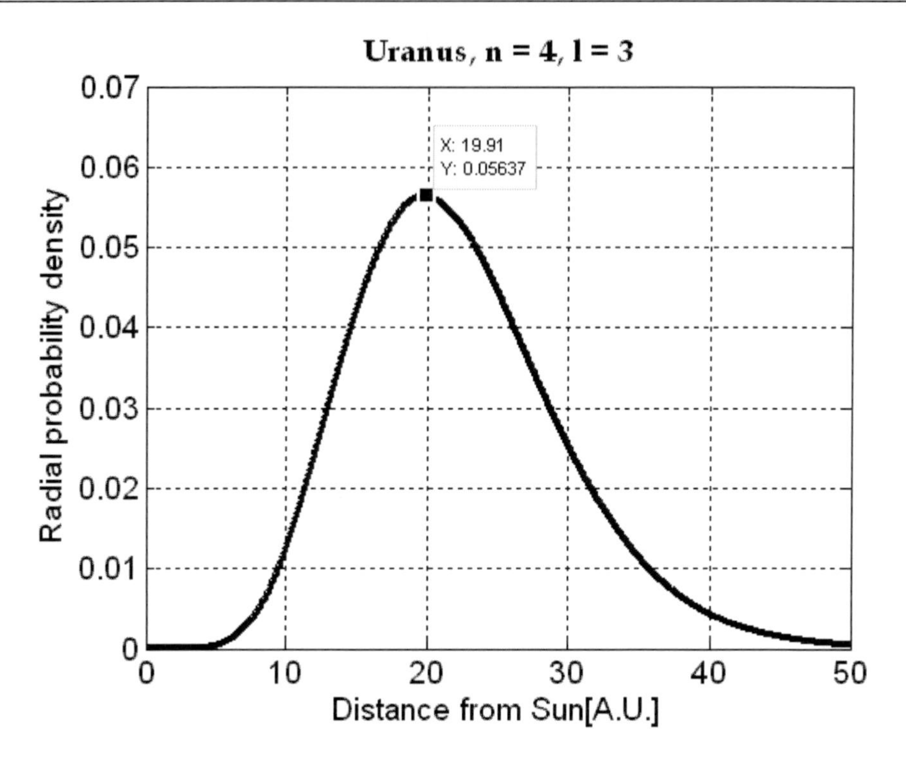

Figure 8. Uranus: radial probability density as a function of the distance from Sun r.

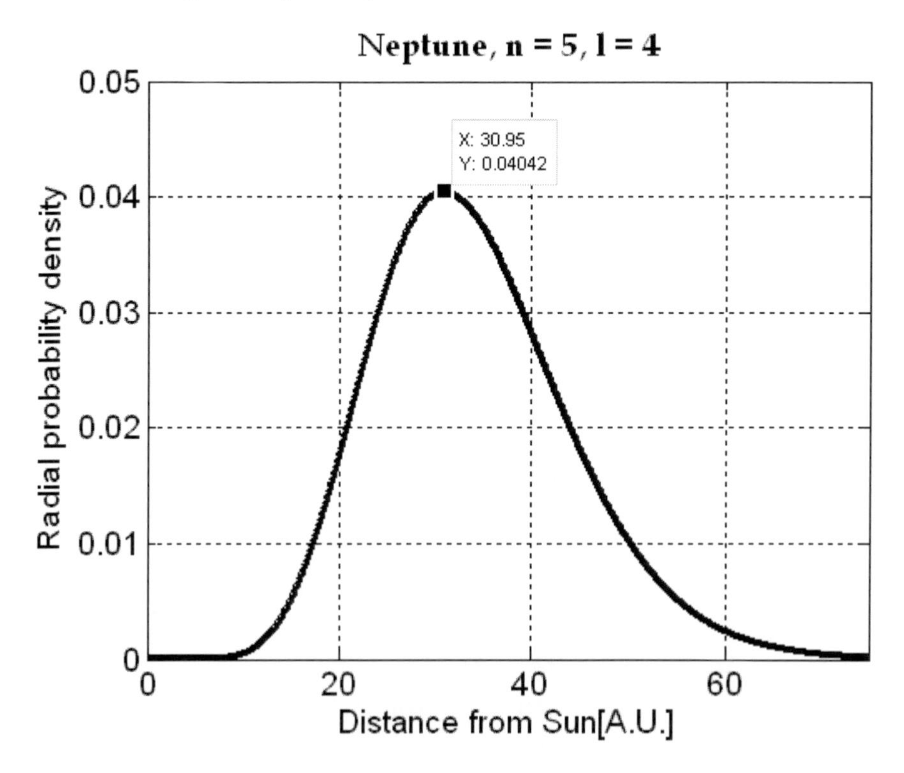

Figure 9. Neptune: radial probability density as a function of the distance from Sun r

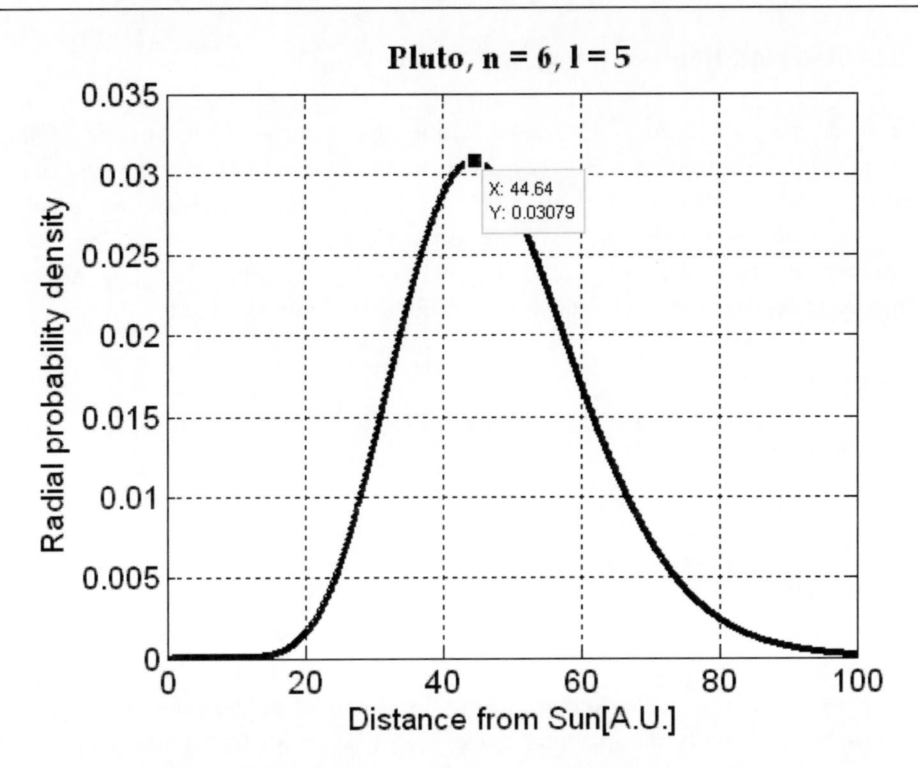

Figure 10. Pluto: radial probability density as a function of the distance from Sun r

Table 4a. Comparison between the semi-major axes values predicted by through the quantum model and the tru observed ones, for inner planets ($\chi = \chi_1$)

	Inner	**planets**	
n	Planetary body	Radius[A.U.] from quantistic model	Observed Mean Radius[A.U.]
3	Mercury	0.38	0.39
4	Venus	0.68	0.72
5	Earth	1.05	1.00
6	Mars	1.52	1.52
8	Caeres	2.72	2.77

Table 4b. Comparison between the semi-major axes values predicted by through the quantum model and the tru observed ones, for inner planets ($\chi = \chi_2$).

	Outer	**planets**	
n	Planetary body	Radius[A.U.] from quantistic model	Observed Mean Radius[A.U.]
2	Jupiter	5.01	5.20
3	Saturn	11.10	9.54
4	Uranus	19.85	19.18
5	Neptune	30.95	30.06
6	Pluto	44.40	39.44

IV.5. The Asteroid Belt as an "Accumulation Zone"

The inner planet sequence terminates within the Asteroid Belt. In this region, the sequence of eigensolutions progressively reduces its spacing and this region becomes notably crowded of possible solutions. It is clear that the larger is the main quantum number, the lower is the probability density peak in the density radial profile.

An analogous trend has been observed beyond the Neptune orbit. This suggests that these zones may be considered as accumulation regions of stable and most-likely orbits.

V. Conclusions and Further Developments

As it has been shown throughout the work, the existence of several planetary distribution laws, as well as the numerical simulations of the Solar System formation, provide the evidence of an underlying criterion driving the planet formation towards the present configuration. Therefore, one can consider the actual planets orbits as the most likely, but being a also the results of long-term resonances, they can be considered also the most stable among all the possible ones.

Such considerations led the Authors to the development of the present work, where a detailed procedure to apply the quantum model to the Solar System dynamics is provided. The quantum model, by providing the system eigensolutions, allows finding the most likely planetary orbits in a stable state, thus meeting both the likelihood and stability requirements of the actual configuration. The results are notably close to the present semi-major axes, revealing an agreement much greater than that of the original Titius-Bode law.

Two major issues deserve a further analysis. First of all, the results obtained depend on the choice of the equivalent Planck's constant χ. The criterion suggested to assess this constant value leads to the impossibility of considering only one constant, because of the different chemical composition of the planets; indeed, even the different dimensions of inner and outer planets suggest that their corresponding sequence should be different. If such a difference is properly taken into account, the final results can provide a better fit of the true observed planetary configuration. In this sense, both the Titius-Bode and the Murray and Dermott rule are nothing but the interpolations of the two planetary sequences.

The second issue to be examined is the meaning of the Asteroid belt, or better, the fact that any of the proposed fitting rule find at least one celestial body (Titius-Bode, Murray and Dermott's rule) or two (the alternative formulation proposed by the Authors), or even an accumulation zone (the quantum model). Now it is well known that the mass of Jupiter prevented the formation of a celestial body through grains collision in the region between Mars and Jupiter itself: the not-aggregated rock material eventually formed what is known as the Asteroid Belt. From this point of view, the fact that some distribution laws find a solution in this region is nothing but a validation of the fact that the underlying criterion to these laws does not take into account the mutual interaction among planets, but only their interaction with the Sun. The quantum model proposed in this Chapter, by providing the evidence of an accumulation zone in the Asteroid region, shows that beside the action of Jupiter, the Asteroid Belt *per se* is a region with a high crowding of orbits. Then, the introduction of a new sequence – triggered in the model by the introduction of the equivalent constant χ_2 - , driven by the temperature radial profile and by the chemical features of the solar nebula, break the inner planets sequence for starting a new one.

As a conclusion, the quantum model proposed in this Chapter provides the best mathematical formulation to describe the present Solar System configuration as the most likely and stable over the long term resonances. The formation process, driven by the Sun and modulated by the nebula chemical composition and by the temperature profile, is pictured through a sequence of eigensolutions, revealing the key feature of the System itself. Beside the encouraging results, whether the quantum approach is fully consistent with the Solar System dynamic, is a point that can be still debated; nevertheless, this approach has the merit to show, through a rigorous mathematical formalism, the presence of an intrinsic order and regularity ruling the Solar System formation, evolution and dynamics.

REFERENCES

[1] V. M. Bakulev, "*The Titius-Bode law of planetary distances: new approaches*". Institute of Physics, St.Petersburg State University, Ulyanovskaya 1, Petergof, 198504 St.Petersburg, Russia.

[2] V. Christianto, "*On the origin of macroquantization in astrophysics and celestial motion*" Annales de la Fondation Louis de Broglie, Volume 31, no 1, 2006 31.

[3] V. Christianto, "*Comparison of predictions of planetary quantization and implications of the Sedna finding*". Apeiron, Vol. 11, No. 3, July 2004.

[4] J. Giné, "*On the origin of gravitational quantization: the Titius-Bode law* ". arXiv:physics/0507072 v3 6 Oct 2005.

[5] P. Lynch, "*On the Significance of the Titius-Bode Law for the Distribution of the Planets*".Mon. Not. R. Astron. Soc. 351, 1174-1178 (2003).

[6] L. Neslusan, "*The significance of the Titius-Bode law and the peculiar locations of the Earth's orbit*". Mon. Not. R. Astron. Soc. 351, 133-136 (2004).

[7] L. Nottale, "*New formulation of stochastic mechanics. Application to Chaos*". CNRS, Observatoire de Paris-Meudon (DAEC) 5, place Janssen, F-92195 Meudon Cedex, France.

[8] M. Pitk¨anen, "*Gravitational Schr¨odinger equation as a quantum model for the formation of astrophysical structures and dark matter?*". September 5, 2004

[9] A. Rubˇci□c and J. Rubˇci□c, "*The quantization of the Solar-like gravitational Systems*". Department of Physics, Faculty of Science, University of Zagreb, Bijeniˇcka cesta 32, HR-10001 Zagreb, Croatia. Received 24 January 1998; Accepted 1 June 1998.

[10] F. Scardigli, "*The Titius-Bode law and a quantum-like description of the planetary systems*".

Cited Books

L. D. Landau and E. M. Lifsitis, "*Meccanica quantistica, teoria non relativistica*", Fisica Teorica 3.

Bertotti, Farinella *"Physics of the Solar System"*.

In: Solar System: Structure, Formation and Exploration ISBN: 978-1-62100-057-0
Editor: Matteo de Rossi

Chapter 8

REGULARITIES IN SYSTEMS OF PLANETS AND MOONS

Vlasta Peřinová,* Antonín Lukš and Pavel Pintr
Joint Laboratory of Optics, Palacký University, RCPTM,
17. listopadu 12, CZ-77146 Olomouc, Czech Republic

Abstract

We deal with regularities of the distances in the solar system. On starting with the Titius–Bode law, these prescriptions include, as "hidden parameters", also the numbering of planets or moons. We reproduce views of mathematicians and physicists of the controversy between the opinions that the distances obey a law and that they are of a random origin. Hence, we pass to theories of the origin of the solar system and demonstrations of the chaotic dynamics and planetary migration, which at present lead to new theories of the origin of the solar system and exoplanets. We provide a review of the quantization on a cosmic scale and its application to derivations of some Bode-like rules.

PACS 96.12-a, 96.15.-g, 96.90.+c.

Keywords: planetology, Bode-like laws.

1. Introduction

The solar system wakes admiration and an attempt at a reasonable argument for this feeling suggests regularities in the systems of planets and moons. In this chapter, we restrict ourselves to the regularities that the distances of secondaries from their primaries indicate. Other parameters leading to the concept of resonances are not treated (cf. [Murray and Dermott (1999)], pages 9, 321). A frequent argument in favor of a formal treatment of the regularities is the failure of the theories of the origin of the solar system, which should be essential at least from the materialist viewpoint. At present, rather new theories are spoken of the successful theories of a new generation and the weight of the usual argument is

*E-mail address: perinova@prfnw.upol.cz

lesser. Some theories seem to confirm the regularity formulas. Therefore, we include also the theories of the origin of the solar system and the extrasolar planets.

Neglecting that both the major sciences, mathematics and physics, have undergone a historical development, we pay attention to the fact that as late as at the times of J. Kepler, astronomy (and astrology) was counted to the mathematics. Kepler's inventions have ushered the establishment of the astronomy as a physical field. Even though till 1781 only six planets of the solar system were known, their distances from the Sun were measured with an appropriate precision.

In 1766, J. Titius von Wittenberg formulated his famous note on planetary distances. J. Bode has published this note and readdressed it, see [Nieto (1972)]. A temporary success of the Bode law may consist also in the fact that it is not quite simple. A geometric progression is obvious only after a subtraction of 0.4 AU (astronomic units) from the distance from the planet to the Sun.

The objection that empirical formulas may be arbitrarily complicated has led to attempts at simple formulas. So, Armellini's law has the form, $r_{n\mathrm{A}} = 1.53^n$, where n assumes the values: -2 for Mercury, -1 for Venus, 0 for the Earth, 1 for Mars, 2 for the asteroid Vesta, 3 for the asteroid Camilla, 4 for Jupiter, 5 for Saturn, 6 for Centaur Chiron, 7 for Uranus, 8 for Neptune and 9 for Pluto [de Oliveira Neto (1996)].

Kant (1755) understood the origin of the solar system as a scientific problem and worded the nebular hypothesis. By his theory, the Sun and the planets became from the gas, which had been located in the volume of the present solar system and had had a high temperature. He assumed that the gravitation could bring about the origin of a proto-Sun and transform the irregular motion into a rotation. The planets originated from the rotating mass.

Independently, Laplace (1796) indicated that the solar system had become from gas and assumed not only a high temperature of the mass, but also its rotation. The nebula rotated as a solid body. His scenario of the evolution includes the cooling, contraction, enhancement of the rotation and flattening. The nebula shed a gaseous ring, which becomes a ball. It repeats as many times as many the planets are. Similarly, the moons of the planets have originated. The Sun has become from the remainder of the nebula. From a single ring more small planets could originate. P. S. Laplace confessed that he was not convinced by his hypothesis.

Maxwell (1859) has provided results confirmed by the flybys by the Voyager spacecraft in the 1980s. In application to the solar nebula, he has remarked that the gaseous ring itself cannot wrap into a spherical body, a planet. It has been stated that the present planetary system and the Sun do not have the total angular momentum that leads to an instability of the rotating nebula. The theories of the origin by the external causes are called catastrophic.

The Chamberlin–Moulton planetesimal hypothesis has been proposed in 1905 by geologist T. C. Chamberlin and astronomer F. R. Moulton [Chamberlin (1905), Moulton (1905)]. The external cause consists in that the star passed close enough to the Sun. Jeans (1914) assumed close encounter between the Sun and a second star. The difference from the previous hypothesis is in that Chamberlin and Moulton assumed separation of some mass on the adjacent and opposite sides of the Sun and the accretion of planetesimals. J. Jeans assumed the separation of the mass only on the adjacent side of the Sun and a direct origin of planets. This hypothesis has been assumed also by the mathematician and astronomer H. Jeffreys, who considered also a collision theory [Jeffreys (1924)]. In the 1920s, H. N. Russell was persuaded by the Jean-Jeffreys tidal hypothesis to affirm that planetary systems are "infrequent" and inhabited planets "matter of pure speculation." Two decades later, however, he gave up this opinion [Russell (1943)]. Russell (1935) measured spectra of binary stars and was interested in the origin of planetary systems. Lyttleton (1936) as an expert on the binary stars assumed that the Sun had been part of such a system.

In contrast, the nebular hypothesis has been resumed. E.g., Nölke (1930) did not derive the shedding of the rings from the assumption of the rotation, but the turbulence. von Weizsäcker (1943) elaborated a similar theory. Kuiper (1951) uses the concept of a protoplanet. The electromagnetic forces have been considered by Alfvén (1942), Dauvillier and Desguins (1942) and Schmidt (1944). In the 1960s, the massive-nebula model [Cameron (1962)] and low-mass-nebula model coexisted [Safronov (1960, 1969)]. The latter has evolved into a "standard" model [Lissauer (1993)].

In section 2, we mention a discussion of the mathematicians, who have not been satisfied with the statement that "the Bode law fits data well enough". They have constructed alternative hypotheses and have found that the likelihood ratio differs significantly from unity in one case [Good (1969)] and it does not differ from unity significantly in the other case [Efron (1971)]. Specialists may pay attention to a hypothesis competing with the statement that the Bode law fits the data well. They are not content with the repetition of a mathematician's idea, but they use the astronomical knowledge. They suggest a random origin of the regularities [Hayes and Tremaine (1998), Murray and Dermott (1999), p. 5]. These imposing analyses are not persuasive, on considering their model dependence [Lynch (2003)].

In section 3, we touch the resumed nebular hypothesis. The topic is estimates of the total mass of the solar nebula and the distribution of its mass [Weidenschilling (1977), Hayashi (1981)] and a modification for the extrasolar nebulas [Kuchner (2004)]. We mention next the 'standard" model of planet formation [Lissauer (1993)]. Finally, we touch dynamical theories of the Titius–Bode law [Graner and Dubrulle (1994), Dubrulle and Graner (1994)]. These theories for the restriction to the Titius–Bode law comprise well intended simplifications. In this framework, Christodoulou and Kazanas (2008) have been able to provide a theory of the dependence of the planetary distance on its ordinary number, which does not express this dependence by a closed formula, but fits the data well. This means even attempts at an application to extrasolar planets. We expand on a method of derivation of the rule involving squares of the ordinary numbers instead of the geometric progression [Krot (2009)]. We provide the dependence of the planetary distances on their ordinary numbers, which is no more based on the squares of the ordinary numbers, but it is satisfactory.

Already in the book [Murray and Dermott (1999), p. 409], a whole chapter is devoted to the chaos and long-term evolution along with appropriate references. In section 4, we pay attention to such reports on numerical integration [Laskar (1989), Sussman and Wisdom (1992)]. The theory of origin of the extrasolar planets meets a difficulty that the Jupiter-mass planets are present on small orbits. This has led to the theory of migrating planets [Murray, Hansen, Holman and Tremaine (1998), Murray, Paskowitz and Holman (2002)]. Long-time scales are not accepted by creationists [Spencer (2007)].

In section 5, we devote attention to the quantization on a cosmic scale. The observed deviations of the absorption lines from the Lyman-α frequency have led to a hypothesis of their origin, which includes the quantization of "megascopic" systems [Greenberger (1983)]. The quantization of microscopic systems has been simulated [de Oliveira Neto (1996), Nottale, Schumacher and Gay (1997), Agnese and Festa (1997), Rubčić and Rubčić (1998), Carneiro (1998), Agnese and Festa (1998, 1999), Nottale, Schumacher and Lefèvre (2000)]. We concentrate ourselves to the approaches, which replace a dynamical theory of the Titius–Bode law by the quantization of orbits and thus derive rather the use of the square of a planet's ordinary number instead of the geometric progression. Next we remember the indirect use of the quantization for the discretization of orbits. First, a wave-function for a planet is chosen and then the expectation value of the distance from the particle to the central body is compared with the observed distance from the planet to the Sun [de Oliveira Neto, Maia and Carneiro (2004)]. Finally, we return to our publications. Pintr and Peřinová (2003–2004) have commented on the proposal of Mohorovičić (1938) positively and have modified it to the moons of giant planets and extrasolar planets. Peřinová, Lukš and Pintr (2007) intended to replace the close relationship of the paper [de Oliveira Neto, Maia and Carneiro (2004)] to the article [de Oliveira Neto (1996)] by a connection with the paper [Agnese and Festa (1997)]. This intention has been realized in part. Pintr, Peřinová and Lukš (2008) have derived a discrete system of orbits using mainly the classical physics. Like an incomplete dynamical theory of the Titius–Bode law, or the quantization of orbits using wave-functions, this theory assigns the distances to the nodal lines of standing waves, even though indirectly, through a transformation.

2. Statistical Decision Making

In the Titius–Bode law doubling occurs, which enables anybody to write down a mathematical formula for the planets Venus, Earth, Mars, Ceres, Jupiter, Saturn, Uranus, Neptune and Pluto. The powers of two may be linearly interpolated,

$$a_{\mathrm{Mercury}} = a,\, a_{\mathrm{Venus}} = a + b,\, a_{\mathrm{Earth}} = a + 2b,\, a_{\mathrm{Mars}} = a + 4b,\, a_{\mathrm{Ceres}} = a + 8b,$$

$$a_{\mathrm{Jupiter}} = a + 16b,\, a_{\mathrm{Saturn}} = a + 32b,\, a_{\mathrm{Uranus}} = a + 64b,\, a_{\mathrm{Neptune}} = a + 96b,$$

$$a_{\mathrm{Pluto}} = a + 128b, \tag{1}$$

where $a = 0.4$, $b = 0.3$, cf. [Christodoulou and Kazanas (2008), Povolotsky (2007)]. The continuation in the formula up to Uranus is obvious.

The Titius–Bode law with slight irregularities provokes to improvement, but also to proposing formulas, whose "validity" is saved by the neglect of critique of numbering of the planets, cf. [Nieto (1972)]. Good (1969) has emended the Bode law to the form

$$b_{\mathrm{Mercury}} = a + b,\, b_{\mathrm{Venus}} = a + 2b,\, b_{\mathrm{Earth}} = a + 4b,\, b_{\mathrm{Mars}} = a + 8b,\, b_{\mathrm{Ceres}} = a + 16b,$$

$$b_{\mathrm{Jupiter}} = a + 32b,\, b_{\mathrm{Saturn}} = a + 64b,\, b_{\mathrm{Uranus}} = a + 128b,\, b_{\mathrm{Neptune}} = a + 256b,$$

$$b_{\mathrm{Pluto}} = a + 512b. \tag{2}$$

Here $a = 0.4$, $b = 0.075$, however.

Just as Efron (1971) has indicated in a footnote of a statistician, it can be expected that he or she practises the "numerology". It seems that the three purposes of his article are given in descending order of importance for the astronomy: (1) The validity of Bode's law, (2) testing whether or not the observed sequence of numbers follows some simple rule, (3) the logical basis of Fisherian significance testing.

The statistical decision making assumes:

(i) A statistical model describing what the statement means that Bode's law is real.

(ii) An alternative statistical model describing the statement that Bode's law is artificial.

The question of the validity of Bode's law is transformed to a problem of hypothesis testing [Efron (1971)].

The statisticians apply Bode's law only to the planets Venus through Uranus. According to [Good (1969)], the statistical model (i) consists in a normal distribution of logarithms of planetary distances. These distances are independent random variables. The means are given by the Titius–Bode law. All the variances are σ^2. It is accepted that the three parameters a, b, and σ^2 are estimated. Model (ii) is a uniform distribution of the logarithms of the planetary distances on some interval $[\log \delta_{\mathrm{l}}, \log \delta_{\mathrm{u}}]$, where the subscript l (u) stands for lower (upper), respecting the observed order of the planets. It is accepted that the parameters $\log \delta_{\mathrm{l}}$, $\log \delta_{\mathrm{u}}$ are estimated. By the use of the Bayesian methods, it has been derived that the data witness for Bode's law. Efron (1971) admits that, in a non-Bayesian framework, the result would be the same, but he adopts the Fisherian methodology along with the model C instead of (ii).

A formulation of the model C demonstrates that, in the mathematical statistics, it is not necessary to specify a joint distribution of the planetary distances completely. Such a distribution, if any, is characterized in terms of the ratios of the planetary spacings to the difference between the distances of Uranus and Venus. For simplicity, we speak of a spacing instead of the difference between the distances from a planet and its successor to the Sun. The planets can be mapped on the interval $[0, 1]$, with 0 corresponding to Venus and 1 to Uranus and the points pertaining to other planets respecting not only the observed order, but also the "law of increasing differences." This way, the model C "impertinently" draws near Bode's law. A classical statistical analysis has utilized a distance statistics, Δ, and it has led to the conclusion that the data do not witness for Bode's law.

It can be expected that a physicist's viewpoint will differ from a mathematician's method. Indeed, Hayes and Tremaine (1998) do not believe in the law of increasing differences. They approach the generation of random planetary systems rather in the sense of the simpler model (ii), with $\delta_{\mathrm{l}} = 0.2$ AU and $\delta_{\mathrm{u}} = 50$ AU. The generation is completed by the rejection of some obviously unstable planetary systems. Nine semi-major axes $r_0, \ldots$, r_8 are generated. Then, a nonlinear least-squares fit of the distances r_i, $i = 0, 1, \ldots, 8$, is performed to the "law" $a + bc^i$, with c being another parameter. Next, a fit is performed on leaving out j out of nine planets, $j = 0, 1, 2, 3$, and, for each j, the best reduction is chosen. The entire procedure is repeated, but, after the generation of semi-major axes, one gap is inserted between two neighboring planets with the largest ratio of $\frac{r_{i+1}}{r_i}$. It means that such planets will have numbers i, $i + 2$, and the numbering ends with number nine.

Nine "reasons" of rejecting have been formulated, first of all, the rejection need not have been attempted at all. Further, such a reason has been a violation of the condition

$$r_{i+1} - r_i > r_{i+1}V_{i+1} + r_iV_i, \tag{3}$$

where, e.g., $V_i = H_{M_i}, 2H_{M_i}, 4H_{M_i}, 8H_{M_i}$, M_i being the mass of the planet i in the solar system, H_{M_i} is the fractional Hill radius,

$$H_{M_i} = \sqrt[3]{\frac{M_i}{3M_{\mathrm{Sun}}}}. \tag{4}$$

When the observed distances in the solar system are processed in the same way as the nine generated semi-major axes, the best fit is obtained in the case, where a gap is added between Mars and Jupiter, whereas Mercury, Neptune and Pluto are ignored.

The view of Lynch (2003) approaches the statement by Efron (1971) that the statistical decision, whether the observed patterns have a physical basis or can be ascribed to chance, depends on the model. He readdresses the geometric progression of orbital periods of five major satellites of Uranus. According to the literature, the model leaves the period of Miranda unchanged and the following satellites have the periods equal to the products of one (Ariel) to four (Oberon) random factors respecting observed ratios of successive periods. He presents a simpler procedure consisting just in random choice of orbital periods in bands covering the values produced by the formula. In the original model, the probability of random origin is about 80 per cent and, in the new model, it is about 20 per cent for the chosen bands.

Since the planetary radii and periods are related by Kepler's third law, Lynch (2003) investigates the solar system in a similar way. First of all, he simplifies the Titius–Bode law to a geometric progression. Further, he repeats the comparison of two models or procedures. The procedure, which is a continuation of the study of major Uranian satellites, indicates that the probability of random origin is only about 40 per cent. In the new model, it is 0.99 for the bands chosen in the similar way as in the new investigation of the Uranian satellites. Even though rigorous mathematical methods used by Efron (1971) may throw new light on these results, Lynch (2003) is right that the possibility of a physical explanation for the observed distributions remains open.

3. Theories of Bode-like Laws

The rejected nebular hypothesis had the advantage that it assumed a mass in the region of the present solar system. After the hypothesis has been resumed, a search could begin for a shortcut of the road leading from the assumptions of the model to the Titius–Bode law. In this section, we first remember an estimate of the total mass of the solar nebula and distribution of its mass [Weidenschilling (1977)] and another one for extrasolar nebulas [Kuchner (2004)]. We mention the model of planet formation, which was standard till recent times [Lissauer (1993)]. We remember some dynamical theories of the Titius–Bode law [Graner and Dubrulle (1994), Dubrulle and Graner (1994), Krot (2009)]. These publications can be criticized, as any of them provides not a unique approach, but at least two different ones. Exceptionally, Christodoulou and Kazanas (2008) have been able to provide a unified approach.

3.1. Resumption of the Nebular Hypothesis

Weidenschilling (1977) remembers theories of cosmogony. Most such theories assume that the planetary system formed from a nebula. It is assumed that the mass fraction of Fe in the solar matter is 1.2×10^{-3}. The mass fractions are at disposal even for the terrestrial planets. To each such a planet, the mass is determined, which has the solar composition. For the planets Jupiter, Saturn, Uranus and Neptune, one proceeds differently, but also in such a way that the appropriate masses are determined.

Weidenschilling (1977) reconstructs the solar nebula by spreading the augmented planetary masses through zones surrounding their orbits. He determines the zones (AU) $(0.22, 0.56)$ for Mercury, $(0.56, 0.86)$ for Venus, $(0.86, 1.26)$ for the Earth, $(1.26, 2.0)$ for Mars, $(2.0, 3.3)$ for asteroids, $(3.3, 7.4)$ for Jupiter, $(7.4, 14.4)$ for Saturn, $(14.4, 24.7)$ for Uranus and $(24.7, 35.5)$ for Neptune in terms of the observed distances.

On the determination of the zones, surface densities can already be found. The surface density is preferred to the volume density, which can be determined only on other assumptions about the vertical structure of the nebula. The surface density proportional to $r^{-\frac{3}{2}}$ has been found, which continues the paper [Cameron and Pine (1973)]. Anomalies (deviations from the power law with the exponent $-\frac{3}{2}$) in Mercury, Mars and asteroids are stated. Processes exist for selective removal of matter from these regions.

So the nebular hypothesis includes the loss of light elements during the planetary formation. The mass of the nebula must be at least between 0.01 and 0.1 solar masses. We may calculate that the mass of the nebula is

$$\begin{aligned} M_{\text{nebula}} &= 2\pi \int_0^{r_{\text{nebula}}} r\sigma_{\text{Earth}} \left(\frac{r}{r_{\text{Earth}}} \right)^{-\frac{3}{2}} \mathrm{d}r \\ &= 4\pi r_{\text{Earth}}^2 \sigma_{\text{Earth}} \sqrt{\frac{r_{\text{nebula}}}{r_{\text{Earth}}}}, \end{aligned} \tag{5}$$

where $\sigma_{\text{Earth}} = 32000\ \mathrm{kgm}^{-2}$, $r_{\text{nebula}} = 35.5$ AU, outer limit of Neptune's zone.

Hayashi (1981) pays attention to the importance of magnetic effects on the origin of the solar system. Nevertheless, he begins with a model of the solar nebula without magnetic effects. He expounds the properties of the nebula, which entail the magnetic effects, and their significance. He takes into account the magnetic and turbulent viscosities. He attempts at an initial condition of a more ancient stage of the evolution.

The model is related to the stages, where according to the theories of planetary formation, the dust sedimented on the equatorial plane. The mass of the nebula is of the order of $0.01\ M_{\text{Sun}}$. It is assumed that $\frac{\mathrm{d}P}{\mathrm{d}r}$, where P is the pressure, has a negligible value. The half-thickness of the nebula, with the temperature dependent on r, is given. The nebula is heated by the Sun and the field temperature is

$$T = 280\sqrt{\frac{r_{\text{Earth}}}{r}} \tag{6}$$

for the luminosity of the Sun identified with the present value.

In the interval $[0.35, 36]$ AU, three kinds of the surface density are considered. Two components of the total density are related only to the dust and gas, but the density of rock increases from 2.7 AU (in the asteroid belt) due to the presence of ice. The gas prevails.

The exponent $-\frac{3}{2}$ is utilized and the surface density $\rho_s(r_{\mathrm{Earth}}) = 17000\ \mathrm{kgm}^{-2}$. As long as the magnetic effects are negligible, the volume density is

$$\rho(r,z) = \rho_0 \left(\frac{r}{r_{\mathrm{Earth}}}\right)^{-\frac{11}{4}} \exp\left(-\frac{z^2}{z_0^2(r)}\right), \tag{7}$$

where $\rho_0 = 1.4 \times 10^{-6}\ \mathrm{kgm}^{-3}$ and $z_0(r) = 0.0472 r_{\mathrm{Earth}} \left(\frac{r}{r_{\mathrm{Earth}}}\right)^{\frac{5}{4}}$. A simple vertical structure is assumed, so that the surface density is

$$\begin{aligned}\rho_s(r) &= \sqrt{\pi}\rho_0 z_0(r) \left(\frac{r}{r_{\mathrm{Earth}}}\right)^{-\frac{11}{4}} \\ &= \rho_s(r_{\mathrm{Earth}}) \left(\frac{r}{r_{\mathrm{Earth}}}\right)^{-\frac{3}{2}}.\end{aligned} \tag{8}$$

Here $\sqrt{\pi}1.4 \times 10^{-6} \times 0.0472 r_{\mathrm{Earth}} = \sqrt{\pi}1.4 \times 10^{-6} \times 0.0472 \times 1.5 \times 10^{11} = 1.7 \times 10^4$ kgm^{-2}.

We note that the exponent $\frac{5}{4} > 1$. Magnitudes of the magnetic fields H_1 and H_2 are determined such that the vertical structure of the nebula is affected for $H > H_1$, with H being the magnitude of a field present in the solar nebula, and a deviation from the Keplerian velocity of rotation occurs for $H > H_2$. The field magnitudes H_1 and H_2 decrease with the increase of r. In a uniformly ionized gas, magnetic fields grow and decay according to a result of magnetohydrodynamics.

The turbulence of an "equilibrium" solar nebula leads to the origin of seed magnetic fields. A possibility of the redistribution of gas density in the solar nebula is studied, which is caused by angular momentum transport due to the presence of magnetic and mechanical turbulent viscosity. The effect of the mechanical viscosity is reduced to the diffusion in the radial direction.

This way the exponent $-\frac{3}{2}$ can be derived. For the isothermal case, the exponent -2 is given. It is admitted that, without further calculations, it is not certain whether the effect of mechanical viscosity alone suffices or magnetic viscosity must contribute. According to an accomplished theory of planetary formation, the planets have been formed except for Uranus and Neptune before the dissipation of the solar nebula, Saturn being formed in an intermediate stage [Hayashi (1981)].

The magnetic effects on the structure of the nebula are negligible in regions of the terrestrial planets, even though it is not valid for its outermost layers. Contrary to this, these effects are significant in regions of the giant planets. Hayashi (1981) recognizes a numerical simulation of the cloud that preceded the nebula. He discusses an initial condition of the collapse. He does not adopt the spherical Jeans condition for it, but properties of the fragment of a rotating isothermal disk.

Kuchner (2004) reviews the mental pictures of the minimum-mass solar nebula. He includes also the papers [Weidenschilling (1977), Hayashi (1981)]. He attempts to take into account the extrasolar planets. He introduces the concept of a minimum-mass extrasolar nebula. He concentrates himself to few-planet, i.e., two-planet and three-planet systems discovered by precise Doppler methods. The astronomer can infer planets with a suitable relation between the orbital period and the ratio of the mass of the planet to the mass of the star corrected by the angle of sight. They detect radial velocity variations of the planet.

Kuchner (2004) chooses 1000 M_{Earth} for the augmented masses of most of the extra-solar planets. The dependence of the surface density on the semi-major axis is obtained by mixing of data of different systems. It emerges that the surface density is proportional to r^{-2}. Separately fitted nebulae are not taken too seriously and their exponents are -2.42 through -1.50. The mixed data force one to admit even the solar exponent $-\frac{3}{2}$.

The minimum-mass solar nebula (exponent $-\frac{3}{2}$) and the uniform-α accretion disk model [Shakura and Sunyaev (1973)] suggest that giant planets do not form at the center of the disk. Migration theories are quite acceptable, if the power law has the exponent $-\beta$, $\beta > 2$. The minimum-mass solar nebula was based on Laplace's concept of the solar nebula that broke up into rings that condensed into planets.

3.2. Protoplanetary Disks

The origin of the solar system is a recognized problem of science [Lissauer (1993)]. Models of planetary formation are developed using the solar system and limited astrophysical observations of star-forming regions and circumstellar disks. Other planetary systems are detected around main sequence stars and pulsars.

A theory of the origin should explain the following facts:

1. Both the orbits of the planets and those of most of asteroids are nearly coplanar and this plane is close to that of the Sun's equator. The orbits of the planets are nearly circular and planets orbit the Sun in the same sense as the Sun rotates.

2. Spaces between the orbits of the planets increase with the distance from the Sun. The orbits of eight planets do not cross. Even though Pluto's orbit crosses that of Neptune, the dwarf planet avoids close encounters with the planet due to $3:2$ resonance.

3. Comets orbit the Sun.

4. Six of the eight planets rotate around their axis in the same direction, in which they revolve around the Sun (cf. point 1), and their obliquities (tilts of their axes) are less than 30°.

5. Most planets have natural satellites.

6. Planetary masses account for less than 0.2 % of the mass of the solar system.

7. Over 98 % of the angular momentum in the solar system is contained in the orbital motions of the Jovian planets.

8. Planets and asteroids have compositions which are rather well known.

9. The size of asteroids and parameters of asteroidal orbits are rather well known.

10. Nearly all meteorites come from the asteroid belt.

11. Ages of meteorites are relatively well known.

12. Isotopic ratios are about the same in all solar system bodies.

13. Meteorites argue for rapid heating and cooling and for magnetic fields of order 1 Gauss (10^{-4} T).

14. Most solid planetary and satellite surfaces are heavily cratered.

Point 1 suggested the hypothesis of planetary formation in a flattened disk [Kant (1755), Laplace (1796)]. In the 1990s, the evidence for the presence of appropriately large disks around pre-main sequence stars increased. Protoplanetary disks contain a mixture of gas and condensed matter. A lower bound on the mass of the protoplanetary disk has been mentioned above [Weidenschilling (1977)]. The planetesimal hypothesis asserts that planets grow within circumstellar disks through pairwise accretion of small solid bodies, the so-called planetesimals. Sufficiently massive planetary bodies embedded in a gas-rich disk can gravitationally capture much gas and produce Jovian-type planets. The absence of a planet in the dynamically stable region inside Mercury's orbit can be attributed to two reasons. Close to the early Sun, nebula temperatures were such high that condensation of material did not take place. Possibly, solid planetesimals felt so strong a gas drag that their decay depleted the region considered of condensed matter [Lissauer (1993)].

Theories considering instabilities of the gas leading to giant gaseous protoplanets fail to explain just compositions of the Jovian planets. Models of planetary growth from small solid bodies do not suffer from such difficulties. Heating in the consequence of the collapse of a molecular cloud core to the solar system dimensions is admitted. Afterwards the disk can cool. Various compounds condense into microscopic grains. The motions of small grains in a protoplanetary disk are strongly coupled to the gas. The vertical component of the star's gravity causes sedimentation onto the midplane of the disk. Models suggest that the volume of the solid material was able to agglomerate into bodies of macroscopic size at least in the terrestrial planet region of the solar nebula [Weidenschilling and Cuzzi (1993)].

With respect to a possible pressure gradient in the radial direction, the gas circles the star less rapidly than the Keplerian rate. Large particles which move at nearly the Keplerian speed experience a headwind, and so the material that survives to form the planets, must accomplish the transition from cm size to km size rather quickly. Further forces upon planetesimals are remembered besides the star's gravity. They are gravitational interactions with other planetesimals and protoplanets. Further mutual inelastic collisions and gas drag.

It is stated that the simplest analytic approach to the evolution of planetesimal velocities is based on methods of the kinetic theory of gases. For the final stages of the evolution, the number of planetesimals becomes small enough that a direct treatment of individual planetesimal orbits is feasible. A numerical n-body integration has been replaced by the alteration of precessing elliptic orbits by close encounters with other planetesimals.

Physical collisions dissipate part or all of the relative kinetic energy of colliding bodies. Models comprise a Boltzmann collision operator for hard spheres modified to allow for inelastic collisions and gravitational interaction. Gas drag damps excentricities and inclinations of planetesimals, especially small ones.

It is accepted that 3-body effects may be neglected. After Safronov (1969), accretion zones and the protoplanet as the largest body in the given zone are introduced. The region, in which 3-body effects are significant, is limited with the radius of protoplanet's Hill Sphere

$$h_S = \sqrt[3]{\frac{M}{3M_*}}a, \tag{9}$$

where M and M_* are the masses of the protoplanet and the star, respectively, and a is the semi-major axis of the protoplanet's orbit.

It is mentioned that the accretion rate of a protoplanet is enhanced by the squared ratio of the escape velocity from the point of contact to the relative velocity of the bodies. It is pointed out that during simulation of early stages, a simultaneous calculation of the velocity evolution and size evolution in the planetesimal swarm is necessary. It was found that the size evolution is of two kinds. The slower evolution exhibits a regular growth of all planetesimals. The more rapid evolution that is related to exceptional planetesimals shows a "runaway" accretion to the largest planetesimal in the local region. The discrete form of the coagulation equation has solutions of two kinds (bifurcation) [Safronov (1969)].

The condition of low velocities of planetesimals for the runaway accretion is remembered. When a protoplanet consumes most of the planetesimals within its gravitational reach, its mass is equal to the so-called isolation mass and the rapid runaway growth may cease. Mechanisms of further growth have been considered. Attention is drawn to the origin of semi-major axes of protoplanets in an arithmetic progression. Such a configuration is not dynamically stable for a long time. Crossing orbits, close gravitational encounters and violent collisions are predicted. It can be referred to, e.g., the paper [Wetherill (1990)].

The necessity of a high-velocity, post-runaway growth phase is emphasized. Times closer to 10^8 are given. Next it is stated that the model of minimum-mass protoplanetary disk does not yield appropriate times for the cores of the giant planets. The exposition of giant planet atmospheres in [Lissauer (1993)] begins with the hypothesis that the atmospheres of the terrestrial planets and other small bodies come from material accreted as solid planetesimals. The view is adopted that first the core of a giant planet was formed. The masses of such cores are of order 10 M_{Earth}. The accretion of the gaseous envelopes of the giant planets long lacked explanation.

The origin of planetary rotation, point 4, is a question, which is difficult to answer. Although part of literature denies a random occurrence, part relies on stochastic factors. The angular momenta brought in by individual planetesimals and hydrodynamic accretion of gas provide planets with rotational angular momentum perpendicular to the midplane of the disk essentially, because the disk is flat.

The analogy of the moons and rings of the giant planets to miniature planetary systems is admitted. Inside Roche's limit, where tidal forces from the planet suffice to disrupt a moon held together only by its gravity, planetary rings dominate. In the outer regions of the satellite systems of Jupiter, Saturn and Neptune, small bodies on highly excentric and inclined orbits occur. This way planetary satellites can be classified as "regular" and "irregular". Regular satellites are related to a circumplanetary disk. Irregular moons may be captured planetoids from the solar nebula. The satellites of Earth and Mars and some other planetary satellites are harder to classify in this manner.

The analogy of the circumplanetary disk to the larger circumsolar disk is nearly imperfect. Satellite systems are much more compact than the planetary system and are evolved dynamically. Most satellite rotations have been locked by planetary tidal forces. Among several moons mean motion commensurabilities persist. Satellites of satellites have not been observed. For the explanation of this simplification, many reasons have been brought in.

The Earth's Moon is different and indicates a stochastic event. Complexity of the satellite systems of the four giant planets in our solar system suggests that stochastic processes participated in satellite formation. It is pointed out to, e.g., little total mass in the asteroid region and their large number. Also the excentricities and inclinations of the orbits of these bodies are higher than the planetary values. We remember also the diversity of the composition. Proximity to Jupiter is an explanation. For dynamical models of the formation of the asteroid belt, one refers to the papers [Wetherill (1989, 1991, 1992)].

The division of comets into two groups is mentioned: short-period comets, with the orbital period shorter than 200 years, and long-period comets, which return after more than 200 years. Hyperbolic orbits are admitted as consequences of perturbations by the planets. The long-period comets come from the Oort cloud. Some short-period comets come from the Oort cloud, but most of them populate the Kuiper belt. The origin of comets, especially the Oort cloud, is connected to the ejection of small planetesimals from the giant planet region. This excludes a restriction of the study to the minimum-mass protoplanetary disk. Also the outer limit of Neptune's zone cannot be the edge of the protoplanetary disk, on considering the formation of the Kuiper belt from planetesimals.

In conclusion, the planetesimal hypothesis is a viable theory of the growth of the terrestrial planets, the cores of the giant planets, and the smaller bodies present in the solar system. The formation of solid bodies of planetary size should be common around young stars, which do not have stellar companions at planetary distances. On certain conditions, planets could form within circumpulsar disks. A summary of the theory of planetary growth from planetesimal accretion within a circumstellar disk is provided in [Lissauer (1993)]. Several different ideas have been added, e.g., the rule that a more massive protoplanetary disk of the same radial extent will probably produce a smaller number of larger planets.

3.3. Dynamical Theories of the Titius–Bode Law

Graner and Dubrulle (1994) point out to the book [Nieto (1972)], in which the notion of a dynamical theory of the Titius–Bode law has been introduced. In spite of the diversity of these theories, in each of the models a Titius–Bode like law arises. Let us note that it means a geometric progression in planetary distances.

Essentially, a description using partial differential equations, which are symmetric, is assumed. The symmetries considered are:

(P1) the invariance with respect to the rotation around an axis z,

(P2) the invariance with respect to the dilatation (scale transformation) in the plane perpendicular to z,

(P3) the time independence.

On denoting $g_j(r, \theta)$ the physical quantities obeying these equations and γ_j the respective exponents, $\Lambda^{\gamma_j} g_j(\Lambda r, \theta)$, where $\Lambda > 0$, satisfy also these equations.

Unfortunately, the exposition comprises imperfections, which we do not evaluate, but must mention them. A physical quantity $g(r, \theta)$ with an exponent γ and its Fourier series decomposition

$$g(r, \theta) \overset{?}{=} \mathrm{Re}\left\{\sum_m a_m(r) \exp\left[i(m\theta + \theta_m)\right]\right\}, \tag{10}$$

where θ_m are real numbers, are considered. However, the closest usual form of this equation is

$$g(r, \theta) = \sum_m a_m(r) \exp(im\theta). \tag{11}$$

We see that both Re and θ_m are superfluous and that the usual statement that $a_0(r)$ is a real number, is substituted with a feeling of $a_0(r)$ being a complex number. But it cannot be defined uniquely.

It is assumed that the coefficient a_m of the decomposition fulfils a symmetric equation

$$h_m\left(\frac{a_m a_m^*}{r^{2\gamma}}, \frac{r}{a_m}\frac{\partial a_m}{\partial r}\right) = 0, \tag{12}$$

where h_m has arisen in a manipulation and the asterisk means the complex conjugation. As a first-order ordinary differential equation, it can be cast to the form

$$\frac{\partial a_m}{\partial r} = \frac{a_m}{r} H_m\left(\frac{|a_m|^2}{r^{2\gamma}}\right), \tag{13}$$

where H_m has arisen in another manipulation, if the dependence of the second argument on the first one can be made explicit. Using the substitution

$$b_m = \frac{a_m}{r^\gamma}, \; x = \log\left(\frac{r}{r_0}\right), \tag{14}$$

where r_0 is a normalizing radius, we obtain that

$$\frac{\partial b_m}{\partial x} = b_m G_m\left(|b_m|^2\right), \tag{15}$$

where $G_m = H_m - \gamma$. As long as $|b_m|^2$ is small, we may utilize the property

$$G_m(|b_m|^2) = \mu_m + ik_m + (\eta_m + i\kappa_m)|b_m|^2 + \mathrm{O}(|b_m|^4), \tag{16}$$

where

$$\mu_m + ik_m = G_m(|b_m|^2)\Big|_{|b_m|^2=0}, \; \eta_m + i\kappa_m = \left.\frac{\mathrm{d}G_m(|b_m|^2)}{\mathrm{d}|b_m|^2}\right|_{|b_m|^2=0}. \tag{17}$$

Let us remark that, in the paper [Graner and Dubrulle (1994)], the subscript m at μ, η, k and κ is omitted. Neglecting the dependence of G_m on $|b_m|^2$ at all, we obtain that

$$b_m = B_0\left(\frac{r}{r_0}\right)^\mu \exp\left[ik\log\left(\frac{r}{r_0}\right)\right], \tag{18}$$

where $\mu \equiv \mu_m$, $k \equiv k_m$. A derivation of a Titius–Bode law is possible in the case of one mode, or in the case of independence of the modal index m. It is also necessary to assume $k \neq 0$. Equal phase cylinders are of the form $r = r_n$, where

$$r_n = r_0 K^n, \quad K = \exp\left(\frac{2\pi}{k}\right). \tag{19}$$

We have criticized that the derivation is only pertinent to $m \neq 0$, so it can hardly be applied to rotationally symmetric solutions, which are important. In spite of this, Graner and Dubrulle (1994) modify the solution (18) to include further the linear term of the Taylor series of the function G_m in $|b_m|^2$. The distances r_n change, do not form a geometric progression and only illustrate the so-called nonlinear Titius–Bode law.

Dubrulle and Graner (1994) utilize the theory from [Graner and Dubrulle (1994), part I] even though, in part I, the assumption (P3) of time independence was made too, which is not made here at first. Therefore, they mention the scale invariance of equations for the hydrodynamic description of the solar nebula. The scalar invariance need not always occur as a consequence of the polytropic gas law

$$P = P_0 \left(\frac{\rho}{\rho_0}\right)^{1+\frac{1}{n}}, \tag{20}$$

where P is the pressure, ρ the volume density, P_0 and ρ_0 are the normalizing pressure and the normalizing density, respectively, and n is the polytropic index, which constrains the validity of the scale invariance to $n = 3$. That is why, Dubrulle and Graner (1994) assume $P = 0$. Another difficulty presents the Poisson equation, in which the second-order partial derivatives occur, while the outline of a theory has included only the first-order derivatives. Therefore, an integral expression of the gravitational potential with the linear density is assumed.

On specifying the equilibrium solution, the assumption of the time independence is abandoned again and the stability of the equilibrium solution is studied by the method of linearization around this solution. The linearized equation is solved by the method of the Fourier series decomposition, which is appropriate to the surface density decreasing like r^{-2}. An option is typical of these problems and their adversity. We obtain

$$\tilde{v}^2 = \left(\frac{\partial}{\partial x} - 1\right)\tilde{\phi}, \tag{21}$$

where $\tilde{v}$ is the scale-invariant tangential component of the velocity and $\tilde{\phi}$ is the scale-invariant gravitational potential of the disk.

The question arises, which mode k (k is a wavenumber of the Fourier series) is suitable for a derivation of the Titius–Bode law. Dubrulle and Graner (1994) argue for $k = k_c$, where k_c separates, which wavenumbers are low and which are high. The low wavenumbers ($< k_c$) are linearly stable and the high wavenumbers ($> k_c$) are linearly unstable. The critical wavenumber k_c depends on $\frac{M_{\rm eg}}{M_C}$, where $M_{\rm eg}$ is an effective gravitating disk mass, substituting for its actual mass, M_D, in a rather complicated manner, and M_C is the mass of the central body.

Christodoulou and Kazanas (2008) have dealt with equilibrium structures of rotating fluids with cylindrical symmetry. They have derived exact results and are convinced that the results are relevant to the location of the planetary orbits. The famous Titius–Bode law also expresses this location, but it is actually opposed to the authors' effort.

The Lane–Emden equation is mentioned, which describes the equilibria of non-rotating fluids. This equation has a spherical symmetry. It can be modified to an equation with cylindrical symmetry easily, what has been utilized by physicists outside astrophysics. In the astrophysics, the paper [Jeans (1914)] may be typically quoted. The results for a finite polytropic index n are known, whereas Christodoulou and Kazanas (2008) deal with the case where $n \to \infty$.

The Titius–Bode "law" is held for the Titius–Bode algorithm. It is also referred to the critiques, which we consider in this review, or to papers, which somewhat underpin this rule. Typically, a pertinent paper expresses both cons and (at least essentially) pros.

Some methodologically related analyzes are mentioned. The simple cylindrical formulation is defended [Jeans (1914)]. Rules are evaluated, which would approximate the observed distances. Only two rules are considered. Such rules rely on a consecutive numbering of planets, to which both asteroid Ceres is counted and the dwarf planet Pluto. The rules do not predict positions of the end bodies. The rule of arithmetic mean, $\frac{1}{2}(a_{i-1}+a_{i+1})$, has errors up to 23.5 and 27.9%. The rule of geometric mean, $(a_{i-1}a_{i+1})^{\frac{1}{2}}$, has errors up to -11.8 and -14.0 %.

In the same vein, we could try another possibility, the rule of the mean with the exponent of $\frac{1}{2}$, $[\frac{1}{2}(a_{i-1}^{\frac{1}{2}} + a_{i+1}^{\frac{1}{2}})]^2$. Having tried it, we obtain the error up to 16.3 % in two cases. A good fit to the rule of the arithmetic mean in the bodies that neighbor the end ones is stated. In the intermediate bodies, the rule of the geometric mean is suitable. The use of the rule, which better predicts the distance to the body, leads to errors up to 5.0 and 9.1 %. We have not improved it using the rule of the mean with exponent $\frac{1}{2}$! It is obvious that the Titius–Bode law begins with a three-term arithmetic progression, which is the shortest nontrivial progression of this kind.

Christodoulou and Kazanas (2008) then mention interesting connections or analogies. In optics, an analogue of the quantity

$$\mathcal{M}_i \equiv \frac{a_{i+1} - a_i}{a_i - a_{i-1}} \tag{22}$$

can be found. Obviously, it is a quantity reducing to the Titius–Bode base two and independent of their coefficients a and b.

We have calculated the ratios $\mathcal{M}_i$ according to equation (22) and obtained results can be found in Table 1 (we insert the error of the rule of the mean with exponent $\frac{1}{2}$). The ratios $\mathcal{M}_i$ can be further rounded to 1 and 2. So the three-term arithmetic progression leads to 1 for single Venus and the Titius–Bode law leads to 2 for the Earth to Saturn. In what preceded, we have presented a continuation of the Titius–Bode rule.

Nevertheless, Christodoulou and Kazanas (2008) have concentrated themselves to the Lane–Emden equation. They have not been attracted by the generous identification of the Titius–Bode sequence with the geometric progression, which has been done by Graner and Dubrulle (1994) and Dubrulle and Graner (1994). Christodoulou and Kazanas (2008) distinguish these notions. They have evaluated also the approach of the cited authors.

The isothermal equilibrium is assumed, i.e., the pressure balances the general gravity.

Table 1. Magnification ratios $\mathcal{M}_i$ of planetary orbits and relative errors of the means with exponent $\frac{1}{2}$

Index i	Planet	$\mathcal{M}_i$	Error (%)
1	Mercury		
2	Venus	0.8	-9.0
3	Earth	1.9	8.7
4	Mars	2.4	16.3
5	Ceres	2.0	11.7
6	Jupiter	1.8	8.5
7	Saturn	2.2	16.3
8	Uranus	1.1	-4.3
9	Neptune	0.9	-5.5
10	Pluto		

The cylindrical symmetry is assumed. It is assumed that the angular velocity has the form

$$\Omega(r) = \Omega_0 f_{\mathrm{CK}}\left(\frac{r}{r_0}\right), \tag{23}$$

where $\Omega_0 = \Omega(0)$ for centrally condensed models and $f_{\mathrm{CK}}(x)$ is an infinite-dimensional parameter such that $f_{\mathrm{CK}}(0) = 1$. It is assumed that an isothermal equation of state for the pressure P and the gas density ρ holds (it is a volume density, but it has much in common with the surface density with respect to the cylindrical symmetry)

$$P = c_0^2 \rho, \tag{24}$$

where c_0 is the isothermal sound velocity. It is declared that for finite polytropic indices, significantly different results are not obtained.

Euler's equation for the unknown functions ρ, Ω, P, and ϕ is presented,

$$\frac{1}{\rho}\frac{\mathrm{d}P}{\mathrm{d}r} + \frac{\mathrm{d}\phi}{\mathrm{d}r} = \Omega^2 r, \tag{25}$$

but we already know that $\Omega \equiv \Omega(r)$ is rather a parameter. Poisson's equation for these functions is also given,

$$\frac{1}{r^{d-1}}\frac{\mathrm{d}}{\mathrm{d}r} r^{d-1}\frac{d\phi}{dr} = 4\pi G\rho, \tag{26}$$

where $d = 2$ with respect to the cylindrical symmetry and G is the gravitational constant. On the elimination of ϕ and with the substitution $x = \frac{r}{r_0}$, an equation is obtained, which may be called the Lane–Emden equation with rotation,

$$\frac{1}{x}\frac{\mathrm{d}}{\mathrm{d}x} x \frac{\mathrm{d}}{\mathrm{d}x}\log\tau + \tau = \frac{\beta_0^2}{2x}\frac{\mathrm{d}}{\mathrm{d}x}(x^2 f_{\mathrm{CK}}^2), \tag{27}$$

where $\tau = \frac{\rho}{\rho_0}$ and $\beta_0^2 = \frac{\Omega_0^2}{2\pi G\rho_0}$. Here ρ_0 is the maximum density.

For $\beta_0 f_{CK} = 0$, equation (27) differs from the usual Lane–Emden equation by the assumed cylindrical symmetry of the matter described and the polytropic index $n \to \infty$. Closed solutions are presented. We ask for which next choice of $\beta_0 f_{CK}$, the solution will be viable and physically meaningful. It is assumed that

$$\tau = \frac{\beta_0^2}{2x}\frac{\mathrm{d}}{\mathrm{d}x}(x^2 f_{CK}^2), \tag{28}$$

then the term with the derivatives on the left-hand side of equation (27) equals zero,

$$\frac{1}{x}\frac{\mathrm{d}}{\mathrm{d}x}x\frac{\mathrm{d}}{\mathrm{d}x}\log\tau = 0. \tag{29}$$

Then the density depends on two integration constants, A and k,

$$\tau(x) = \frac{\beta_0^2}{2}Ax^{k-1} \tag{30}$$

and $\beta_0 f_{CK}$ also depends on the integration constant B,

$$\beta_0 f_{CK}(x) = \frac{\sqrt{Ag(x) + B}}{x}, \tag{31}$$

where

$$g(x) = \begin{cases} \frac{x^{k+1}}{k+1}, & \text{if } k \neq -1, \\ \log x, & \text{if } k = -1, \end{cases} \tag{32}$$

implying that $\frac{\mathrm{d}g}{\mathrm{d}x} = x^k$ for all values of k. Five points are added to this result. Point 4 on the composite profiles is important. These profiles can be utilized to the predictions we have mentioned above and we will also return to them below. The striking assumption (29) is connected with the enthalpy gradient being constant.

Even though the surface density corresponds to $k = -\frac{1}{2}$, also $k = 1$ is interesting, for which we obtain a finite value at the axis of the cylinder, $\tau(0) = \beta_0^2$, which cannot be changed on a choice of A, since necessarily $A = 2$ with respect to the boundary condition $f(0) = 1$. The boundary condition $\tau(0) = 1$ is mentioned, which contradicts the property $\tau(0) = \beta_0^2$, in general. The boundary conditions are "imposed" to the desired function at the axis of the cylinder. It is stated that it has been performed using a numerical integration.

The solution for $\tau(0) = \beta_0^2$ and $\left.\frac{\mathrm{d}\tau}{\mathrm{d}x}\right|_{x=0} = 0$ represents a "baseline" solution, about which the solution oscillates that fulfils the proper initial condition. The oscillatory behavior of the density profiles is utilized for fitting the observed planetary distances to the density peaks. A construction of the density profile and rotation law in three regions, $x \leq x_1$, $x_1 < x < x_2$, $x \geq x_2$ is described. The model depends on at least three parameters, $x_1 > 0$, $x_2 > x_1$, and $1 - k \equiv \delta > 2$, with k being related to the region $x_1 < x < x_2$. The fourth parameter of the model is β_0. The mentioned four parameters have been chosen such that they predict the observed planetary distances. In the discussions, many further interesting connections with research of planet formation are presented. The model has been applied to the solar nebula. An application to the satellites of Jupiter and the five planets of 55 Cancri is possible.

In the paper [Christodoulou and Kazanas (2008)], solutions have been found successfully, which exhibit a pronounced oscillatory behavior. The adaptable slope of the profile of the differential rotation leads to arithmetic partial or geometric partial progressions of the mass density peaks. It is a clear explanation of the Titius–Bode "law" of planetary distances. A criticism of the explanations of the order invoking new dynamical laws or "universal" constants and solar-system "quantizations" is indicated. At this place, we could object that the choices of slopes of the profile of the differential rotation have only been verified using a numerical integration. It is admitted that the currently available observations have not supported the idea that our solar system is a representant of the planetary systems around stars. The most similar systems such as 55 Cancri and HD37124 are mentioned.

Krot (2009) mentions his statistical theory of gravity. He promises to expound the gravitational condensation. He derives an anti-diffusion equation. To our opinion, this derivation is faulty. His equation (48) is the beginning, but it already expresses what one has intended to derive. After a certain time, the maximum value of the density from the reference time t_0 is present in a point at a distance Δr from the origin, where the maximum was attained at the time t_0. Thus, the density is closer to the origin than it was at the time t_0. It does not seem to be an error, but equation (48) is not so evident. Then the anti-diffusion equation is combined with the Euler equation. The uniform rotation around the axis z is assumed. For example, a distribution with the shape of oblate ellipsoid, but still the Laplacian–Gaussian one, is derived.

The derivation of the distribution of specific angular momentum λ, $\lambda = \Omega h^2$, where Ω is the angular velocity of the uniform rotation around the axis z and h is the distance from the axis z, is promised. Let us note that it means the double of the areal velocity. This distribution is

$$f(\lambda) = \frac{\alpha(1-\varepsilon_0^2)}{2\Omega} \exp\left(-\frac{\alpha(1-\varepsilon_0^2)}{2\Omega}\lambda\right), \tag{33}$$

where $\alpha \equiv \alpha(t)$ is a positively defined monotonically increasing function and ε_0 is a constant, ε_0^2 is a squared eccentricity of ellipse [Krot (2009)].

Schmidt's hypothesis is a picture that particles or planetesimals, which have sufficiently close values of the specific angular momentum or those of the areal velocity, condense or accrete [Schmidt 1944]. The hypothesis may include just cutting of the distribution of the areal velocities to equal parts.

A solution of this problem is described. We look for specific angular momenta μ_n, λ_n, such that

$$\mu_n = \frac{\lambda_n + \lambda_{n+1}}{2}, \tag{34}$$

$$\lambda_n = \frac{\int_{\mu_{n-1}}^{\mu_n} \lambda f(\lambda)\mathrm{d}\lambda}{\int_{\mu_{n-1}}^{\mu_n} f(\lambda)\mathrm{d}\lambda}. \tag{35}$$

Using the approximation

$$\lambda_n \approx \frac{\lambda_{n+1} + \lambda_{n-1}}{2}, \tag{36}$$

we obtain λ_n in the form of an arithmetic progression. This way, Schmidt (1944) has derived his law for planetary distances, r_n. Then he has encountered the fact that it is not accurate enough for all the planets together. He has recommended to use it separately for the terrestrial planets and the Jovian planets.

The approximation (36) is based on the possibility of considering $f(\lambda)$ a constant in the interval $[\mu_{n-1}, \mu_n]$. Krot (2009) performs at least the linear approximation of the function $f(\lambda)$ (his equation (101)),

$$f(\lambda) \approx \frac{\alpha(1-\epsilon_0^2)}{2\Omega}\left[1 - \frac{\alpha(1-\epsilon_0^2)}{2\Omega}\lambda\right]. \tag{37}$$

He adds a generalization of the Schmidt law using the inverted equation

$$\lambda_n = \sqrt{2GM_{\mathrm{Sun}} r_n}. \tag{38}$$

We will return to the derivation below.

The proposed theory is applied to the formation of the solar system. As usual, the whole enterprize with the planets is based on merely eight planets and on the dwarf planet Pluto. We criticize the fitting of data to the theoretic dependence. An excessive number of parameters, namely seven, only the number of planets minus one, have been proposed. So we are very close to a mere data transformation, which is not recommended. Every further comparison then results in favor of this *incorrect* procedure, whose inappropriateness may not be conceded by the author.

Now, the planets are successively numbered, although the asteroid belt might deserve its number. The asteroid belt is equated one planet usually. Krot (2009) simply neglects this belt. We criticize a large number of parameters for expressing a relatively small number of observed distances. Let us see, how the author has arrived at this solution. The author has processed various assumptions to the "hypothesis" that the cloud of particles has an exponential distribution of specific angular momentum λ with the probability density (33). Later he has sacrificed this assumption to the approximation (37). On substituting into equation (35), he has obtained that

$$\lambda_n \approx \frac{\frac{\mu_n+\mu_{n-1}}{2} - \frac{\alpha(1-\varepsilon_0^2)}{6\Omega}\left(\mu_n^2 + \mu_n\mu_{n-1} + \mu_{n-1}^2\right)}{1 - \frac{\alpha(1-\varepsilon_0^2)}{4\Omega}\left(\mu_n + \mu_{n-1}\right)}, \tag{39}$$

where μ_n is defined by equation (34). On substituting (34) into (39) and using the property

$$\bar{\lambda} = \frac{2\Omega}{\alpha(1-\varepsilon_0^2)}, \tag{40}$$

the author obtains the finite difference equation

$$\lambda_n = \frac{\bar{\lambda}(C_n + D_n) - \frac{1}{3}\left(C_n^2 + C_nD_n + D_n^2\right)}{4\bar{\lambda} - (C_n + D_n)}, \tag{41}$$

where $C_n = \lambda_{n+1} + \lambda_n$, $D_n = \lambda_n + \lambda_{n-1}$ and where, for readability, we have already written the equality sign instead of the more correct $\approx$.

We mind that this finite difference equation is not completed with initial or boundary conditions. Before we complete it, we will return to the original formulation. The sequence to be found is not denoted by a single letter, but it has the form

$$(\lambda_1, \mu_1, \lambda_2, \mu_2, \lambda_3, \mu_3, \ldots). \tag{42}$$

So the finite difference equations (34) and (35) are equations for a single sequence, even though it is denoted by two letters. But it is justified by the alternation of the equations. We begin with equation (35) for $n = 1$, where we put $\mu_0 = 0$, hopefully. The author has not explained it. We proceed with equation (34) for $n = 1$, etc. Another assumption to be considered is the hypothesis that the sequence is finite. In the model of formation of the solar system, it follows from the finite number of the planets.

Krot (2009) does not recommend a length of the sequence. The author demonstrates that the longer sequence that includes Pluto does not fit well. Although we have not made numerical calculations, we could also exclude Pluto. Then we obtain

$$(\lambda_1, \mu_1, \lambda_2, \mu_2, \ldots, \lambda_7, \mu_7, \lambda_8). \tag{43}$$

To each term of this sequence, an equation exists. We simplify the last equation with the assumption $\mu_8 = +\infty$. This assumption is reasonable. In combination with the approximation (37) it should rather be assumed that $\mu_8 = \bar{\lambda}$. So we have arrived at fifteen equations for fifteen unknowns, which depend only on the parameter of the distribution $f(\lambda)$. It is unique for the time being. It is far from the result by Krot (2009), who asserts something else.

To assess the equations, which are solved in an obviously incorrect way, we note that somewhere a "usual" mistake occurs and the numbering is shifted by one. Certainly, $\lambda_0 = a_0$ for $n = 0$ in equation

$$\lambda_n = a_0 + dn, \tag{44}$$

where d is the difference and a_0 is the first (the zeroth) term of the arithmetic progression. Later in the misleading substitution,

$$\lambda_n = Z^n, \quad n = 1, 2, 3, \ldots, \tag{45}$$

$n \geq 1$ already.

Since, in equation (41), the variable n is comprised only as the subscript, we can utilize our note. In this equation, we put $n = 2, \ldots, 7$, considering the boundary conditions

$$\lambda_1 = \frac{\bar{\lambda}(\lambda_2 + \lambda_1) - \frac{1}{3}(\lambda_2 + \lambda_1)^2}{4\bar{\lambda} - (\lambda_2 + \lambda_1)}, \tag{46}$$

$$\lambda_8 = \frac{\bar{\lambda}(2\bar{\lambda} + \lambda_8 + \lambda_7) - \frac{1}{3}\left[(2\bar{\lambda} + \lambda_8 + \lambda_7)^2 - 2\bar{\lambda}(\lambda_8 + \lambda_7)\right]}{4\bar{\lambda} - (2\bar{\lambda} + \lambda_8 + \lambda_7)}. \tag{47}$$

Although the problem is formulated, it is meaningless to expound its solution. One of many reasons is that Krot (2009) proceeds in a completely different way. We will explain the essence of our method using a different class of distributions of the specific angular momentum than (33).

Let us assume that the random variable $\underline{\lambda}$ has the log uniform distribution, i.e., that $\log \underline{\lambda}$ has the uniform distribution on the interval $[\log \lambda_{\min}, \log \lambda_{\max}]$. Then

$$f(\lambda) = \frac{1}{\lambda \log\left(\frac{\lambda_{\max}}{\lambda_{\min}}\right)}. \tag{48}$$

Then the finite difference equation (35) becomes

$$\lambda_n = \frac{\mu_n - \mu_{n-1}}{\log\left(\frac{\mu_n}{\mu_{n-1}}\right)}, n = 1, \ldots, 8, \tag{49}$$

where $\mu_0 = \lambda_{\min}$, $\mu_8 = \lambda_{\max}$. Given the differential rotation of the form $\Omega_0 \left(\frac{h}{h_0}\right)^{\frac{k-1}{2}}$, with $h_0 \equiv r_0$, the specific angular momentum will be

$$\lambda = \lambda_0 \left(\frac{h}{h_0}\right)^{\frac{k+3}{2}}, \quad -3 < k < 1, \tag{50}$$

where λ_0 is the specific angular momentum for $h = h_0$. Our distribution of distances has the probability density

$$l(h) = \frac{k+1}{h_{\max}^{k+1} - h_{\min}^{k+1}} h^k, k \neq -1, \tag{51}$$

where $h_{\min}$ and $h_{\max}$ are the least and greatest possible distances, respectively. For $k = -1$, a limiting formula with the logarithm holds. Since

$$\frac{h}{h_0} = \left(\frac{\lambda}{\lambda_0}\right)^{\frac{2}{k+3}}, \tag{52}$$

it holds that

$$\begin{aligned} f(\lambda) &= l(h) \left|\frac{\mathrm{d}h}{\mathrm{d}\lambda}\right| \\ &= \frac{k+1}{\lambda_{\max}^{\frac{2(k+1)}{k+3}} - \lambda_{\min}^{\frac{2(k+1)}{k+3}}} \lambda^{\frac{k-1}{k+3}}, k \neq -1. \end{aligned} \tag{53}$$

For $k = -1$, a limiting formula with the logarithm holds, see above.

We have solved this problem with a small change, on leaving out the case $n = 1$, namely Mercury. Contrary to [Krot (2009)], we include the asteroid Ceres with number $n = 5$. Here $n = 8$ means Uranus. But the optimal choice of μ_1 and μ_8 leads to relative errors up to forty per cent. Then the difference equation (35) has also the form

$$\lambda_n = \frac{2k+2}{3k+5} \frac{\mu_n^{\frac{3k+5}{k+3}} - \mu_{n-1}^{\frac{3k+5}{k+3}}}{\mu_n^{\frac{2k+2}{k+3}} - \mu_{n-1}^{\frac{2k+2}{k+3}}}, n = 2, 3, 4, 5, 6, 7, 8, k \neq -\frac{5}{3}, -1. \tag{54}$$

For $k = -\frac{5}{3}, -1$, a limiting formula with the logarithm holds. We search, moreover, for the optimum k. We have found that the best fit is achieved for $k = -2.228$, min=0.017387.

For the numerical illustration, we assume that

$$\begin{aligned} a_1 &= 0.3871, a_2 = 0.7233, a_3 = 1.0000, a_4 = 1.5237, \\ a_5 &= 2.765, a_6 = 5.2028, a_7 = 9.580, a_8 = 19.141, \qquad (55) \\ r_{\text{Earth}} &= 1.49597870691 \times 10^{11}, M_{\text{Sun}} = 1.9891 \times 10^{30}, G = 6.67428 \times 10^{-11}, \end{aligned}$$

(56)

Table 2. The radii d_n and their relative errors $\frac{d_n - a_n}{a_n}$ 100 %

Index n	Planet	d_n	Error $(\%)$
2	Venus	0.702972	-2.81
3	Earth	0.994763	-0.52
4	Mars	1.578346	3.59
5	Ceres	2.745513	-0.70
6	Jupiter	5.079846	-2.36
7	Saturn	9.748512	1.76
8	Uranus	19.085846	-0.29

where a_j, $j = 1, \ldots, 8$, are the main half-axes of planet orbits beginning the Mercury and ending Uranus in the astronomical units, r_{Earth} is the astronomical unit, M_{Sun} is the mass of the Sun and G is the gravitational constant. The values a_j for $j \neq 5$ are according to [Krot (2009)], a_5 corresponds to the asteroid Ceres and it has been involved according to [Christodoulou and Kazanas (2008)].

We choose $k \neq -1$, but in the surroundings of the value $k = -1$. We search for 15 unknowns creating the sequence

$$(\mu_1, \lambda_2, \mu_2, \lambda_3, \mu_3, \lambda_4, \mu_4, \lambda_5, \mu_5, \lambda_6, \mu_6, \lambda_7, \mu_7, \lambda_8, \mu_8), \tag{57}$$

which obey the equation

$$\sum_{n=2}^{8} \left(\frac{d_n - a_n}{a_n} \right)^2 = \min, \tag{58}$$

where

$$d_n = \frac{\lambda_n^2}{2GM_{\mathrm{Sun}}} \frac{1}{r_{\mathrm{Earth}}}, \tag{59}$$

$$\mu_n = \frac{\lambda_n + \lambda_{n+1}}{2}, n = 2, 3, 4, 5, 6, 7, \tag{60}$$

$$\lambda_n = \frac{2k+2}{3k+5} \frac{\mu_n^{\frac{3k+5}{k+3}} - \mu_{n-1}^{\frac{3k+5}{k+3}}}{\mu_n^{\frac{2k+2}{k+3}} - \mu_{n-1}^{\frac{2k+2}{k+3}}}, n = 2, 3, 4, 5, 6, 7, 8. \tag{61}$$

The probability density function $f(\lambda)$ of specific angular momentum λ is proportional to $\lambda^{\frac{k-1}{k+3}}$, $k = -2.228$. The nebula is divided formally into rings suitable for assignment of the angular momentum to circular loops of the radii d_n. The values of d_n and errors $\frac{d_n - a_n}{a_n} 100\%$ can be found in Table 2.

4. Chaotic Behavior and Migration

Laskar (1989) has reported on an extensive analytic system of averaged differential equations describing secular evolution of the orbits of eight main planets, accurate to the second order in the planetary masses and to the fifth order in eccentricity and inclination. His effort reminds one of the numerical integration of the long-term evolution of the solar system, e.g., [Sussman and Wisdom (1988)]. Through the analysis of results, it has been decided, whether the initial conditions of the solar system lead to a quasiperiodic or a chaotic solution. In the time of the publication, direct numerical integration was not yet able to take into account the inner planets and their orbital motion was held for too rapid. Laskar (1989) has estimated the maximum Lyapunov exponent to be about $\frac{1}{5}$ Myr^{-1}. It indicates a chaotic motion. This conclusion has been left for next evaluation.

Sussman and Wisdom (1992) have numerically integrated the evolution of the whole planetary system for a time span of nearly 100 million years. They have estimated the Lyapunov exponent to be about $\frac{1}{4}$ Myr^{-1}. They have confirmed the characteristics of Pluto's motion [Sussman and Wisdom (1988)]. They have encountered the complication that the subsystem of the Jovian planets is chaotic or quasi-periodic in the dependence on the initial values. The method of integration resembles an approach intended for a provocation of the chaos, but it can be used in a regime, which excludes that the exponential divergence is a numerical artifact.

Murray, Hansen, Holman and Tremaine (1998) have elaborated an explanation of the presence of Jupiter-mass planets in small orbits. E.g., a planet orbits τ Bootis at a distance of 0.0462 AU (Wikipedia gives a more recent value of 0.0481 AU). They have assumed that the giant planets may form at orbital radii of several AU and then migrate inwards. Such a planet originates in a planetesimal disk and it decreases the angular momentum of the planetesimals with resonant interactions. The planetesimals either collide with or are ejected by the planet. The ejection process removes orbital energy from the planet, which moves closer to the star. Other consequences of a close encounter may be a collision of the planetesimal with the star or a long-term capture into a mean-motion resonance.

Murray, Paskowitz and Holman (2002) have concentrated themselves to the resonant migration of planets, which produces large eccentricities. In their study, they have distinguished the migration due to ejection of planetesimals and that by tidal torques. These possibilities rely respectively on the assumptions of planetesimal and gas disks. The resonant migration in the gas disk is related to two bodies of roughly Jupiter mass.

Spencer (2007) reviews the migrating planets from a creationist's standpoint. He remembers the Nebular Hypothesis, which comprised the idea that all the planets in our solar system formed in the regions, where they are now located. Extrasolar planetary systems require rather pictures of a migration. As the long accepted "naturalistic" origins explanations for our solar system do not operate for extrasolar planetary systems, planetary scientists modify past theories for our solar system. It is welcome, when the nebular models have had problems with the formation of Uranus and Neptune. The Nice model is much appealing. The theory has been published as the paper [Tsiganis, Gomes, Morbidelli and Levinson (2005)]. The initial order of the giant planets JSNU has changed to the actual order JSUN after 6.6 Myr. It can be described also as a "rapid inclination of the orbits of Uranus and Neptune and exchange of their orders in distance from the Sun". The time axis is drawn in the length of over 80 Myr. Spencer (2007) does not accept naturalistic origins scenarios that conflict with the biblical time scale.

It can be expected that long time spans, not only 10^9 yr, but also 10^{18} yr, lead to considerations of the inclusion of the quantum gravity and the uncertainty relations. For them, we refer to the essay [Page (2010)].

5. Quantization on a Cosmic Scale

In this section, we provide an account of our publications. Pintr and Peřinová (2003–2004) have reported on the proposal of Mohorovičić (1938). The then popular Bohr–Sommerfeld quantum theory has convinced him of the occurrence of "allowed" orbits in planetary science. His "magnification ratios" cf., [Christodoulou and Kazanas (2008)]) are lesser than one (reduction ratios). Another peculiar hypothesis is the connection of the allowed orbits with moons, which have "subjected" themselves to the primaries later. Peřinová, Lukš and Pintr (2007) have tried to use the concept of wave-function instead of or together with the notion of an allowed orbit. Pintr, Peřinová and Lukš (2008) have returned to the allowed orbits using mainly the classical physics. Some modern physics concepts are applied as well.

Before this account, we review ideas which influenced us. Greenberger (1983) admitted that quantization of "megascopic" systems exists, which has a macroscopic limit in common with the familiar quantization of microscopic systems. de Oliveira Neto (1996) and Agnese and Festa (1998) have constructed Bode-like laws for our solar system. Later, de Oliveira Neto, Maia and Carneiro (2004) have provided a continuation of the paper [de Oliveira Neto (1996)], using the concept of a wave-function. Interesting coincidences can be found even in the application of these ideas to extrasolar planets.

5.1. Quantization of Megascopic Systems

Greenberger (1983) has concerned himself with absorption lines in quasars with a high redshift parameter z. He assumes that a hydrogen atom is attracted gravitationally to a quasar. The world is quantized on a microscopic scale and is quantized on a cosmic scale as well by the model theory. He restricts himself to the absorption of the Lyman-α frequency of ultraviolet radiation by an atom, i.e., the transition from the level $n = 1$ to $n = 2$. When the atom also passes over into a higher gravitational state, the absorption lines different from the Lyman-α line will be present in the analogy with the Raman effect.

Greenberger (1983) has expressed the quantized gravitational energy as

$$E_n = -\frac{E_0}{n^{\frac{2}{3}}}, \tag{62}$$

where E_0 can only be derived theoretically in a complicated way and it is not as yet observable. In observing the absorption lines with the wavelengths λ_k, $k = 1, 2$, we must consider jumps from the gravitational levels n_{k1} to n_{k2} and the equation

$$\frac{hc}{\lambda_k} - \frac{hc}{\lambda_\alpha} = \frac{E_0}{1+z}\left(n_{k1}^{-\frac{2}{3}} - n_{k2}^{-\frac{2}{3}}\right), k = 1, 2, \tag{63}$$

where h is Planck's constant and λ_α is the wavelength of the Lyman-α line seen on the Earth. From this

$$\frac{\frac{\lambda_\alpha}{\lambda_1} - 1}{\frac{\lambda_\alpha}{\lambda_2} - 1} = \frac{n_{11}^{-\frac{2}{3}} - n_{12}^{-\frac{2}{3}}}{n_{21}^{-\frac{2}{3}} - n_{22}^{-\frac{2}{3}}} \tag{64}$$

and E_0 is not needed.

An analysis of absorption spectra of four quasars has provided good fit for two of them and the third comparison has been also meaningful. The fourth case has provided an illustration of the method at least. This complicated procedure is not similar to the Titius–Bode law. Considerations of elliptical rings around the normal elliptical galaxies in [Malin and Carter (1980)] remind us of the Titius–Bode law. The major half-axes very closely fit the formula

$$r_n = r_0 n^{\frac{2}{3}}. \tag{65}$$

Greenberger (1983) distinguishes the "kinetic" momentum from the canonical momentum and remembers that

$$[x, p_c] = i\hbar, \tag{66}$$

where x is a position coordinate and p_c the canonical momentum. We assume that the physical system is described by a Hamiltonian H. Then

$$\dot{x} \equiv \frac{\mathrm{d}x}{\mathrm{d}t} = -\frac{i}{\hbar}[x, H] = \frac{\partial H}{\partial p_c} \tag{67}$$

and the kinetic momentum is $p = m\dot{x}$, when the "system" means a particle of the mass m. We have understood that the author considers equations for $\dot{x}$ of the form

$$[x, \dot{x}] = C(\dot{x}), \tag{68}$$

where the commutator C may depend on $\dot{x}$ again. We obtain an equation for the Hamiltonian H of the form

$$\frac{\partial^2 H}{\partial p_c^2} = -\frac{i}{\hbar} C\left(\frac{\partial H}{\partial p_c}\right). \tag{69}$$

For the original treatment, we refer to [Greenberger (1983)].

de Oliveira Neto (1996) starts with an imaginary observer, who is compelled to apply the quantum theory to our solar system by his/her cosmic size. He represents the mean planetary distances by the formula

$$r_{nm} = \frac{n^2 + m^2}{2} r_0, 0 \leq m \leq n, \tag{70}$$

where n and m are integers and r_0 is the mean distance of Mercury to the Sun. In the dependence on a planet, we note that

$$n(\text{Mercury}) = 1,\ \ n(\text{Venus}) = n(\text{Earth}) = n(\text{Mars}) = 2,\ \ n(\text{Jupiter}) = 4,$$

$$n(\text{Saturn}) = 5,\ \ n(\text{Uranus}) = 7,\ \ n(\text{Neptune}) = 9,\ \ n(\text{Pluto}) = 10. \tag{71}$$

Next, $m(p) = n(p)$ for $p =$ Mercury, Mars, Jupiter, Saturn, Uranus, Neptune, Pluto. Three of the inner planets are described only by a variation of m: $m(\text{Venus}) = 0$, $m(\text{Earth}) = 1$, $m(\text{Mars}) = 2$. Besides the planets and Pluto, the dwarf planet, some asteroids have been represented.

Nottale, Schumacher and Gay (1997) continue the approach in the book [Nottale (1993)]. They assume that at very large time-scales, the solar system can be described in terms of fractal trajectories governed by a Schrödinger-like equation. They let the fractal motion depend on a universal constant w_0 having the dimension of a velocity. According to an illustrative example by Greenberger (1983), a connection of w_0 with the fundamental length λ may be

$$w_0\lambda = \frac{GM}{c}, \tag{72}$$

where $M = M_{\rm Sun}$. The average distance to the Sun is given in the form

$$\langle r_{n\ell}\rangle = \frac{1}{2}[3n^2 - \ell(\ell+1)]\frac{GM}{w_0^2}, \tag{73}$$

where n is the principal number of the quantized orbit and ℓ is the number of quanta of the angular momentum, $\ell = 0, 1, \ldots, n-1$. It has been found that the appropriate value for the inner system is $w_0 = (144.3 \pm 1.2)$ km s^{-1} and $w_0 \to w_{\rm out}$ for the outer system, $w_{\rm out} = \frac{w_0}{5}$, where $w_0 = (140 \pm 3)$ km s^{-1}.

Agnese and Festa (1997) remind the paper [Greenberger (1983)], especially his equation (8) where, without loss of generality, they write $\bar{\lambda}f$ instead of λf, $\bar{\lambda} = \frac{\lambda}{2\pi}$. We have understood that these authors propose the commutator

$$[x, m\dot{x}] = i(\hbar + mc\bar{\lambda}), \tag{74}$$

where $\hbar = \frac{h}{2\pi}$ is the reduced Planck constant and

$$\dot{x} = -\frac{i}{\hbar}[x, H]. \tag{75}$$

As above, we obtain that

$$\frac{\partial^2 H}{\partial p_c^2} = \frac{1}{m} + \frac{c\bar{\lambda}}{\hbar}, \tag{76}$$

or

$$H = \left(\frac{1}{m} + \frac{c\bar{\lambda}}{\hbar}\right)\frac{p_c^2}{2} + \text{potential energy}. \tag{77}$$

It is just the direction, in which Greenberger (1983) did not intend to proceed. It is obvious that, sometimes, an alternative expression for the kinetic energy may be useful.

Agnese and Festa (1997) do not utilize the full quantum mechanics, but the Bohr–Sommerfeld discretization rules for (multiply) periodic motions

$$\oint p_j \mathrm{d}q_j = n_j 2\pi\hbar, \tag{78}$$

where q_j, p_j are generalized coordinates and canonically conjugate momenta, respectively, and n_j are integers. Using the equation

$$\hbar = \frac{\tilde{e}^2}{\alpha_c c}, \tag{79}$$

where c is the velocity of light and the charge of the electron $\tilde{e}$,

$$\tilde{e} = \frac{e}{\sqrt{4\pi}}\epsilon_0, \tag{80}$$

with the elementary charge $e = 1.602 \times 10^{-19}$C, the permittivity of the vacuum ϵ_0, the constant of fine structure α_c, we obtain that the periodic plane motions of a particle should fulfil the equations

$$\oint p_r \mathrm{d}r = k2\pi\frac{\tilde{e}^2}{\alpha_c c}, \tag{81}$$

$$\oint p_\varphi \mathrm{d}\varphi = l2\pi\frac{\tilde{e}^2}{\alpha_c c}, \tag{82}$$

where k and l are the radial and azimuthal numbers, respectively. This way the particle-mass independence of the orbits is not obtained. Hypothetically, the authors replace

$$\frac{\tilde{e}^2}{\alpha_c} \rightarrow \frac{GMm}{\alpha_g}, \tag{83}$$

which gives

$$\oint p_r \mathrm{d}r = k2\pi\frac{GMm}{\alpha_g c}, \tag{84}$$

$$\oint p_\varphi \mathrm{d}\varphi = l2\pi\frac{GMm}{\alpha_g c}, \tag{85}$$

where α_g is the gravitational constant. This assumption entails the particle-mass, m, the independence of orbits. Nevertheless, a new quantization rule is not needed, when one correctly writes p_c instead of p and uses the new kinetic energy. In application to the solar system, we put $M = M_{\mathrm{Sun}}$. The principal number $n = k + l$ is introduced.

The major half-axes of the elliptic orbits have the form

$$a_n = a_1^{\mathrm{Sun}} n^2, \tag{86}$$

where

$$a_1^{\mathrm{Sun}} = \frac{1}{\alpha_g^2}\frac{GM_{\mathrm{Sun}}}{c^2}. \tag{87}$$

For circular orbits, the canonical momentum $p_r = 0$, $k = 0$. The estimation of a_1^{Sun} has been carried out and, simultaneously, the numbers $n(p)$ for the planets have been searched for. It has been found that $a_1^{\mathrm{Sun}} = (0.0439 \pm 0.0004)$ AU and

$$n(\mathrm{Mercury}) = 3,\ \ n(\mathrm{Venus}) = 4,\ \ n(\mathrm{Earth}) = 5,\ \ n(\mathrm{Mars}) = 6,\ \ n(\mathrm{Ceres}) = 8,$$
$$n(\mathrm{Jupiter}) = 11,\ \ n(\mathrm{Saturn}) = 15,\ \ n(\mathrm{Uranus}) = 21,\ \ n(\mathrm{Neptune}) = 26,$$
$$n(\mathrm{Pluto}) = 30. \tag{88}$$

From this

$$\frac{1}{\alpha_g} = 2113 \pm 15. \tag{89}$$

It can be objected that the number of parameters is greater by one than the number of observations, so that the solution of the problem is too easy. But it would be an oversimplified opinion, since most of the unknowns are integers. Such parameters cannot be counted as "full-blown" degrees of freedom. Intuitively, a quasi-continuum of consecutive integers (here $3, 4, 5, 6$) would be counted as a value of a single parameter.

The assumption that α_g is a universal constant is tested by satellites of the planets. The first allowed orbit, which lies out of the planet, has a relatively high (≥ 19) principal number for Mercury, Venus, the Earth, and Mars. For the giant planets, such orbits have lesser numbers. But the "universality" of the constant α_g has not been confirmed.

Agnese and Festa (1997) indicate that the spin of a purely gravitationally bounded celestial body may be proportional to the square of its mass,

$$J = \frac{1}{2}\frac{Gm^2}{\alpha_g c}. \tag{90}$$

In fact, the law $J = pM^2$, where p is a proportionality constant, has been proposed, with $p = 8 \times 10^{-17}\ \mathrm{m^2kg^{-1}s^{-1}}$ [Wesson (1981)]. In continuation, these authors have calculated $G/(2\alpha_g c) = 2.35 \times 10^{-16}\ \mathrm{m^2kg^{-1}s^{-1}}$.

Rubčić and Rubčić (1998) have criticized the vacant orbits in the quadratic law in [Agnese and Festa (1997)]. They have been aware that it is advantageous to analyze the terrestrial planets and the Jovian planets as separate systems. The quadratic law governs even the satellites of Jupiter, Saturn and Uranus [Rubčić and Rubčić (1995, 1996)].

They relate the quantization in the cosmic world to the effect of chaos [Nottale (1993, 1996)]. Rubčić and Rubčić (1998) begin with the "quantization" of the specific angular momentum $\frac{J_n}{m_n}$, where J_n is the angular momentum of the planet and m_n is its mass. Then $\frac{J_n}{m_n} = nH'$, with $H' = fAM$, where M is the mass of the central body,

$$A = 2\pi\frac{G}{\alpha_c c}, \tag{91}$$

and f indicates the impossibility of an invariable quantization of the systems considered. Particularly, $f = 2.41 \pm 0.03$ for the terrestrial planets and $f = 12.61 \pm 0.16$ for the Jovian ones. The principle of consecutive numeration is respected. Especially, Mercury, Venus, the Earth, Mars, Ceres are numbered by 3, 4, 5, 6, 8 and, e.g., Jupiter, Saturn and Uranus are labeled by 2, 3 and 4, respectively. The asteroid Ceres only represents the main belt, which is related to numbers 7 to 9.

As an approximation, the radius of the nth orbit is

$$r_n = \frac{1}{G}(fA)^2Mn^2. \tag{92}$$

Through an analysis of the behavior of f^{-1}, a rougher approximation (cf., equation (13) in [Rubčić and Rubčić (1998)]) has been derived,

$$r_n = \frac{GM}{v_0^2k^2}n^2, \tag{93}$$

where $v_0 = (25.0 \pm 0.7)\ \mathrm{kms^{-1}}$ and $k = 1$ for the Jovian planets and the satellites of Uranus, $k = 2, 4$ for the satellites of Jupiter and Saturn, respectively, and $k = 6$ for the terrestrial planets. The value $v_0 \approx 24\ \mathrm{kms^{-1}}$ is one of increments of galactic red-shifts [Arp and Sulentic (1985), Rubčić and Rubčić (1998)], cf., also [Arp (1998)].

The universal constant A can be written in the form

$$A = \frac{h}{m_0^2}, \tag{94}$$

where $m_0^2 = \alpha_c m_\mathrm{P}^2$, m_P is Planck's mass.

After the observation of quantized red-shifts, Dersakissian (1984) formulated a cosmic form of quantum theory. The studies of the red-shift of galaxies reverberate also in [Carvalho (1985)]. In that paper, the quantized gravitational energy (62) has been obtained using a model, which we understand on paying attention to an illustrative example in [Greenberger (1983)]. Indeed, the equation for p, $[x,p] = i\hbar g(x)$, where $g(x)$ is a potential function, can be written as an equation for H,

$$\frac{\partial^2 H}{\partial p_c^2} = \frac{1}{m} g(x), \tag{95}$$

which leads to the Hamiltonian by Carvalho (1985) on generalization to three dimensions, where, e.g., $g(\mathbf{r})$ must replace $g(x)$ and be proportional to $\frac{1}{|\mathbf{r}|^2}$.

5.2. Tentative "Universal" Constants

Carneiro (1998) remembers Dirac's hypothesis that cosmological large numbers, as mass M, radius R and age T of our Universe can be related to the typical values of mass m, size r and life time t appearing in particle physics, by a scale factor $\Lambda \sim 10^{38}$–10^{41} [Dirac (1937, 1938)]

$$\frac{T}{t} = \frac{R}{r} = \left(\frac{M}{m}\right)^{\frac{1}{2}} = \Lambda. \tag{96}$$

A simple one-dimensional analysis is utilized for introducing new concepts as a scaled quantum of action H,

$$\frac{H}{h} = \Lambda^3. \tag{97}$$

If Universe rotates, its spin must be of the order $\frac{H}{2\pi} \sim 10^8$ Js. Another, intermediate, scale of quantization related to the angular momenta of stars is proposed, whose values concentrate around $\frac{H'}{2\pi} \sim 10^{42}$ Js,

$$\frac{T'}{t} = \frac{R'}{r} = \left(\frac{M'}{m}\right)^{\frac{1}{3}} = \lambda. \tag{98}$$

Here $\lambda = \sqrt{\Lambda}$, $T' \sim 10^4$ s, $R' \sim 10^4$ m and $M' \sim 10^{30}$ kg. For the astrophysical meaning of these values, we refer to the paper [Carneiro (1998)]. Gravitation constants of three kinds are considered, $G = G_2$ on the Λ scale, $G' = G_3$ on the λ scale and g on the microscopic scale,

$$G_n = g\left(\frac{M_n}{m}\right)^{\frac{1}{n}-1}, n = 2, 3, \tag{99}$$

where $M_2 = M$ and $M_3 = M'$. Here g can be named the strong gravity constant [Recami, Raciti, Rodrigues, Jr., and Zanchin (1994)]. It has been guaranteed that $G_3 = G_2$, or $G' = G$.

In a continuation of [Agnese and Festa (1997)], it has been chosen

$$H' = 2\pi\sqrt{GMm^2 r_1}, \tag{100}$$

where m is the average mass of planets of the solar system and $r_1 \equiv a_1^{\rm Sun}$, and calculated that $H' = 1.2 \times 10^{42}$ Js for $m = 2.10 \times 10^{26}$ kg [Carneiro (1998)].

Agnese and Festa (1998) have considered the universal constant of the form

$$v_* = \alpha_g c \tag{101}$$

and have found that $v_* \sim 143.7$ km s^{-1}. An elliptical orbit has been tried for Pluto, $(n, l) = (30, 29)$, and for 1996TL$_{66}$, $(n, l) = (44, 36)$, which is not much valued even by themselves. They have analyzed about twenty star–planet pairs. The extrasolar planets can be discovered near the star. The mass of these stars was not known, but it can be at least roughly estimated from the star type. From the theory, it follows that the ratio of the orbital periods of planets orbiting different stars at the level $n = 1$ equals the ratio between the masses of the respective stars. The predictions have agreed fairly for (stars) HD187123, τ Bootis and HD75289 and their companions.

Agnese and Festa (1999) have refined the analysis in the paper [Agnese and Festa (1997)] with the assumption that the distance of Jupiter from the Sun has least changed since the formation. They have obtained that

$$a_1^{\rm Sun} = 0.04297 \text{ AU}, \quad \frac{1}{a_g^{\rm Sun}} = 2086 \pm 14. \tag{102}$$

The planetary system of υ Andromedae with three planets was paid attention to. These planets are called b, c, d and the planet c is the most massive. From the orbital period and the principal number of the planet, the mass of the star can be found. As the mass of υ Andromedae is known with some accuracy, it has been feasible to number the planet c with $n = 4$. The authors then have corrected the mass of the star. Similarly, the numbers $n = 1$ for the planet b and $n = 7$ for the planet d have been determined, but the value of the stellar mass have no more been changed. The authors are aware that their scheme contradicts the theories of the planet migration. Since the numbers $n = 1, 4, 7$ are not consecutive, it is natural to assume the existence of smaller, not yet observed, planets.

Nottale, Schumacher and Lefèvre (2000) have continued the study in [Nottale (1996)]. They have accepted the gravitation coupling (not the "structure") constant $\alpha_g = \frac{w_0}{c}$. Planets in the inner solar system and exoplanets have been treated altogether. The mean velocity has been calculated from the parent star mass, M, and the planet period, P, as $v = \sqrt[3]{\frac{M}{P}}$. The histogram of the values of $|\delta n| = \left|\frac{144}{v} - n\right|$, with n being the nearest integer of $\frac{144}{v}$, has been plotted. The distribution has differed from the uniform one significantly. Rather a condensation of δn about $\delta n = 0$ has been observed.

de Oliveira Neto, Maia and Carneiro (2004) apply full quantum mechanics in contrast to [Agnese and Festa (1997)]. Such an approach is peculiar in that the indeterminism of the complete quantum mechanics does not permit to find the planetary distances with certainty. As in many similar cases, the predictions are formulated in terms of the planetary mean distances. In the application of the full quantum mechanics, wave functions are associated with the planets. For the calculation of the planetary mean distances, squares of moduli of these wave functions are used. Next comments may resemble the philosophy of quantum mechanics. A superposition of planetary wave functions is a wave function again. But this function does not describe all the planets simultaneously. At least in quantum mechanics, Schrödinger's cat paradox is not related to a description of two cats, but to that of two *states* of a single cat. In application to the planetary formation, we need a nonstandard interpretation of quantum mechanics indeed. Let us consider a weighted average of squares of moduli of wave functions, which differs from a square of a modulus of an appropriate superposition of these functions in the neglect of the interference terms. The weighted average respecting the masses of protoplanets much resembles the solar nebula according to the paper [Christodoulou and Kazanas (2008)].

de Oliveira Neto, Maia and Carneiro (2004) remember the paper [Nelson (1966)] based on the idea from the paper [Fényes (1952)]. A cosmic Brownian motion could lead to the validity of a Schrödinger-type diffusion equation. In the framework of this model, we can hardly imagine a Brownian motion, whose complexity would depend on the number of the planets to be formed. A theory of planet formation must then content with the superposition of wave functions, which "paradoxically" do not correspond to states of a single protoplanet, but directly to the protoplanets.

It is proper to remember the book [Nottale (1993)], where the inner and outer planetary systems are treated separately. de Oliveira Neto, Maia and Carneiro (2004) solve the time-independent Schrödinger equation with Newton's potential in two dimensions. In order a relationship to the planetary system to be established, also the component of the Hamiltonian, which is the kinetic energy, must be changed, as we have mentioned above. As Peřinová, Lukš and Pintr (2007) proceed almost in the same way, deviating only in detail, we do not give the theory twice, we present it only in what follows. Using the numbers k, $k = n - \frac{1}{2}$, where n is the principal quantum number and ℓ, $\ell = l - 1$, the following correspondences have been established in Table 3.

It is obvious that, only in several cases, the assumption of circular orbits [Agnese and Festa (1997)] $l = n$, or $\ell + 1 = k + \frac{1}{2}$, or $\ell = k - \frac{1}{2}$ is used.

5.3. Membrane Model

Mohorovičić (1938) returned to the matter of distances of the bodies of the solar system. He established a law, which can be expressed as follows: The distance of every body of the solar system from the Sun is given in the astronomic units by the formula

$$r_k^{\mp} = 3.363(1 \mp 0.88638^k), \tag{103}$$

where the minus sign corresponds to inner parts of the solar system and the plus sign describes the more remote regions of the solar system. The number k can assume positive integer values for the minus sign, it can take both positive and negative integer values for the plus sign in the formula (103). If other real numbers are substituted, the result corresponds to unstable orbits. This formula encompasses all inner planets and asteroids, giant planets, and even the distances of comets. It cannot be recommended without tables, in which the numeration of planets is striking. In comparison with the Titius–Bode series, we have $k = 1, 2, 3, 5, 14$ for the bodies that are respected by the Titius–Bode series. But the

Table 3. Predicted distances of bodies from the Sun

Body	k	ℓ	$r_{k\ell}$ [AU]	Error [%]
—	$\frac{1}{2}$	0	0.055	—
Mercury	$\frac{3}{2}$	1	0.332	-15
Mercury	$\frac{3}{2}$	0	0.387	-1
Venus	$\frac{5}{2}$	2	0.829	15
Earth	$\frac{5}{2}$	1	0.995	-0.5
Earth	$\frac{5}{2}$	0	1.050	5
Mars	$\frac{7}{2}$	3	1.548	2
Hungaria	$\frac{7}{2}$	2	1.824	-6
Hungaria	$\frac{7}{2}$	1	1.990	2.5
Hungaria	$\frac{7}{2}$	0	2.046	5.5
Vesta	$\frac{9}{2}$	4	2.488	5.5
Ceres	$\frac{9}{2}$	3	2.875	9
Hygeia	$\frac{9}{2}$	2	3.151	-0.5
Camilla	$\frac{9}{2}$	1	3.317	-4.5
Camilla	$\frac{9}{2}$	0	3.372	-3
Jupiter	$\frac{11}{2}$	0	5.031	-3
—	$\frac{13}{2}$	0	7.021	—
Saturn	$\frac{15}{2}$	0	9.343	-2
Chiron	$\frac{17}{2}$	0	11.997	-12.5
Chiron	$\frac{19}{2}$	0	14.982	9.5
Uranus	$\frac{21}{2}$	0	18.300	-4.5
—	$\frac{23}{2}$	0	21.948	—
HA2 (1992), DW2 (1995)	$\frac{25}{2}$	0	25.929	4.5
Neptune	$\frac{27}{2}$	0	30.241	0.5
—	$\frac{29}{2}$	0	34.885	—
Pluto	$\frac{31}{2}$	0	39.861	1

Mohorovičić formula is double similar as proposals of some other authors. Perhaps, the split is peculiar. So far we have mentioned the terrestrial planets and the asteroid Ceres, the formula with the upper sign. We have $k = 5, -5, -13, -17, -20$ for farther bodies, which are described by the Titius–Bode series well or wrong. The Titius–Bode series would give good predictions to the bodies with $k = 5, -5, -13$ and the lower sign. These chosen values evidence the attention paid to the asteroids and comets. Typical is the condensation of orbits in the asteroid belt, where at 3.363 AU there is also the interface.

Pintr and Peřinová (2003–2004) generalize the Mohorovičić formula for stars of different masses, noting that the formula (103) can be rewritten in the form

$$a_k = a_{\mathrm{lim}}(1 \mp 0.88638^k), \tag{104}$$

where $a_{\rm lim} = 3.363$ AU for our solar system. Substituting

$$a_{\rm lim} = \frac{GM^{(\rm star)}}{\mu_g^2 c^2 1.5 \times 10^{11}}, \tag{105}$$

where $M^{(\rm star)}$ is the stellar mass, c is the speed of light and μ_g is a Mohorovičić constant with the property

$$\frac{1}{\mu_g} = 18448.1, \tag{106}$$

they obtain the calculated distances of the bodies in AU in the form

$$a_k = \frac{GM^{(\rm star)}(1 \mp 0.88638^k)}{\mu_g^2 c^2 1.5 \times 10^{11}}. \tag{107}$$

In a comparison with the paper [Agnese and Festa (1997)], it can be stated that the interface, $a_{\rm lim}$, is obtained for $n = 10$, but the correspondence is very rough.

The relation (107) has been modified and for distances of moons of the giant planets, a similar relation has been obtained

$$a_k = \frac{GM^{(\rm planet)}(1 \mp 0.88638^k)}{\mu_g^2 c^2}, \tag{108}$$

where the distance is measured in meters. Tables for moons and rings of Jupiter, Saturn, Uranus and Neptune are presented, in which they observe that:
(i) In the case of Jupiter, $k = 18$ for an inner moon and $k = 8$ for an outer moon is utilized.
(ii) In the case of Saturn, $k = 33$ for an inner ring and $k = 14$ for an outer ring is utilized.
(iii) In the case of Uranus, the interface is not utilized, the ring and moons are outer.
(iv) In the case of Neptune, the interface is not utilized, the moons are outer.
When the interface is utilized, the tables lack indication of the sign in the formula (108).

The formula (107) has been applied to the systems of υ Andromedae and 47 Ursae Majoris. The interface has not been utilized. The planets are inner.

Peřinová, Lukš and Pintr (2007) have solved the Schrödinger equation with appropriately modified Hamiltonian deviating in detail from the paper [de Oliveira Neto, Maia and Carneiro (2004)]. They formulate the problem as follows. They consider a body of the mass $M_{\rm p}$, which orbits a central body of the mass M and has the potential energy $V(x, y, z)$ in its gravitational field. As planets and moons of the giant planets revolve approximately in the same plane, they consider $z = 0$. As they revolve in the same direction, they choose directions of the axes x, y and z such that the planets or moons of giant planets revolve anticlockwise. Then they write the modified Schrödinger equation for the wave function $\psi = \psi(x, y)$ from the part of the Hilbert space $L_2(R^2) \cap C^2(R^2)$ and the eigenvalue $0 > E \in R$ in the form

$$-\frac{\hbar_M^2}{2M_{\rm p}} \left(\frac{\partial^2}{\partial x^2} + \frac{\partial^2}{\partial y^2} \right) \psi + V(x, y)\psi = E\psi, \tag{109}$$

where $\hbar_M \approx 1.48 \times 10^{15} M_{\rm p}$, $V(x, y) = V(x, y, z)$ and E is the total energy. Negative values of E classically correspond to the elliptic Kepler orbits and the localization property (bound state) is also conserved in the quantum mechanics for such total energies E. The factor 1.48×10^{15} is not a dimensionless number, but the unit of its measurement is $\mathrm{m}^2\mathrm{s}^{-1}$. With respect to the unusual unit, they do not wonder that Agnese and Festa (1997) consider this factor in the form of a product, such that $\hbar_M = \bar{\lambda}_M c M_\mathrm{p}$, where $\bar{\lambda}_M \approx 4.94 \times 10^6$ m.

They transform equation (109) into the polar coordinates,

$$-\frac{\hbar_M^2}{2M_\mathrm{p}}\left(\frac{\partial^2\tilde{\psi}}{\partial r^2}+\frac{1}{r}\frac{\partial\tilde{\psi}}{\partial r}+\frac{1}{r^2}\frac{\partial^2\tilde{\psi}}{\partial\theta^2}\right)+\tilde{V}(r)\tilde{\psi}=E\tilde{\psi}, \tag{110}$$

where $\tilde{\psi} \equiv \tilde{\psi}(r,\theta) = \psi(r\cos\theta, r\sin\theta)$, $\tilde{V}(r) = V(r\cos\theta, r\sin\theta)$ does not depend on θ. Particularly, they choose

$$\tilde{V}(r) = -\frac{GM_\mathrm{p}M}{r}. \tag{111}$$

With respect to the Fourier method, they assume a solution of equation (111) in the form

$$\tilde{\psi}(r,\theta) = R(r)\Theta(\theta). \tag{112}$$

The original eigenvalue problem is transformed equivalently to two eigenvalue problems

$$\Theta''(\theta) = -\Lambda\Theta(\theta), \tag{113}$$

$$\Theta(0) = \Theta(2\pi) \tag{114}$$

and

$$R''(r)+\frac{1}{r}R'(r)+\left\{-\frac{\Lambda}{r^2}+\left[E-\tilde{V}(r)\frac{2M_\mathrm{p}}{\hbar_M^2}\right]\right\}R(r)=0, \tag{115}$$

$$\lim_{r\to 0+}[\sqrt{r}R(r)]=0,\quad \sqrt{r}R(r)\in L_2((0,\infty)). \tag{116}$$

The solution of the problem (113)–(114) has the form

$$\Theta_\ell(\theta) = \frac{1}{\sqrt{2\pi}}\exp(i\ell\theta) \tag{117}$$

for $\ell = \pm\sqrt{\Lambda} \in Z$.

Here $\ell = 0$ should mean a body, which does not revolve at all. In the classical mechanics, such a body moves close to a line segment ending at the central body, and it spends a short time in the vicinity of this body. Peřinová, Lukš and Pintr (2007) utilize some – not all – of the concepts of quantum mechanics and will not avoid the case $\ell = 0$ [de Oliveira Neto, Maia and Carneiro (2004)]. In the formula (117), $\ell = 1, 2, \ldots, \infty$ corresponds to the anticlockwise revolution.

On respecting (111), equation (115) becomes

$$R''(r)+\frac{1}{r}R'(r)+\left\{-\frac{\ell^2}{r^2}-B-\frac{2M_\mathrm{p}}{\hbar_M^2}\left(-\frac{GM_\mathrm{p}M}{r}\right)\right\}R(r)=0, \tag{118}$$

where

$$B = -\frac{2M_\mathrm{p}E}{\hbar_M^2} = -\frac{2}{(\bar{\lambda}_M c)^2}\frac{E}{M_\mathrm{p}}. \tag{119}$$

It holds that

$$\frac{M_{\mathrm{p}}GM_{\mathrm{p}}M}{\hbar_M^2} = \frac{GM}{(\bar{\lambda}_M c)^2}. \tag{120}$$

On substituting $r = \frac{\rho}{2\sqrt{B}}$ and introducing

$$\tilde{R}(\rho) = R\left(\frac{\rho}{2\sqrt{B}}\right), \tag{121}$$

equation (118) becomes

$$\tilde{R}''(\rho) + \frac{1}{\rho}\tilde{R}'(\rho) + \left(-\frac{1}{4} + \frac{k}{\rho} - \frac{\ell^2}{\rho^2}\right)\tilde{R}(\rho) = 0, \tag{122}$$

where

$$k = \frac{GM}{(\bar{\lambda}_M c)^2\sqrt{B}}. \tag{123}$$

For later reference, it holds inversely that

$$\sqrt{B} = \frac{GM}{(\bar{\lambda}_M c)^2 k}, \tag{124}$$

$$\frac{-E}{M_{\mathrm{p}}} = \frac{(\bar{\lambda}_M c)^2}{2}B = \frac{(GM)^2}{2(\bar{\lambda}_M c)^2 k^2}. \tag{125}$$

Expressing $\tilde{R}(\rho)$ in the form

$$\tilde{R}(\rho) = \frac{1}{\sqrt{\rho}}u(\rho), \tag{126}$$

they obtain an equation for $u(\rho)$,

$$u''(\rho) + \left[-\frac{1}{4} + \frac{k}{\rho} - \left(\ell'^2 - \frac{1}{4}\right)\frac{1}{\rho^2}\right]u(\rho) = 0, \tag{127}$$

where $\ell' = \ell$. It is familiar that this equation has two linear independent solutions $M_{k,\ell'}(\rho)$, $M_{k,-\ell'}(\rho)$, if ℓ' is not an integer number. When ℓ' is integer, the solution $M_{k,-\ell'}(\rho)$ must be replaced with a more complicated solution. It can be proven that the other solution is not regular for $\rho = 0$ (it diverges as $\ln\rho$ for $\rho \to 0$). The remaining solution $M_{k,\ell}(\rho)$ can be transformed to a wave function from the space $L_2((0,\infty))$ if and only if $k - \ell - \frac{1}{2} = n_{\mathrm{r}}$ is any nonnegative integer number. They choose this function to be

$$u_{k\ell}(\rho) = C_{k\ell}M_{k,\ell}(\rho), \tag{128}$$

where $C_{k\ell}$ is an appropriate normalization constant and $M_{k,\ell}(\rho)$ is a Whittaker function, namely

$$M_{k,\ell}(\rho) = \rho^{\ell+\frac{1}{2}}\exp\left(-\frac{\rho}{2}\right)\Phi\left(\ell - k + \frac{1}{2}, 2\ell + 1; \rho\right), \tag{129}$$

where Φ is the confluent (or degenerate) hypergeometric function. In the formula (128), the constant $C_{k\ell}$ has the property

$$\int_0^\infty r[R_{k\ell}(r)]^2\,\mathrm{d}r = 1, \tag{130}$$

or it is

$$C_{k\ell} = 2\sqrt{B}\frac{1}{(2\ell)!}\sqrt{\frac{(n+\ell-1)!}{2k(n-\ell-1)!}}. \qquad (131)$$

Then

$$R_{k\ell}(r) = 2\sqrt{B}\sqrt{\frac{(n-\ell-1)!}{2k\Gamma(n+\ell)}}\exp(-r\sqrt{B})(2r\sqrt{B})^{\ell}L_{n-\ell-1}^{2\ell}(2r\sqrt{B}), \qquad (132)$$

where $n = k + \frac{1}{2}$, $L_{n-\ell-1}^{2\ell}(2r\sqrt{B})$ is a Laguerre polynomial and the relation (124) holds.

Having solved the modified Schrödinger equation, they address an interpretation of the formulas derived. The probability density $P_{k\ell}(r)$ of the revolving body occurring at the distance r from the central body is

$$P_{k\ell}(r) = r[R_{k\ell}(r)]^2, r \in [0,\infty). \qquad (133)$$

Mean distances of the planets are given by the relation

$$r_{k\ell} = \int_0^{\infty} rP_{k\ell}(r)\,\mathrm{d}r \qquad (134)$$

$$= \frac{(\bar{\lambda}_M c)^2}{4GM}\left[(2k-n_{\mathrm{r}})(2k-n_{\mathrm{r}}+1) + 4n_{\mathrm{r}}(2k-n_{\mathrm{r}}) + n_{\mathrm{r}}(n_{\mathrm{r}}-1)\right], \qquad (135)$$

where $n_{\mathrm{r}} = n - \ell - 1$, $k = \frac{1}{2}, \frac{3}{2}, \frac{5}{2}, \ldots, \infty$ and $\ell = 0, 1, 2, \ldots, n$. In fact, it is a particular case $d = 2$ of a formula depending on the dimension d. It reduces to the familiar formula for $d = 3$. The derivations in the framework of the results of [Nouri (1999)] are easy.

Peřinová, Lukš and Pintr (2007) recall that de Oliveira Neto, Maia and Carneiro (2004) have defined the Bohr radius of the solar system $r_{\frac{1}{2}0} = 0.055$ AU. As Agnese and Festa (1997) have preferred circular orbits, they expect an approach with $\ell = k - \frac{1}{2}$, which has not been adopted by de Oliveira Neto, Maia and Carneiro (2004). As Agnese and Festa (1997) have demonstrated, such an approach may require a different Bohr radius.

The orbits, on which big bodies – planets – may originate, are listed in Table 4. It emerges that for every number k, there exists only one stable orbit, on which a big body – a planet – may originate. Then they can interpret the number k as the principal quantum number and ℓ as the orbital quantum number equal to the number of possible orbits, but only for the greatest ℓ there exists a stable orbit of a future body. A planet which does not confirm this theory, is the Earth. Since the description based on the modified Schrödinger equation for the planetary system is not fundamental, it could not fit all the stable orbits. Other deviations are likely to be incurred by collisions of the bodies in early stages of the origin of the planets, thus nowadays it is already possible to observe elliptical orbits, which are very close to circular orbits.

Using the graphs of the probability densities that they have plotted for every predicted orbit of this system, they have obtained expected results. The graphs of the probability densities for each orbit with $k \leq \frac{9}{2}$ and with $\frac{11}{2} \leq k \leq \frac{39}{2}$ are contained in Figure 1 and in Figure 2, respectively. The vertical axis denotes the probability density $P_{k\ell}(r)$ and the longitudinal axis designates the planetary distance r from the Sun. In Figure 1, the graph

Table 4. Bodies with stable circular orbits

Body	k	ℓ	$r_{k\ell}$ [AU]	Error [%]
—	$\frac{1}{2}$	0	0.055	—
Mercury	$\frac{3}{2}$	1	0.332	−15
Venus	$\frac{5}{2}$	2	0.83	15.5
Mars	$\frac{7}{2}$	3	1.54	1.5
Vesta	$\frac{9}{2}$	4	2.49	5.5
Fayet comet	$\frac{11}{2}$	5	3.64	−3.5
Jupiter	$\frac{13}{2}$	6	5.03	−3.5
Neujmin comet	$\frac{15}{2}$	7	6.636	−2.5
—	$\frac{17}{2}$	8	8.46	—
Saturn	$\frac{19}{2}$	9	10.5	10
—	$\frac{21}{2}$	10	12.77	—
Westphal comet	$\frac{23}{2}$	11	15.26	−2.5
Pons–Brooks comet	$\frac{25}{2}$	12	17.97	4
Uranus	$\frac{27}{2}$	13	20.9	9
—	$\frac{29}{2}$	14	24.055	—
—	$\frac{31}{2}$	15	27.43	—
Neptune	$\frac{33}{2}$	16	31.02	3
—	$\frac{35}{2}$	17	34.84	—
Pluto	$\frac{37}{2}$	18	38.88	−1.5
—	$\frac{39}{2}$	19	43.134	—

for $n = 2$ is interpreted such that the highest probability density is assigned to the orbit of the radius of 0.332 AU and from the calm shape of the graph they infer that an ideal circular orbit is tested.

The previous procedure has been applied to moons of giant planets [Peřinová, Lukš and Pintr (2007)]. It emerges that the moons of giant planets are also fitted by the modified Schrödinger equation and appropriate expectation values. Especially, the predicted stable circular orbits of Jupiter's moons are presented in Table 5. For Jupiter it holds that $M = M^{\text{Jup}}$ and the Bohr radius of this system is $r_1 = 6287$ km. It emerges that the predicted lunar orbits fit the measured orbits of the moons orbiting Jupiter.

Pintr, Peřinová and Lukš (2008) have paid attention to the following event. In the year 2004, a new theory emerged, which assumes basing on a study of chemical compounds in meteorites and a study of astronomical objects of the distant universe that, like most low-mass stars, the Sun formed in a high-mass star-forming region, where some stars went supernova [Desch, Healy and Leshin (2004)]. The radiation from massive stars carves out ionized cavities in the dense clouds, within which the stars formed. Such regions are called H II (ionized hydrogen regions). Examples of these regions of star formations are the Orion Nebula, the Eagle Nebula, and many other nebulas. The decisive argument of this new theory is the presence of isotope ^{60}Fe in meteorites. This isotope is unstable and it may originate only in cores of high-mass stars. The presence of the isotope ^{60}Fe in our solar

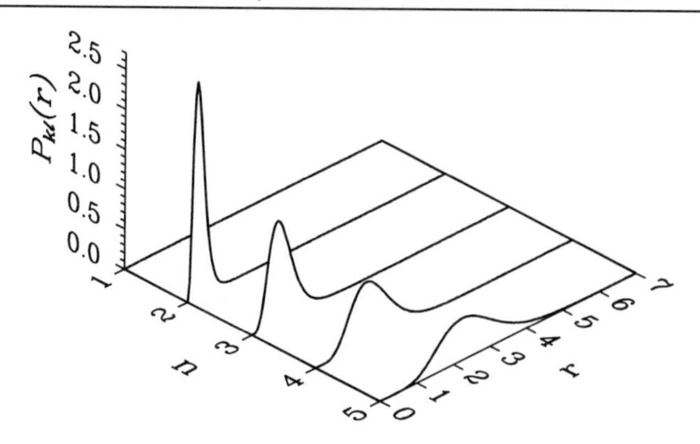

Figure 1. Probability densities for a particle in states with quantum numbers k, ℓ, which correspond, respectively ($n = k + \frac{1}{2}$), to Mercury ($n = 2, \ell = 1$), Venus ($n = 3, \ell = 2$), Mars ($n = 4, \ell = 3$) and asteroid Vesta ($n = 5, \ell = 4$). Here $k \in \{\frac{3}{2}, \frac{5}{2}, \ldots, \frac{9}{2}\}$ and r is measured in AU.

Table 5. Moons of Jupiter with stable circular orbits

Body	k	ℓ	$r_{k\ell}$ [km]	Error [%]
—	$\frac{1}{2}$	0	6287	—
—	$\frac{3}{2}$	1	37722	—
Halo ring	$\frac{5}{2}$	2	94305	5.5−−23.5
Outer ring	$\frac{7}{2}$	3	176036	43−−27.5
—	$\frac{9}{2}$	4	282915	—
Io	$\frac{11}{2}$	5	414942	−1.5
Europa	$\frac{13}{2}$	6	572117	−14.5
—	$\frac{15}{2}$	7	754440	—
—	$\frac{17}{2}$	8	961911	—
Ganymede	$\frac{19}{2}$	9	1.19×10^6	11
—	$\frac{21}{2}$	10	1.452×10^6	—
Callisto	$\frac{23}{2}$	11	1.735×10^6	−8

system favours the hypothesis that the Sun was born near high-mass stars in ionized clouds of gas and dust. The protagonist of the new theory is Hester and the group of astronomers in the Arizona State University [Hester, Desch, Healy and Leshin (2004)]. The origin of the solar system happened probably as follows: A shock wave, which compresses molecular gas into dense cores, is driven in advance of an ionization front. After the cores emerge into the H II region interior, they evaporate. Some of the cores contain a star and a circumstellar disk. The disk evaporates. The massive stars pelt the low-mass young stellar objects and the protoplanetary disks, which surround them, with ejecta.

Assuming this to be a probable mechanism of the origin of stellar systems, Pintr, Peřinová and Lukš (2008) attempt to find answers to some basic questions about the traits of the solar system. They consider our solar system on some simplifications:

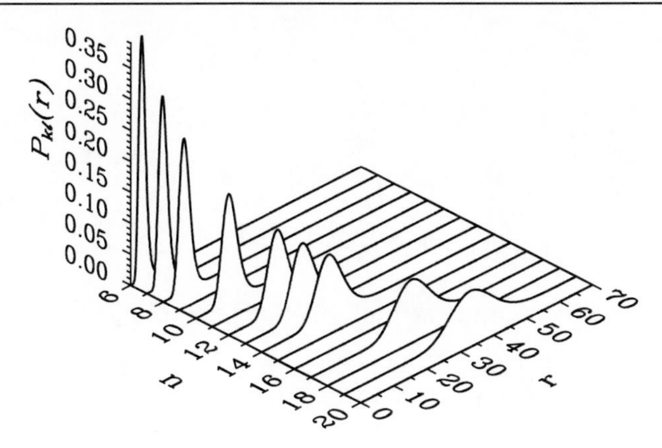

Figure 2. Probability densities for a particle in states with quantum numbers k, ℓ, which correspond, respectively ($n = k + \frac{1}{2}$), to Fayet comet ($n = 6$), Jupiter ($n = 7$), Neujmin comet ($n = 8$), Saturn ($n = 10$), Westphal comet ($n = 12$), Pons-Brooks comet ($n = 13$), Uranus ($n = 14$), Neptune ($n = 17$) and Pluto ($p = 19$), $k \in \{\frac{11}{2}, \frac{13}{2}, \ldots, \frac{39}{2}\}$, $\ell = n - 1$. Here r is measured in AU.

(i) The orbits of the bodies about the Sun are considered to be circular.

(ii) These orbits lay in one plane with respect to the Sun.

They do not explain the elliptical orbits and the inclinations of these orbits. Then they can define a two-dimensional planetary model of a radius r_0.

Pintr, Peřinová and Lukš (2008) argue that an ionized nebula is the medium, where hydromagnetic waves may propagate at a velocity, v_A (Alfvén's velocity). For understanding possible evolution of the solar system, they introduce the circular membrane model (two-dimensional model) of the radius r_0. It describes a perfectly flexible circular membrane of a constant thickness [Brepta, Půst and Turek (1994)]. It is pulled with a force F_l over a unit of length acting at the distance r_0 from the center of the membrane. In formulating the equation of motion, they take into account only the transverse displacements w, assuming that they are small on comparison with the size of the circular membrane. They let μ_A denote the mass of a unit of area.

Introducing polar coordinates r and φ with the relations $x = r\cos\varphi$ and $y = r\sin\varphi$, they solve the appropriate problem for the transverse displacement $\tilde{w}$ using the method of separation of variables. This method is based on solutions in the form

$$\tilde{w}(r, \varphi, t) = R(r, \varphi)T(t), \tag{136}$$

where $T(t)$ is a solution of a simpler problem and

$$R(r, \varphi) = R_1(r)\Phi(\varphi), \tag{137}$$

with $R_1(r)$ and $\Phi(\varphi)$ solutions of simpler problems. Below, $J_n(z)$ will be the Bessel function of the order n, $\mu_m^{(n)}$ will be positive roots of the equation

$$J_n(\mu) = 0, \tag{138}$$

and

$$\lambda_m^{(n)} = \pm \frac{\mu_m^{(n)}}{r_0}, \quad n = 0, 1, \ldots, \infty, \quad m = 1, 2, \ldots, \infty. \tag{139}$$

Omitting the standard developments, they state that

$$\tilde{w}_{nm}(r, \varphi, t) = R_{1nm}(r)\Phi_n(\varphi)T_{nm}(t), \, n = 0, 1, \ldots, \infty, \, m = 1, 2, \ldots, \infty, \tag{140}$$

where they have taken into account the multiplicity of the solutions and

$$R_{1nm}(r) = C_n J_n \left(\mu_m^{(n)} \frac{r}{r_0} \right), \tag{141}$$

with $\lambda_m^{(n)} \geq 0$,

$$\Phi_n(\varphi) = A_n \cos(n\varphi) + B_n \sin(n\varphi), \tag{142}$$

$$T_{nm}(t) = E_{nm} \exp(i\Omega_{nm} t), \tag{143}$$

where

$$\Omega_{nm} = \lambda_m^{(n)} \sqrt{\frac{F_l}{\mu_{\mathrm{A}}}}. \tag{144}$$

The superposition form of a solution suggests that, without a loss of generality, they may specify

$$A_{nm} = 1, B_{nm} = -i, C_{nm} = 1, E_{nm} = 1. \tag{145}$$

The superposition solution (140) is valid for any circular membrane. For $n = 0$, the membrane has a nonzero amplitude at the point with $r_0 = 0$. For other n, $J_n(0) = 0$. The subscripts n and m control the number of nodal lines. The membrane has $2n$ radial nodal segments and m nodal circles including the circumference.

The membrane model considered must take into account the mass $M^{(\mathrm{star})}$ of a future central body and the membrane radius r_0 must also be connected with this mass. A multiplicity has been conceded. A distance a_r is defined in the form

$$a_r = \frac{GM^{(\mathrm{star})}}{c^2}, \tag{146}$$

which is known as half the Schwarzschild radius. On using the fine structure constant α_e, the radii $r_0^{(a)}$ are defined in the form

$$r_0^{(a)} = \frac{a_r}{\alpha_e^a}, \tag{147}$$

where α_e has the property $\frac{1}{\alpha_e} \simeq 137.0$ and a is an exponent, $a = 0, 1, 2, ..., \infty$.

Substituting $M^{(\mathrm{star})} = M^{(\mathrm{Sun})}$, Pintr, Peřinová and Lukš (2008) have observed the necessity of the following radii: $r_0^{(0)} = 1.48$ km, half the Schwarzschild gravitational radius of the Sun; $r_0^{(4)} = 3.481$ AU, the membrane for terrestrial planets (the center of the asteroid belt); $r_0^{(5)} = 476.897$ AU, the membrane for giant planets and $r_0^{(6)} = 65334.9$ AU, the membrane for formation of the Oort cloud of comets.

Pintr, Peřinová and Lukš (2008) replace a mechanical velocity v with the Alfvén velocity v_{A}. They rewrite the angular frequency of the membrane (144) in the form

$$\Omega_{nm}^{(a)} = \frac{\mu_m^{(n)}}{r_0^{(a)}} v_{\mathrm{A}}. \tag{148}$$

They may calculate the time of revolution of the ionized parts,

$$T_{nm}^{(a)} = 2\pi \frac{r_0^{(a)}}{\mu_m^{(n)}} \frac{1}{v_{\mathrm{A}}}. \tag{149}$$

This period depends on the physical properties of the membrane. It is assumed that a future body revolving on a circular orbit about born central body at a distance $r_{nm}^{(a)}$ inherits this period from the wave. The velocity of such a body can be written in two forms,

$$v_{nm}^{(a)} = \frac{2\pi r_{nm}^{(a)}}{T_{nm}^{(a)}} = \sqrt{\frac{GM}{r_{nm}^{(a)}}}, \tag{150}$$

or (cf., [Murray and Dermott (1999)])

$$(2\pi)^2 r_{nm}^{(a)3} = GMT_{nm}^{(a)2}. \tag{151}$$

This equation for $r_{nm}^{(a)}$ has the solution

$$r_{nm}^{(a)} = \sqrt[3]{\frac{GMT_{nm}^{(a)2}}{(2\pi)^2}} = \sqrt[3]{\frac{GMr_0^{(a)2}}{\mu_m^{(n)2} v_{\mathrm{A}}^2}}, \tag{152}$$

where (149) have been utilized.

The membrane model admits that, under conditions present at the origin of stellar systems, a hydro-magnetic wave originates, whose nodal lines provide stable orbits of future planets. For the calculation of possible orbits of planets, it has been assumed that $n = 1$ and the Alfvén velocity for the inner parts of the solar system was $v_{\mathrm{A}}^{(4)} = 16500$ ms^{-1} (Table 6) and for the formation of the outer planets, it was $v_{\mathrm{A}}^{(5)} = 7000$ ms^{-1}, while $n = 0$ (Table 7).

In application to giant planets, it holds that $M^{(\mathrm{star})} = M^{(\mathrm{planet})}$. A membrane model of Jupiter's system has been formulated with $r_0^{(0)} = 0.0014$ km, half the Schwarzschild gravitational radius of Jupiter, $r_0^{(4)} = 496043$ km, the membrane for small moons (rings), $r_0^{(5)} = 6.796 \times 10^7$ km, the membrane for natural moons, and $r_0^{(6)} = 9.310 \times 10^9$ km, the membrane for the remotest parts of the system.

A similar model of Saturn's system has been reported with $r_0^{(0)} = 0.00042$ km, half the Schwarzschild gravitational radius of Saturn, $r_0^{(4)} = 14803$ km, the membrane for small moons (center of rings), $r_0^{(5)} = 2.028 \times 10^7$ km, the membrane for natural moons, and $r_0^{(6)} = 2.778 \times 10^9$ km, the membrane for the remotest parts.

In application to planets of υ Andromedae, it holds that $M^{(\mathrm{star})} = M^{(\upsilon\ \mathrm{And})}$. A membrane model of the υ-Andromedae system has been formulated with $r_0^{(0)} = 1.705$ km, half the Schwarzschild gravitational radius of υ Andromedae, $r_0^{(4)} = 4.003$ AU, the membrane

Table 6. Distances of formation of inner bodies of the solar system up to 0.389 AU

m	$\mu_m^{(1)}$	r [AU]	Body
1	3.832	1.39	Mars
2	7.016	0.930	Earth
3	10.173	0.725	Venus
4	13.323	0.606	asteroid 1999 MN
5	16.471	0.526	—
6	19.616	0.468	—
7	22.760	0.424	—
8	25.903	0.389	Mercury

Table 7. Distances of formation of outer bodies of the solar system up to 9.59 AU

m	$\mu_m^{(0)}$	r [AU]	Body
1	2.405	89.364	planet X
2	5.52	51.359	—
3	8.654	38.05	Pluto – Charon
4	11.792	30.963	Neptune
5	14.931	26.455	—
6	18.071	23.2944	—
7	21.212	20.934	—
8	24.353	19.090	Uranus
9	27.494	17.610	Halley comet
10	30.635	16.384	—
11	33.776	15.352	—
12	36.917	14.468	—
13	40.058	13.701	—
14	43.2	13.029	—
15	46.341	12.434	—
16	49.483	11.901	—
17	52.624	11.423	—
18	55.766	10.989	—
19	58.907	10.596	—
20	62.049	10.234	—
21	65.19	9.903	—
22	68.331	9.59	Saturn
..	...	...	...
55	$\mu_{55}^{(0)}$	5.203	Jupiter

for terrestrial planets (probable center of planetoids), $r_0^{(5)} = 548.432$ AU, the membrane for Jovian planets, and $r_0^{(6)} = 75135.2$ AU, the membrane for the remotest parts of the system.

The last calculations have been related to the star GJ876. Such a model has been presented with $r_0^{(0)} = 0.474$ km, half the Schwarzschild gravitational radius of GJ876, $r_0^{(4)} = 1.114$ AU, the membrane for terrestrial planets (probable center of planetoids), $r_0^{(5)} = 152.607$ AU, the membrane for Jovian planets, and $r_0^{(6)} = 20907.2$ AU, the membrane for the remotest parts of the system.

6. Conclusion

We have concerned with a specific astronomical theme, the regularities in the distances of the planets and moons from the respective central bodies. Historically, this topic is restricted to appropriate mentions of the Titius–Bode law. We have taken into account that this narrow theme has been elaborated on in recent years due to the advances in the discoveries of the extrasolar planetary systems. The doubt that the regularity has emerged accidentally is an alternative to the hypothesis that some law is in force in conjunction with causes of some deviations from it. A relative certainty can be achieved by mathematical methods of the statistical decision making. But it entails to try to define a complicated random sample.

Stochasticity elements can be expected in the dynamics of the rotating nebula. We have not dealt with such scenarios, but with some acceptable estimates of the mass distribution in the primordial nebula. The present dynamical theories of the origin of the planetary system assume a mixture of gas and condensed matter. We have admitted that a number of authors are searching for a shortcut of the pathway leading from the assumptions of the model to the regularities of the planetary orbits. We have considered a formal division of the nebula into rings defined by optimum, "convenient", assignment of the angular momentum to circular loops to be easy to describe.

Even though we have not aimed just at detailed calculations, we have taken into account that they are feasible at present, also due to the competition with the research demonstrating the chaotic behavior and migration of the planets. The possibility of the chaotic behavior has already been admitted in the past and it seems to be a reason why other researchers moved away the quantization of a theory with an unusual kinetic energy somewhere to the frontier of the observable universe. We have included a formal use of the quantization to the discretization of orbits and various efforts for finding a physical underpinning of such an advantage. In extremity, the vibration modes of a circular membrane with orbits of planets or moons can be compared. The mechanical vibrations model a specific magneto-hydrodynamical behavior, which is expected in the region of the birth of stars and their planetary systems.

Acknowledgments

The authors acknowledge the financial support by the Ministry of Education of the Czech Republic in the framework of the project No. 1M06002. They are obliged to Ing. J. Křepelka for the careful preparation of figures.

References

Agnese, A. G.; Festa, R. *Phys. Lett. A* 1997, 227, 165-171.

Agnese, A. G.; Festa, R. *Hadronic Journal* 1998, 8, 237-254.

Agnese, A. G.; Festa, R. (1999). Discretizing υ-Andromedae planetary system. arXiv: astro-ph/9910534, v2.

Alfvén, H. *Stockholms Observatorium Annales* 1942, 14, 2.1-2.33.

Arp, H. *Seeing Red: Redshifts, Cosmology and Academic Science*; Apeiron: Montreal, Canada, 1998.

Arp, H.; Sulentic, J. W. *Astrophys. J.* 1985, 291, 88-111.

Brepta, R.; Půst, L.; Turek, F. *Mechanical Vibration*; Sobotáles: Prague, CZ, 1994, p. 268 (in Czech).

Cameron A. G. W. *Icarus* 1962, 1, 13–69.

Cameron, A. G. W.; Pine, M. R. *Icarus* 1973, 18, 377–406.

Carneiro, S. *Found. Phys. Lett.* 1998, 11, 95-102.

Carvalho, J. C. *Lett. Nuovo Cimento* 1985, 6, 337-342.

Chamberlin, T. C. *Carnegie Institution Year Book # 3 for 1904*; Carnegie Inst.: Washington, DC, 1905, pp. 208-233.

Christodoulou, D. M.; Kazanas, D. (2008). Exact solutions of the isothermal Lane–Emden equation with rotation and implications for the formation of planets and satellites, arXiv:astro-ph/0706.3205 v2.

Dauvillier, A.; Desguins, E. *La genèse de la vie*; Hermann: Paris, 1942.

de Oliveira Neto, M. *Ciência e Cultura* (*J. Brazil. Assoc. Advance. Sci.*) 1996, 48, 166-171.

de Oliveira Neto, M.; Maia, L. A.; Carneiro, S. *Chaos, Solitons & Fractals* 2004, 21, 21-28.

Dersarkissian, M. *Lett. Nuovo Cimento* 1984, 13, 390-394.

Desch, S.; Healy, K. R.; Leshin, L. (2004). New theory proposal for solar system formation. *Universe Today*, http://www.universetoday.com.

Dirac, P. A. M. *Nature* 1937, 139, 323-323.

Dirac, P. A. M. *Proc. Roy. Soc. London* 1938, 165, 199-208.

Dubrulle, B.; Graner, F. *Astron. Astrophys.* 1994, 282, 269-276.

Efron, B. *J. Am. Statist. Assoc.* 1971, 66, 552-559.

Fényes, I. *Z. Physik* 1952, 132, 81-106.

Good, I. J. with discussion by H. O. Hartley, I. J. Bross, H. A. David, M. Zelen, R. E. Bargmann, F. J. Anscombe, M. Davis and R. L. Anderson, *J. Am. Statist. Assoc.* 1969, 64, 23-66.

Graner, F.; Dubrulle, B. *Astron. Astrophys.* 1994, 282, 262-268.

Greenberger, D. M. *Found. Phys. Lett.* 1983, 13, 903-951.

Hayashi, Ch. *Supplement of the Progress of Theoretical Physics* 1981, 70, 35-53.

Hayes, W.; Tremaine, S. *Icarus* 1998, 135, 549-557.

Hester, J. J.; Desch, S. J.; Healy, K. R.; Leshin, L. A. *Science* 2004, 304, 5674-5675.

Jeans, J. H. *Phil. Trans. Royal Soc. London* 1914, Series A, 213, 457-485.

Jeffreys, H. *The Earth, Its Origin, History and Physical Constitution* ; Cambridge University Press: Cambridge, 1924.

Kant, I. *Allgemeine Naturgeschichte und Theorie des Himmels* ; J. F. Petersen: Königsburg, Leipzig, GE, 1755.

Kuiper, G. P. In: *Astrophysics: a topical symposium* ; Hynek, J. A.; Ed.; McGraw-Hill: New York, NY, 1951, pp 357-424.

Krot, A. M. *Chaos, Solitons & Fractals* 2009, 41(3), 1481-1500.

Kuchner, M. J. *Astrophys. J.* 2004, 612, 1147-1151.

Laplace, P. S. *Exposition du système du monde* ; De l'Imprimerie du Cercle-Social: Paris, FR, 1796.

Laskar, J. *Nature* 1989, 338, 237-238.

Lissauer, J. J. *Annu. Rev. Astron. Astrophys.* 1993, 31, 129-174.

Lynch, P. *Mon. Not. R. Astron. Soc.* 2003, 341, 1174-1178.

Lyttleton, R. A. *Monthly Notices of the R. A. S.* 1936, 96, 559–568.

Malin, D. F.; Carter, D. *Nature* 1980, 285, 643-645.

Maxwell, J. C. *On the Stability of the Motions of Saturn's Rings* ; MacMillan: London, 1859.

Mohorovičić, S. *Astronomische Nachrichten* 1938, 266, 149-160.

Moulton, F. R. *Astrophys. J.* 1905, 165–181.

Murray, C. D.; Dermott, S. F. *Solar System Dynamics*; Cambridge University Press: Cambridge, GB,1999.

Murray, N.; Hansen, B.; Holman, M.; Tremaine, S. *Science* 1998, 279, 69-72.

Murray, N.; Paskowitz, M.; Holman, M. *Astrophys. J.* 2002, 565, 608-620.

Nelson, E. *Phys. Rev.* 1966, 4, 1079-1085.

Nieto, M. M. *The Titius–Bode Law of Planetary Distances: Its History and Theory* ; Pergamon Press: Oxford, MA, 1972.

Nölke, F. *Der Entwicklungsgang unseres Planetensystems* ; Dümmler Verlh.: Berlin, 1930.

Nottale, L. *Fractal Space–Time and Microphysics: Towards a Theory of Scale Relativity* ; World Scientific: Singapore, 1993, p. 311.

Nottale, L. *Astron. Astrophys.* 1996, 315, L9-L12.

Nottale, L.; Schumacher, G.; Gay, J. *Astron. Astrophys.* 1997, 322, 1018-1025.

Nottale, L.; Schumacher, G.; Lefèvre, E. T. *Astron. Astrophys.* 2000, 361, 379-387.

Nouri, S. *Phys. Rev. A*, 1999, 60, 1702-1705.

Page, D. N. *Int. J. Mod. Phys. D* 2010, 19, 2271–2274.

Peřinová, V.; Lukš, A.; Pintr, P. *Chaos, Solitons & Fractals* 2007, 34, 669-676.

Pintr, P.; Peřinová, V. *Acta Universitatis Palackianae* 2003–2004, Physica 42–43, 195-209.

Pintr, P.; Peřinová, V.; Lukš, A. *Chaos, Solitons & Fractals* 2008, 36, 1273-1282.

Povolotsky, A. (2007) (personal communication).

Recami, E.; Raciti, F.; Rodrigues, Jr., W. A.; Zanchin, V. T. In *Gravitation: The Space-Time Structure. (Proceedings of Silarg-VIII)* ; Latelier, P. S.; Rodrigues, Jr., W. A.; Eds.; World Scientific: Singapore, Singapore, 1994, p. 355-372.

Rubčić, A.; Rubčić, J. *Fizika B* 1995, 4, 11-28.

Rubčić, A.; Rubčić, J. *Fizika B* 1996, 5, 85-92.

Rubčić, A.; Rubčić, J. *Fizika B* 1998, 7, 1-13.

Russell, H. N. *The Solar System and Its Origin*; Macmillan: New York, NY, 1935.

Russell, H. N. *Scientific American* 1943, 169, 18-19.

Safronov, V.S. *Ann. Astrophys.* 1960, 23, 901–904.

Safronov, V. S. *Evolution of the Protoplanetary Cloud and Formation of the Earth and Planets*; Nauka: Moscow, 1969. Engl. trans. NASA TTF-677, 1972.

Schmidt, O. Y. *Dokl. Akad. Nauk SSSR* 1944, 45, 245-249 (in Russian).

Shakura, N. I.; Sunyaev, R. A. *Astron. Astrophys.* 1973, 24, 337-355.

Spencer, W. *J. Creat.* 2007, 21(3), 12-14.

Sussman, G. J.; Wisdom, J. *Science* 1988, 241, 433-437.

Sussman, G. J.; Wisdom, J. *Science* 1992, 257, 56-62.

Tsiganis, K.; Gomes, R.; Morbidelli, A.; Levinson, H. F. *Nature* 2005, 435, 459–461.

von Weizsäcker, C. F. *Z. Astrophysik* 1943, 22, 319-355.

Weidenschilling, S. J. *Astrophys. Space Sci.* 1977, 51, 153-158.

Weidenschilling, S. J.; Cuzzi, J. N. In *Protostars and Planets III*; Levy, E. H.; Inine, J. I.; Matthews, M. S.; Eds.; Univ. Ariz. Press: Tucson, AR, 1993, 1031-1060.

Wesson, P. S. *Phys. Rev. D* 1981, 23, 1730-1734.

Wetherill, G. W. In *Asteroids II*; Binzel, R. P.; Gehrels, T.; Matthews, M. S.; Eds.; Univ. Ariz. Press: Tucson, AR, 1989, 661-680.

Wetherill, G. W. *Annu. Rev. Earth Planet. Sci.* 1990, 18, 205-256.

Wetherill, G. W. *Science* 1991, 253, 535-538.

Wetherill, G. W. *Icarus* 1992, 100, 307-325.

Reviewed by Professor RNDr. Jan Novotný. Ph.D., Faculty of Science, Masaryk University, Brno, Czech Republic.

In: Solar System: Structure, Formation and Exploration
Editor: Matteo de Rossi
ISBN: 978-1-62100-057-0

Chapter 9

A MODELS OF FORMING PLANETS AND DISTRIBUTION OF PLANETARY DISTANCES AND ORBITS IN THE SOLAR SYSTEM BASED ON THE STATISTICAL THEORY OF SPHEROIDAL BODIES

Alexander M. Krot[*]
Laboratory of Self-Organization System Modeling, United Institute of Informatics Problems, National Academy of Sciences of Belarus, Minsk, Belarus

ABSTRACT

In this chapter, we consider a statistical theory of gravitating spheroidal bodies to explore and develop a model of forming and self-organizing the Solar system. It has been proposed the statistical theory for a cosmological body forming (so-called spheroidal body) by means of numerous gravitational interactions of its parts (particles). The proposed theory starts from the conception for forming a spheroidal body inside a gas-dust protoplanetary nebula; it permits us to derive the form of distribution functions, mass density, gravitational potentials and strengths both for immovable and rotating spheroidal bodies as well as to find the distribution function of specific angular momentum. As the specific angular momentums are averaged during conglomeration process, the specific angular momentum for a planet of the Solar system (as well as a planetary distance) can be found by means of such procedure. This work considers a new law for the Solar system planetary distances which generalizes the well-known Schmidt law. Moreover, unlike the well-known planetary distances laws the proposed law is established by a physical dependence of planetary distances from the value of the specific angular momentum for the Solar system.

The proposed simple statistical approach to investigation of our Solar system forming describes only a natural self-evolution inner process of development of protoplanets from a dust-gas cloud. Naturally, this approach however does not include any dynamics like collisions and giant impacts of protoplanets with large cosmic bodies. Henceforth, the presented statistical theory will only be able to predict surely the

* Laboratory of Self-Organization System Modeling, United Institute of Informatics Problems, National Academy of Sciences of Belarus, Surganov Str. 6, 220012 Minsk, Belarus, E-mail address: alxkrot@newman.bas-net.by

protoplanet's positions according to the proposed ent[n/2] rule, i.e. the findings in this work are useful to predict if today's position or orbit of a considered planet coincides with its protoplanet's location or not.

Though orbits of moving particles into the flattened rotating spheroidal body are circular ones initially, however, they could be distorted by collisions with planetesimals and gravitational interactions with neighboring originating protoplanets during evolutionary process of protoplanetary formation. Really, at first the process of evolution of gravitating and rotating spheroidal body leads to its flattening, after that the evolutionary process results in its decay into forming protoplanets. This work shows that the orbits of moving particles are formed by action of centrally-symmetrical gravitational field mainly on the later stages of evolution of gravitating and rotating spheroidal body, i.e. when the particle orbits become Keplerian ones. In this connection, this work investigates the orbits of moving planets and bodies in centrally-symmetrical gravitational field of gravitating and rotating spheroidal body during the planetary stage of its evolution (in particular, the angular shift of the Mercury's perihelion is estimated). This work shows that according to the proposed statistical theory of gravitating spheroidal bodies the turn of perihelion of Mercury' orbit is equal to 43.93" per century that well is consistent with conclusions of the general relativity theory of Einstein (this estimation is equal to 43.03") and astronomical observation data ($43.11 \pm 0.45''$).

Keywords: gas-dust protoplanetary nebula, initial gravitational interactions, statistical theory, spheroidal bodies, distribution functions, specific angular momentum, planetary distances law, forming planetary orbits, angular shift of perihelion of orbit

INTRODUCTION

In spite of great succeses and achievements in astrophysics and geophysics of the last decades, the problems of the Solar system forming remain to be actual and important at least because now there is not a just general and noncontradictory scenario of formation of proto-Sun and protoplanetary system from a protosolar nebula (a molecular cloud).

Beginning from the 40-s of the previous century a great number of woks has been carried out in cosmogony: in particular, nebular theories for exploring evolution processes in the Solar system forming have been proposed by von Weizsäcker [1, 2], Schmidt [3, 4], Kuiper [5], Hoyle [6, 7], ter Haar [8, 9] and Cameron [8, 10] et al. As a result, different theories were proposed for the explanation of as structure, formation and exploration of the Solar System (see, for example, [11]). These theories, overlapping each other, can be divided into 5 categories:

- electromagnetic,
- gravitational,
- nebular,
- quantum mechanical,
- statistical theories.

In particular, the *electromagnetic* theory of Alfvén is found to be very important because it points to a damping action of the magnetic field at transferring an angular momentum from the Sun to an ionizing gas around the Sun [12].

The *gravitational* Schmidt's theory [3, 4] is based on the idea of the capture of a dust-gas cloud by the Sun as well as a conglomeration process. This process leads to "scooping out" a cloud substance by neighboring proto-planets owing to the difference between specific angular momentums for these proto-planets. Schmidt's theory was continued by his colleagues actively: Gurevich and Lebedinsky [13, 14], Safronov [15], Vityazev et al. [16]. Viktor S.Safronov was the first who suggested so-called planetesimal hypothesis stating that planets form out of dust grains that collide and stick to form larger and larger bodies. Dole [17] also developed a dust accumulation process describing absorption of gas by planets with "critical sizes" and, hence, leading to origin the giant-planets.

The *nebular* von Weizsäcker's theory [1, 2] revealed the greatest interest among scientists. It pointed to the importance of turbulent processes in forming protocloud. It was shown by von Weizsäcker that the turbulence leads to vortex cells forming and then a substance condensation origin. The initial von Weizsäcker's hypothesis was modified by many scientists [11]. The origin of modern theories can be traced to Kuiper [5] who not only suggested that the protoplanetary nebula was significantly more massive than the present-day sum of planetary masses, perhaps exceeding 0.1 of Sun's mass M_S but also proposed that the gas giant planets are the result of gaseous accretion of solid protoplanetary cores (see [18], p.495).

Some recent works have been devoted to investigating the possibility of describing planetary orbits based on *quantum mechanical approaches* [19–26]. A brief review of the set of these theoretical studies was made by De Oliveira Neto and co-workers [25]. One of the interesting results found refers to the prediction of a fundamental radius given by $r = 0.05$AU, also predicted by Nottale [20, 21] and Agnese and Festa [22, 23] in their studies. Agnese and Festa [22] described the Solar system as a gravitational atom (these authors also proposed a formula describing the distances of the bodies of the Solar system). Following the Bohr–Sommerfeld atomic theory foundations, quantum mechanics emerged with its Schrödinger equation. A derivation of the Schrödinger equation from the Newtonian mechanics was given in the works [27, 28]. The important point in Nelson's works is that a diffusion process can be described in terms of a Schrödinger-type equation, with help of the hypothesis that any particle in the empty space, under the influence of any interaction field, is also subject to a universal Brownian motion [29] based on the quantum nature of space–time in quantum gravity theories or on quantum fluctuations on cosmic scale [21, 30, 31]. In this context, the possibility of describing a classical process like the formation of Solar system in terms of quantum mechanics can be seriously considered [25]. As for macroscopic bodies, the chaotic behaviour of the Solar system during its formation and evolution [32, 33] indicates a diffusion process to be described in terms of a Schrödinger-type equation. The description of the planetary system using a Schrödinger-type diffusion equation has been presented in [25]. El Naschie has obtained the dimensionless Newton gravity constant by using descriptive set theory [34]. An analysis of the relationship to the electrostatic and gravitational forces has revealed the utility of the Newton gravity constant for estimation of planetary orbits [26].

In spite of great number of works aimed to the formation of the solar system exploring and significant efforts of many brilliant scientists, however, the mentioned theories were not able to explain all phenomena completely occurring in the Solar system. In this connection, it has been proposed the *statistical* theory [35–55] for a cosmological body forming (so-called *spheroidal body*) by means of numerous gravitational interactions of its parts (particles). The domain of investigations within framework of the proposed statistical theory of gravity includes Newtonian gravity and Newtonian quantum gravity.

The proposed theory starts from the conception for forming a spheroidal body inside a gas-dust protoplanetary nebula; it permits us to derive the form of distribution functions, mass density, gravitational potentials and strengths for an immovable spheroidal bodies [35–40, 54, 55] and rotating ones [41–45, 48–53] as well as to find the distribution function of specific angular momentum [45–49, 52, 53] for a rotating uniformly spheroidal body based on the proposed distribution function of particles in a space.

In this work, we consider a statistical theory of gravity to explore and develop a model of forming the Solar system. This paper investigates a gas-dust protoplanetary cloud as rotating and gravitating spheroidal bodies having the specific angular momentum distribution functions. As the specific angular momentums are averaged during conglomeration process, the specific angular momentum for a planet of the solar system (as well as a planetary distance) can be found by means of such procedure. The problem of gravitational condensation of a gas-dust protoplanetary cloud with a view to protoplanet formation in its own gravitational field is also considered here. This paper presents a new law for the Solar system planetary distances which generalizes the well-known Schmidt law.

On the other hand, this work shows that the orbits of moving particles are formed by action of centrally-symmetrical gravitational field mainly on the later stages of evolution of gravitating and rotating spheroidal body, i.e. when the particle orbits become Keplerian ones. In this connection, this work investigates also the orbits of moving planets and bodies in centrally-symmetrical gravitational field of gravitating and rotating spheroidal body during the planetary stage of its evolution.

1. The Derivation of Distribution Function of Particles in a Uniformly Rotating Spheroidal Body and Calculation of its Gravitational Potential in a Remote Zone

Let us consider the statistical theory of formation of cosmological bodies [35–55] beginning from the derivation of distribution function of particles in a space filled in a homogeneous gaseous (dust-like) nebula. In other words, the question is about the distribution of particles in space. The statistical aspect of the problem results from the fact that the considered body consisting of gaseous matter is a system containing a large number of particles interacting among themselves by gravitational interactions only. Numerous fluctuations of particle concentration caused by their local gravitational interactions do not allow to predict surely the behavior of the system. First of all, we represent the protoplanetary nebula as a self-gravitating body satisfying the following assumptions [35–38, 52]:

1. The gaseous (dust-like) body is considered in a homogeneous space;
2. The gaseous gravitating body under consideration is homogeneous in its chemical structure, i.e. it consists of N identical particles with the mass m_0;
3. The gravitating body is isolated one, i.e. it is not subjected to influence of external fields and bodies;

4. The gaseous gravitating body is isothermal and has a low temperature T, besides $T_{\deg} < T$, where $T_{\deg} = (h^2 / m_0 k_B) n^{2/3}$ is a degeneration temperature [56], n is a concentration of particles, h is the Planck constant, k_B is the Boltzmann constant;
5. The process of gravitational interaction of particles is slow-flowing in time: both immovable and rotating gravitating bodies.

To describe the stage of evolution from the protoplanetary nebula (the solar nebula) to a forming protoplanetary gas-dust disk, let us consider a spheroidal body consisting of N particles with mass m_0 and rotating uniformly with angular velocity $\vec{\Omega}$. We choose a plane of rotation as (x, y) and Oz as an axis of rotation of spheroidal body. To obtain a distribution function for particles in a space we use a cylindrical frame of reference (h, ε, z):

$$x = h \cdot \cos\varepsilon; \quad y = h \cdot \sin\varepsilon; \quad z = z, \tag{55}$$

where $h = \sqrt{x^2 + y^2}$, $0 \le h < \infty, -\infty \le z \le \infty, 0 \le \varepsilon \le 2\pi$. The choice of the cylindrical rotating frame of reference gives us an advantage because a rotating spheroidal body remains relatively immovable in this system.

Analogously to the case of an immovable frame of reference (x, y, z) [35–40, 54, 55], we introduce a distribution for all three coordinates h, ε and z in the rotating frame of reference [43, 52]:

$$dp_{h,\varepsilon,z} = \Phi(h, \varepsilon, z) h dh d\varepsilon dz, \tag{56}$$

where $\Phi(h, \varepsilon, z)$ is a probability volume density function to locate a particle in a rotating frame of reference. As far as particles into a rotating spheroidal body do not move relatively rotating frame of reference, the derivation of distribution function can be done according to the non-rotational case (see [35–40, 52]) though there are some peculiarities connected with the differential equation and integrating process with respect to the *angular* coordinate ε. Analogously to this non-rotational case, in a cylindrical rotating frame of reference (h, ε, z) the probabilities of any particle having coordinates in the intervals $[h, h + dh]$, $[\varepsilon, \varepsilon + d\varepsilon]$, $[z, z + dz]$ are respectively equal to:

$$dp_h = \varphi(h) dh = \frac{dN_h}{N}; \; dp_\varepsilon = \varphi(\varepsilon) h d\varepsilon = \frac{dN_\varepsilon}{N}; \; dp_z = \varphi(z) dz = \frac{dN_z}{N}, \tag{10.1.3}$$

where $\varphi(h), \varphi(\varepsilon), \varphi(z)$ are one-dimensional probability densities. As for an immovable frame of reference (x, y, z), we introduce a distribution for all three coordinates h, ε and z in a rotating frame of reference [43, 52]:

$$dp_{h,\varepsilon,z} = \Phi(h, \varepsilon, z) h dh d\varepsilon dz, \tag{10.1.4}$$

where $\Phi(h,\varepsilon,z)$ is a probability volume density function to locate a particle in a rotating frame of reference.

We suppose that a particle into a rotating spheroidal body does not move relatively rotating frame of reference, therefore, such particle has all three coordinates independently each other. Then according to the theorem of complex event probability we have:

$$dp_{h,\varepsilon,z} = \varphi(h)\varphi(\varepsilon)\varphi(z)hdhd\varepsilon dz\ , \tag{10.1.5}$$

Comparing (10.1.4) with (10.1.5) we obtain:

$$\Phi(h,\varepsilon,z) = \varphi(h)\varphi(\varepsilon)\varphi(z) \tag{10.1.6}$$

As the distance r from the center of spheroidal body to an observed particle is equal to $r = \sqrt{x^2+y^2+z^2} = \sqrt{h^2\cos^2\varepsilon + h^2\sin^2\varepsilon + z^2} = \sqrt{h^2+z^2}$, then the equation is true:

$$\Phi(r) = \varphi(h)\varphi(\varepsilon)\varphi(z)\,. \tag{10.1.7}$$

Starting from the functional equation (10.1.7) let us try to define the form of functions φ and Φ [43, 52]. Differentiating Φ as a composite function with respect to h, let us represent eq. (10.1.7) in the differential equation form:

$$\dot{\Phi}_r(r)\frac{\partial r}{\partial h} = \dot{\varphi}_h(h)\varphi(\varepsilon)\varphi(z). \tag{10.1.8}$$

Let us calculate the partial derivative:

$$\frac{\partial r}{\partial h} = \frac{\partial}{\partial h}\left(\sqrt{h^2+z^2}\right) = \frac{1}{2}\cdot\frac{2h}{\sqrt{h^2+z^2}} = \frac{h}{r}. \tag{10.1.9}$$

With provision for eq. (10.1.9) the equation (10.1.8) becomes

$$\dot{\Phi}_r(r)\frac{h}{r} = \dot{\varphi}_h(h)\varphi(\varepsilon)\varphi(z). \tag{10.1.10}$$

Dividing eq. (10.1.10) by eq. (10.1.7), we have:

$$\frac{\dot{\varphi}_h(h)}{h\varphi(h)} = \frac{\dot{\Phi}_r(r)}{r\Phi(r)}\,. \tag{10.1.11}$$

Analogously we can obtain a differential equation relatively z as the following:

$$\frac{\dot{\varphi}_z(z)}{z\varphi(z)} = \frac{\dot{\Phi}_r(r)}{r\Phi(r)}. \qquad (66)\ (10.1.12)$$

Now from eq. (10.1.7) let us find the differential equation with respect to ε :

$$\dot{\Phi}_r(r)\frac{\partial r}{\partial \varepsilon} = \varphi(h)\dot{\varphi}_\varepsilon(\varepsilon)\varphi(z). \qquad (10.1.13)$$

From the beginning a non-uniform rotation with an angular velocity $f(\vec{\Omega})$ of gravitating spheroidal body, inside it a particle begins to move in the opposite direction to this rotation owing to an inertia force action. In this connection, if a core of spheroidal body turns on a small angle ε then external particles rotate on $-\varepsilon$ [52]. Because the angle ε is enough small then distance r (from the center of spheroidal body to an moving particle) is equal to

$$r = \sqrt{h^2 + z^2 + h^2(-\varepsilon)^2}. \qquad (10.1.14)$$

According to eq. (10.1.14) the desired partial derivative is equal to the following [52]:

$$\frac{\partial r}{\partial \varepsilon} = \frac{\frac{1}{2}h^2 2\varepsilon}{\sqrt{h^2 + z^2 + h^2\varepsilon^2}} = \frac{h^2\varepsilon}{r}. \qquad (10.1.15)$$

Substituting eq. (10.1.15) in eq. (10.1.13) we obtain:

$$\frac{\dot{\varphi}_\varepsilon(\varepsilon)}{h^2\varepsilon\cdot\varphi(\varepsilon)} = \frac{\dot{\Phi}_r(r)}{r\Phi(r)}. \qquad (10.1.16)$$

Since in eqs.(10.1.11), (10.1.12), (10.1.16) the right parts are the same, and the left parts are functions of either h or z, or h and ε, then according to the independence of these coordinates it takes place identity in the case of constancy of these left parts individually [43–45, 52]:

$$\dot{\varphi}_h(h)/h\varphi(h) = \dot{\varphi}_z(z)/z\varphi(z) = (1/h^2)\dot{\varphi}_\varepsilon(\varepsilon)/\varepsilon\varphi(\varepsilon) = -\alpha, \qquad (10.1.17)$$

where α is some constant being *the parameter of gravitational compression* [37, 52] and obviously $\alpha > 0$. Integrating eqs. (10.1.17) we obtain:

$$\int \dot{\varphi}_h(h)dh/\varphi(h) = -\alpha\int hdh;$$

$$\int \dot{\varphi}_z(z)dz/\varphi(z) = -\alpha\int zdz;\quad \int_{\varphi(\varepsilon)}^{\varphi_0} \dot{\varphi}_\varepsilon(\varepsilon)d\varepsilon/\varphi(\varepsilon) = -\alpha h^2\int_{\varepsilon}^{\varepsilon_0} \varepsilon d\varepsilon, \qquad (10.1.18)$$

whence

$$\varphi(h) = c_1 e^{-\alpha h^2/2};\ \varphi(z) = c_2 e^{-\alpha z^2/2};\ \varphi(\varepsilon) = \varphi_0 e^{\alpha h^2 \varepsilon_0^2/2}\ (\varepsilon^2 \to 0). \tag{10.1.19}$$

The third relation in eq. (10.1.19) has been obtained under the condition of stabilization of the coordinate ε to a constant value ε_0. Because ε tends to zero we should take into account only the upper limit value ε_0.

According to eqs. (10.1.7) and (10.1.19) now it is not difficult to write down a general expression for a probability volume density function:

$$\Phi(h, z, \varepsilon_0) = \varphi(h)\varphi(z)\varphi(\varepsilon_0) = C\, e^{-\alpha(h^2(1-\varepsilon_0^2)+z^2)/2}, \tag{10.1.20}$$

where C is a constant of integration. Starting from the normalization condition for the probability density (10.1.20):

$$\int_V \Phi dV = 1$$

we can find a probability volume density function describing a particle distribution into a rotating and gravitating spheroidal body in the cylindrical coordinates [43–45, 52]:

$$\Phi(h, z) = (\alpha/2\pi)^{3/2}(1-\varepsilon_0^2)e^{-\alpha(h^2(1-\varepsilon_0^2)+z^2)/2}, \tag{10.1.21a}$$

as well as in the Cartesian coordinates:

$$\Phi(x, y, z) = (\alpha/2\pi)^{3/2}(1-\varepsilon_0^2)e^{-\alpha(x^2(1-\varepsilon_0^2)+y^2(1-\varepsilon_0^2)+z^2)/2} \tag{10.1.21b}$$

and in the spherical coordinate coordinates:

$$\Phi(r, \theta) = (\alpha/2\pi)^{3/2}(1-\varepsilon_0^2)e^{-\alpha r^2(1-\varepsilon_0^2 \sin^2\theta)/2}, \tag{10.1.21c}$$

where ε_0 is a constant of stabilization of the variable ε. Obviously, when $\varepsilon_0^2 \to 0$ then the equation (10.1.21c) goes to the relevant equation for the non-rotational case of spheroidal body [35–40, 52]. By analogy with this non-rotational case we also can find the mass density for rotating and gravitating spheroidal body in the cylindrical, Cartesian, and spherical coordinate system respectively [43–45, 52, 53]:

$$\rho(h, z) = M(\alpha/2\pi)^{3/2}(1-\varepsilon_0^2)e^{-\alpha(h^2(1-\varepsilon_0^2)+z^2)/2} = \rho_0(1-\varepsilon_0^2)e^{-\alpha(h^2(1-\varepsilon_0^2)+z^2)/2}, \tag{10.1.22a}$$

$$\rho(x, y, z) = \rho_0(1-\varepsilon_0^2)e^{-\alpha(x^2(1-\varepsilon_0^2)+y^2(1-\varepsilon_0^2)+z^2)/2}, \tag{10.1.22b}$$

$$\rho(r,\theta) = \rho_0(1-\varepsilon_0^2)e^{-\alpha r^2(1-\varepsilon_0^2 \sin^2\theta)/2}, \tag{10.1.22c}$$

where $M = m_0 N$ is the mass of rotating and gravitating spheroidal body. Obviously, the iso-surfaces of the mass density (10.1.22a-c) are flattened ellipsoidal ones, and ε_0^2 is a parameter of their flatness (a squared eccentricity of ellipse). Really, if even the squared eccentricity $\varepsilon_0^2 \to 1$ then the equation (10.1.22a) can describe a mass density of a protoplanetary flattened gaseous (gas-dust) disk [52]:

$$\rho(h, z) = \rho_c(h)e^{-\alpha z^2/2}, \tag{10.1.23}$$

where $\rho_c(h) = \lim\limits_{\substack{\varepsilon_0^2 \to 1 \\ M\to\infty}} M(\alpha/2\pi)^{3/2}(1-\varepsilon_0^2)e^{-\alpha h^2(1-\varepsilon_0^2)/2}$ is a value of mass density in a *central* flat of this gaseous disk, M is a mass of star (the Sun) plus mass of gaseous disk, i.e. the total mass of an initial presolar molecular cloud. It is interesting to note that this equation (10.1.23) coincides completely the known *barometric formula* (for a flat rotating disk) obtained with the usage of hydrostatic mechanical equilibrium condition [14 p.769, 24 p.36] or the same formula of mass density distribution in the disk "standard" model derived on the basis of the hydrostatic equilibrium condition jointly with the ideal gas state equation [16 p.19]. Besides, Gurevich and Lebedinsky used designation $\rho_{max}(h)$ instead of $\rho_c(h)$.

Thus, the function of mass density (10.1.22a-c) characterizes a flatness process: from initial spherical forms (for a non-rotational spheroidal body case) through flattened ellipsoidal forms (for a rotating and gravitating spheroidal body) to fuzzy contour disks when the squared eccentricity ε_0^2 varies from 0 till 1.This means that the derived function (76a-c) is appropriate to describe evolution of a protoplanetary gaseous (gas-dust) disk around a star (in particular, the Sun).

As it has been shown in the previous works [35–40, 52, 53], there is a *threshold value* α_g that if $\alpha \ge \alpha_g$ then a weak gravitational field with a gravitational potential φ_g arises in the spheroidal body. According to (10.1.22a-c) we can try to seek a solution φ_g bearing in mind the Poisson equation.

As it well known from the theory of potential [57], the general solution of the Poisson equation

$$\nabla^2\varphi_g = 4\pi\gamma\rho \tag{10.1.24}$$

has the form:

$$\varphi_g(\vec{r}) = -\gamma\int_V \frac{\rho(\vec{r}\,')}{|\vec{r}-\vec{r}\,'|}dV', \tag{10.1.25}$$

where $\rho(\vec{r}')$ is a mass density of a gravitating body, $\vec{r}'$ is a radius vector of volume element of body (distance from the center of mass of the body to a given volume element), $\vec{r}$ is a radius-vector of observations of the gravitational field.

Initially, we apply (10.1.25) to the calculation of potential of a *non-rotating* spheroidal body with the following mass density [35–40, 52]:

$$\rho(\vec{r}) = \rho_0 \cdot e^{-\alpha \vec{r}^2/2}. \tag{10.1.26}$$

where $\rho_0 = M(\alpha/2\pi)^{3/2}$ .To do this, first note (see [57, 58]) that

$$\left|\vec{r} - \vec{r}'\right| = \sqrt{r^2 + r'^2 - 2rr'\cos\psi} \ , \tag{10.1.27b}$$

where

$$\cos\psi = \cos\theta \cdot \cos\theta' + \sin\theta \cdot \sin\theta' \cdot \cos(\varepsilon - \varepsilon') \ . \tag{10.1.27c}$$

Let us note that if $\varepsilon' \to \varepsilon$ then we obtain $\cos\psi = \cos\theta \cdot \cos\theta' + \sin\theta \cdot \sin\theta' = \cos(\theta - \theta')$, i.e. $\psi = \theta - \theta'$.

$$\varphi_g(\vec{r}) = -\gamma \int_0^r \int_0^{\pi} \int_0^{2\pi} \frac{\rho_0 e^{-\alpha r'^2/2} r'^2 \sin\theta \, dr' d\theta d\varepsilon'}{\sqrt{r^2 + r'^2 - 2rr'\cos\psi}} = -\gamma\rho_0 \int_0^r r'^2 e^{-\alpha r'^2/2} \int_0^{\pi} \int_0^{2\pi} \frac{\sin\theta \, d\theta d\varepsilon'}{\sqrt{r^2 + r'^2 - 2rr'\cos\psi}} dr'$$

. (10.1.28)

Let us consider initially the case $r' << r$ by selecting *a spherical volume of radius r'* around the origin of coordinates. To estimate the gravitational potential of the spherical volume:

$$\varphi_g(\vec{r})\Big|_{r >> r'} = \lim_{r'/r \to 0} -\gamma\rho_0 \int_0^{r'} r'^2 e^{-\alpha r'^2/2} \int_0^{\pi} \int_0^{2\pi} \frac{\sin\theta' \, d\theta' d\varepsilon'}{\sqrt{r^2 + r'^2 - 2rr'\cos\psi}} dr' , \tag{10.1.29a}$$

let us compute separately the inner integral in this expression:

$$\lim_{r'/r \to 0} \int_0^{\pi} \int_0^{2\pi} \frac{\sin\theta' \, d\theta' d\varepsilon'}{\sqrt{r^2 + r'^2 - 2rr'\cos\psi}} = \lim_{r'/r \to 0} \frac{1}{r} \int_0^{\pi} \int_0^{2\pi} \frac{\sin\theta' \, d\theta' d\varepsilon'}{\sqrt{1 + \left(\frac{r'}{r}\right)^2 - 2\left(\frac{r'}{r}\right)\cos\psi}} = \lim_{r'/r \to 0} \frac{1}{r} \int_0^{\pi} \int_0^{2\pi} \sin\theta' \, d\theta' d\varepsilon' = \frac{4\pi}{r}$$

(10.1.29b)

Substituting (10.1.29b) in (10.1.29a) we obtain:

$$\varphi_g(\vec{r})\Big|_{r>>r'} = \lim_{r'/r\to 0}\left\{-\gamma\rho_0\cdot\frac{4\pi}{r}\int_0^{r'} r'^2 e^{-\alpha\frac{r'^2}{2}}dr'\right\} = -4\pi\gamma\rho_0 \lim_{r'/r\to 0}\frac{1}{r}\left\{\frac{1}{\alpha}\left[\int_0^{r'} e^{-\alpha\frac{r'^2}{2}}dr' - r'\cdot e^{-\alpha\frac{r'^2}{2}}\Bigg|_0^{r'}\right]\right\} =$$

$$= -\frac{4\pi\gamma\rho_0}{\alpha}\lim_{r'/r\to 0}\left\{\frac{1}{r}\int_0^{r'} e^{-\alpha\frac{r'^2}{2}}dr' - \frac{r'}{r}\cdot e^{-\alpha\frac{r'^2}{2}}\Bigg|_0^{r'}\right\} = -\frac{4\pi\gamma\rho_0}{\alpha}\cdot\frac{1}{r}\int_0^{r'} e^{-\alpha\frac{r'^2}{2}}dr'. \quad (10.1.30)$$

So, if $r >> r'$ then (10.1.30) becomes the known formula for the gravitational potential of a non-rotating spheroidal body [35, 36, 52]. The second case $r' \propto r >> r_*$ leads to the same result if instead of the limit $r'/r \to 0$ in (10.1.30) we shall consider the limit $r' \to r$ under the condition $r >> r_*$.

Thus, the formula (10.1.30) coincides with the expression for the gravitational potential of a non-rotating spheroidal body, besides the magnitude of the potential $\varphi_g(r)$ of spheroidal body is determined by mass of a inner ball with the radius r. Similarly, we attempt to derive the potential of the gravitational field in a remote zone for the case of a uniformly rotating spheroidal body on the basis of the general solution (10.1.25) of the Poisson equation.

More exactly, using the general solution (10.1.25) of the Poisson equation (10.1.24) let us calculate the estimation of the gravitational potential of a *uniformly rotating* spheroidal body under the following conditions [50], [51]:

1. a distance r' from the center of mass of the spheroidal body to a volume element dV in the process of integration in this area V is not greater than the distance r from the center to the point of observation of the gravitational field (the test body in the point M):

$$r' \le r; \quad (10.1.31a)$$

2. the distance r from the center of mass O to the observation point M is much larger than the distance r_* from the center to the point of density inflection (a conditional shell) of spheroidal body or the extremum point of strength of gravitational field:

$r_* << r$ (the condition of the remote zone); (10.1.31b)

3. an evaluation of the gravitational potential φ_g is carried out by accounting the terms of first order relative to small quantity ε_0^2, i.e.

$$\varphi_g = O\left(\varepsilon_0^2\right). \quad (10.1.31c)$$

Now we find the gravitational potential of a uniformly rotating spheroidal body in a remote zone of the gravitational field based on the conditions (10.1.31a-c); to this end let us extract initially some ellipsoidal volume around the origin of coordinates (Figure 10.1).

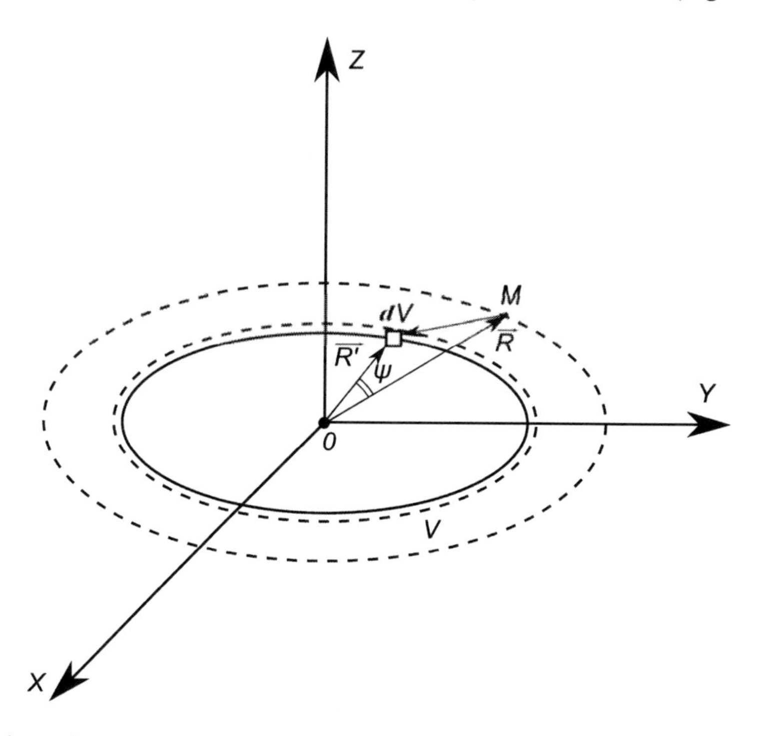

Figure10.1. Scheme for calculating the gravitational potential of a uniformly rotating spheroidal body in a remote zone of the gravitational field.

Then the integration over the volume V should be carried out on the elements dV that are disposed concentrically on oblate ellipsoidal surfaces. These confocal oblate ellipsoidal surfaces are *isosteres* of uniformly rotating spheroidal body with a mass density of the kind (10.1.22b):

$$\rho(x,y,z)=M\left(\frac{\alpha}{2\pi}\right)^{3/2}\left(1-\varepsilon_0{}^2\right)\cdot\exp\left[-\frac{\alpha}{2}\left\{\left(1-\varepsilon_0{}^2\right)x^2+\left(1-\varepsilon_0{}^2\right)y^2+z^2\right\}\right]. \quad (10.1.32a)$$

As a result of such constructions, the radius vector $\vec{R}'$ of a volume element dV of the body is a hodograph whose endpoint moves on a given ellipsoidal surface. Similarly, we assume that the endpoint of the radius vector $\vec{R}$ of the observation point of the field belongs to an ellipsoidal surface limiting the volume V of a rotating spheroidal body. So, $\vec{R}'$ are the radius vectors of points of the ellipsoidal surfaces in a Cartesian coordinate system:

$$\frac{x^2}{2/\left(1-\varepsilon_0{}^2\right)\cdot\alpha}+\frac{y^2}{2/\left(1-\varepsilon_0{}^2\right)\cdot\alpha}+\frac{z^2}{2/\alpha}=1, \quad (10.1.32b)$$

besides $\overrightarrow{R'}$ is a radius vector of a point on the surface of the ellipsoid bounding the volume V. Note that according to the Newton's theorem [57] a homogeneous body, bounded by two similar and similarly situated concentric ellipsoids, has no attraction at points inside the inner cavity V. In view of the fact that the mass density $\rho(x,y,z)$ of a spheroidal body (under carrying out the condition (10.1.31b)) is insignificant and can be considered homogeneous, then in accordance with Newton's theorem an external to the volume V area of spheroidal body does not have practically any influence on the magnitude of potential φ_g that also follows from the derivation of formula (10.1.30) for the case of non-rotating spheroidal body. In a spherical coordinate system

$$\begin{cases} x = r\sin\theta\cos\varepsilon; \\ y = r\sin\theta\sin\varepsilon; \\ z = r\cos\theta \end{cases} \tag{10.1.33a}$$

the argument of the function of mass density (10.1.32a) takes the form:

$$\left(1-\varepsilon_0{}^2\right)\cdot x^2 + \left(1-\varepsilon_0{}^2\right)\cdot y^2 + z^2 = \left(1-\varepsilon_0{}^2\right)\cdot r^2\sin^2\theta + r^2\cdot\cos^2\theta = r^2\cdot(1-\varepsilon_0{}^2\sin^2\theta) \tag{10.1.33b}$$

.

If, however, we shall use the elliptical, more exactly, the spheroidal (in the flattened spherical coordinates) coordinate system

$$\begin{cases} x = \dfrac{1}{\sqrt{1-\varepsilon_0{}^2}} R\sin\Theta\cos E; \\ y = \dfrac{1}{\sqrt{1-\varepsilon_0{}^2}} R\sin\Theta\sin E; \\ z = R\cos\Theta, \end{cases} \tag{10.1.34a}$$

then the mentioned above argument of the function of the mass density (10.1.32a) can be written more simply [50, 51]:

$$\left(1-\varepsilon_0{}^2\right)\cdot x^2 + \left(1-\varepsilon_0{}^2\right)\cdot y^2 + z^2 = R^2\sin^2\Theta + R^2\cos^2\Theta = R^2. \tag{10.1.34b}$$

Comparing (10.1.33b) and (10.1.34b) it is not difficult to see that

$$R^2 = r^2\cdot\left(1-\varepsilon_0{}^2\sin^2\theta\right). \tag{10.1.35}$$

In the flattened spherical coordinates the length of vector $\vec{R}$ according to (10.1.34b) is defined by relation:

$$\left|\vec{R}\right| = \sqrt{\left(1-\varepsilon_0{}^2\right)\cdot x^2 + \left(1-\varepsilon_0{}^2\right)\cdot y^2 + z^2}\ , \tag{10.1.36a}$$

and the distance between two vectors $\vec{R}$ and $\vec{R'}$ is found by the cosine theorem:

$$\left|\vec{R}-\vec{R'}\right| = \sqrt{R^2 + R'^2 - 2RR'\cos\psi} = R\sqrt{1+\left(\frac{R'}{R}\right)^2 - 2\left(\frac{R'}{R}\right)\cos\psi}\ . \quad (6.6.16b)\ (10.1.36b)$$

Moreover, in view of (10.1.34b) the function of mass density (10.1.32a) in the flattened spherical coordinates has a very simple form [50, 51]:

$$\rho(R) = M\left(\frac{\alpha}{2\pi}\right)^{3/2}\left(1-\varepsilon_0{}^2\right)\cdot e^{-\alpha R^2/2} \tag{10.1.37}$$

reminding the mass density function (10.1.26) for a non-rotating spheroidal body (up to a factor $1-\varepsilon_0{}^2$). In this regard, the gravitational potential of a uniformly rotating spheroidal body in the remote zone can be calculated similarly to the mentioned above approach (10.1.28) - (10.1.30) for a non-rotating spheroidal body.

Now let us calculate the Lamé coefficients for the flattened spherical coordinates in the general form (10.1.34a):

$$H_R = \sqrt{\left(\frac{\partial x}{\partial R}\right)^2 + \left(\frac{\partial y}{\partial R}\right)^2 + \left(\frac{\partial z}{\partial R}\right)^2} = \sqrt{\frac{1-\varepsilon_0{}^2\cos^2\Theta}{1-\varepsilon_0{}^2}}\ ; \tag{10.1.38a}$$

$$H_\Theta = \sqrt{\left(\frac{\partial x}{\partial \Theta}\right)^2 + \left(\frac{\partial y}{\partial \Theta}\right)^2 + \left(\frac{\partial z}{\partial \Theta}\right)^2} = R\sqrt{\frac{1-\varepsilon_0{}^2\sin^2\Theta}{1-\varepsilon_0{}^2}}\ ; \tag{10.1.38b}$$

$$H_E = \sqrt{\left(\frac{\partial x}{\partial E}\right)^2 + \left(\frac{\partial y}{\partial E}\right)^2 + \left(\frac{\partial z}{\partial E}\right)^2} = \frac{R\sin\Theta}{\sqrt{1-\varepsilon_0{}^2}}\ . \tag{10.1.38c}$$

Using (10.1.38a-c) let us find the volume element dV in the flattened spherical coordinates [50, 51]:

$$dV = H_R H_\Theta H_E dR d\Theta dE = \frac{R^2 \sin\Theta}{(1-\varepsilon_0{}^2)^{3/2}} \cdot \sqrt{(1-\varepsilon_0{}^2 \sin^2\Theta)\cdot(1-\varepsilon_0{}^2\cos^2\Theta)} dR d\Theta dE$$

(10.1.39)

Finally, in view of formulas (10.1.36b), (10.1.37), (10.1.39) and conditions (10.1.31a) - (10.1.31c) let us calculate an estimation of the gravitational potential (10.1.25) in the remote zone of a uniformly rotating spheroidal body in the flattened spherical coordinates [50, 51]:

$$\varphi_g(\vec{R})\Big|_{R>>R*} = -\gamma \int_V \frac{\rho(R')dV'}{R\sqrt{1+(R'/R)^2-2(R'/R)\cos\psi}} =$$

$$= -\frac{\gamma M(\alpha/2\pi)^{3/2}\left(1-\varepsilon_0{}^2\right)}{\left(1-\varepsilon_0{}^2\right)^{3/2}} \int_0^R\int_0^\pi\int_0^{2\pi} \frac{R'^2 e^{-\alpha R'^2/2}}{R\sqrt{1+(R'/R)^2-2(R'/R)\cos\psi}} \cdot \sin\Theta' \sqrt{\left(1-\varepsilon_0{}^2\sin^2\Theta'\right)\left(1-\varepsilon_0{}^2\cos^2\Theta'\right)} dR' d\Theta' dE' =$$

$$= -\frac{\gamma M(\alpha/2\pi)^{3/2}}{\sqrt{1-\varepsilon_0{}^2}} \int_0^R \frac{R'^2 e^{-\alpha R'^2/2} dR'}{R\sqrt{1+(R'/R)^2-2(R'/R)\cos\psi}} \cdot 2\pi \int_0^\pi \sqrt{\left(1-\varepsilon_0{}^2\sin^2\Theta\right)\left(1-\varepsilon_0{}^2\cos^2\Theta\right)} \sin\Theta\, d\Theta$$

(10.1.40)

Let us calculate separately the integrals entering in (10.1.40). Similarly to the case of a non-rotating spheroidal body, i.e. according to derivations (10.1.29a, b), (10.1.30), to calculate the integral by R' first we *select an ellipsoidal volume* with the radius vector $\vec{R'}$ around the origin of coordinates (see Figure 10.1) and then apply the limiting condition for the denominator:

$$\lim_{R'/R\to 0} \int_0^{R'} \frac{R'^2 e^{-\alpha R'^2/2} dR'}{R\sqrt{1+(R'/R)^2-2(R'/R)\cos\psi}} = \lim_{R'/R\to 0} \frac{1}{R}\int_0^{R'} R'^2 e^{-\alpha R'^2/2} dR' =$$

$$= \lim_{R'/R\to 0} \frac{1}{\alpha}\left\{\frac{1}{R}\int_0^{R'} e^{-\alpha R'^2/2} dR' - \left(\frac{R'}{R}\right)\cdot e^{-\alpha R'^2/2}\Big|_0^{R'}\right\} = \frac{1}{\alpha R}\int_0^{R'} e^{-\alpha R'^2/2} dR' \quad (10.1.41)$$

Since the relation (10.1.41) holds for all concentric ellipsoidal volumes with the lenght of the radius vector $\left|\vec{R'}\right| << \left|\vec{R}\right|$, then according to (10.1.31a) it is also suitable for the considered ellipsoidal area of volume V when $\left|\vec{R'}\right| \propto \left|\vec{R}\right|$, i.e. $\left|\vec{R'}\right| < \left|\vec{R}\right|$ and even $\left|\vec{R'}\right| = \left|\vec{R}\right|$. Indeed, at $R' = R$ the function $e^{-\alpha R^2/2} \to 0$ because $R >> R_*$ according to the condition of remote zone (10.1.31b). Thus, starting from (10.1.31a) and (10.1.31b) the required integral in (10.1.40) is equal [50, 51]:

$$\lim_{\substack{R_*/R\to 0\\ R'<R}} \int_0^R \frac{R'^2 e^{-\alpha R'^2/2} dR'}{R\sqrt{1+(R'/R)^2-2(R'/R)\cos\psi}} = \frac{1}{\alpha R}\int_0^R e^{-\alpha R'^2/2} dR' . \quad (10.1.42)$$

To calculate the second integral by Θ', entering into (10.1.40), we use the substitution $s=\varepsilon_0\cos\Theta'$:

$$\int_0^\pi \sqrt{(1-\varepsilon_0^2\sin^2\Theta')\cdot(1-\varepsilon_0^2\cos^2\Theta')}\sin\Theta'\, d\Theta' = \frac{1}{\varepsilon_0}\int_{-\varepsilon_0}^{\varepsilon_0}\sqrt{[(1-\varepsilon_0^2)+s^2]\cdot[1-s^2]}ds = \frac{2}{\varepsilon_0}\int_0^{\varepsilon_0}\sqrt{[(1-\varepsilon_0^2)+s^2]\cdot[1-s^2]}ds$$

(10.1.43)

The integral (10.1.43) can be expressed in terms of elliptic integrals of 1st and 2nd kind [59 p.262]. However, taking into account the third condition (10.1.31c) the desired integral by Θ' (10.1.43) can be calculated much easier [50, 51]:

$$\int_0^\pi \sqrt{\left(1-\varepsilon_0^2\sin^2\Theta'\right)\cdot\left(1-\varepsilon_0^2\cos^2\Theta'\right)}\sin\Theta'\, d\Theta' \approx \int_0^\pi \sqrt{1-\varepsilon_0^2\cdot\left(\sin^2\Theta'+\cos^2\Theta'\right)}\sin\Theta'\, d\Theta' = 2\sqrt{1-\varepsilon_0^2}$$

(10.1.44)

Obviously, the terms of fourth order of smallness $O(\varepsilon_0^4)$ are neglected under the derivation of (10.1.44).

Thus, in view of the integrals (10.1.42) and (10.1.44) calculated under the conditions (10.1.31a) - (10.1.31c), the estimation of the gravitational potential in the remote zone (10.1.40) for an uniformly rotating spheroidal body is equal [50, 51]:

$$\varphi_g(R)\Big|_{R>>R_*} = -\frac{\gamma M\alpha^{3/2}}{\sqrt{2\pi(1-\varepsilon_0^2)}}\cdot 2\sqrt{1-\varepsilon_0^2}\cdot\frac{1}{\alpha R}\int_0^R e^{-\alpha R'^2/2}dR' = -\sqrt{\frac{2\alpha}{\pi}}\cdot\gamma M\cdot\frac{1}{R}\int_0^R e^{-\alpha R'^2/2}dR'$$

. (10.1.45)

Under derivation (10.1.30) and (10.1.45), an idea about maximal exclusion of dependences on spatial coordinates on equipotential surfaces has been exploited (in particular, on the spheres in the first case and on the flattened ellipsoids in the second case). If we use an *eigen coordinate system* (for example, spherical or ellipsoidal) then canonical coordinate is to be r for a sphere or R for an oblate ellipsoid (instead of r and θ in accordance with (10.1.35)). In this connection, using (10.1.35) we can estimate the gravitational potential in a remote zone of uniformly rotating spheroidal body in the *spherical coordinates* [50, 51]:

$$\varphi_g(r,\theta)\Big|_{r>>r_*} = -\sqrt{\frac{2\alpha}{\pi}}\cdot\gamma M\cdot\frac{1}{r\sqrt{1-\varepsilon_0^2\sin^2\theta}}\cdot\int_0^{r\sqrt{1-\varepsilon_0^2\sin^2\theta}} e^{-\alpha r'^2/2}dr' . \quad (10.1.46)$$

Now we can verify the accuracy of the derived estimation (10.1.46) by substituting (10.1.46) into the left side and (10.1.22c) into the right side of the Poisson equation (10.1.24). As a result, the absolute error of estimation of the Laplacian of the gravitational potential (10.1.46) of uniformly rotating spheroidal body is expressed by the following relation:

$$\Delta^{abs}(\theta,\varepsilon_0,\rho)=\frac{\varepsilon_0^2\cos^2\theta}{1-\varepsilon_0^2\sin^2\theta}\cdot\frac{4\pi\gamma\rho(r,\theta)}{1-\varepsilon_0^2}. \qquad (10.1.47a)$$

The relative error in estimating the Laplacian of the gravitational potential (10.1.46) is given by

$$\Delta^{rel}(\theta,\varepsilon_0)=\frac{\varepsilon_0^2}{1-\varepsilon_0^2}\cdot\frac{\cos^2\theta}{1-\varepsilon_0^2\sin^2\theta}. \qquad (10.1.47b)$$

Obviously, absolute and relative errors depend on the angular coordinate θ and the value of oblateness ε_0^2. In particular, if $\varepsilon_0^2\to 0$ then there are no errors: $\Delta^{abs}(\theta,\varepsilon_0,\rho)\to 0$, $\Delta^{rel}(\theta,\varepsilon_0)\to 0$.

Thus, for the considering case of a *weakly flattened* spheroidal body ($\varepsilon_0^2 << 1$) the obtained formula (10.1.46) of estimation of the gravitational potential in the remote zone is exact enough since the maximum relative error in calculating the Laplacian of the gravitational potential tends to zero: $\Delta_{\max}^{rel}(0,\varepsilon_0)=\varepsilon_0^2/(1-\varepsilon_0^2)<<1$.

2 A Statistical Model of Evolution for Rotating and Gravitating Spheroidal Bodies and Its Application to the Problem of Planets Forming and of Distribution of Planetary Distances in the Solar System

The obtained results in the previous sections of this chapter permit us to pass the next stage of evolution from a protoplanetary flattened gaseous (gas-dust) disk to originating protoplanets. Let us note the formulas (10.1.22a-c), (10.1.23) point to a *common scenario* of formation both a star and a protoplanetary gas-dust disk around it (in particular, the Sun and the solar protoplanetary gas-dust disk) because M is considered as a total mass of star (the Sun) and mass of gas-dust disk [52, 53]. Really, it has been mentioned in the work [60]: “Measurements of the composition of the Earth, Moon, and meteorites support a common origin for the Sun and planets (e.g., Harris 1976; Anders & Grevesse 1989)”. In this connection let us consider a statistical model of origin of protoplanets embedded in a flattened gaseous (gas-dust) protoplanetary disk based on specific angular momentum value distribution function.

A distribution function of a specific angular momentum value for an uniform rotating spheroidal body are obtained simply using the derived relation for the probability volume density in the cylindrical coordinate system. First of all, let us use the probability volume density (10.1.21a) for particles being at distances close to h from the axis of rotation Oz in a rotating spheroidal body in *the relative mechanical equilibrium state* (Figure 10.2). The probability of value h being belonged an interval between h and $h+dh$ for a considered particle is equal to:

$$dp_h = \frac{dN_h}{N} = f(h)dh, \quad h \in [h, h+dh], \tag{78}$$

where dN_h / N is a share of particles being at distances (from the axis of rotation Oz) close to h, and $f(h)$ is an one-dimensional probability density to locate a particle at distance h from the axis of rotation.

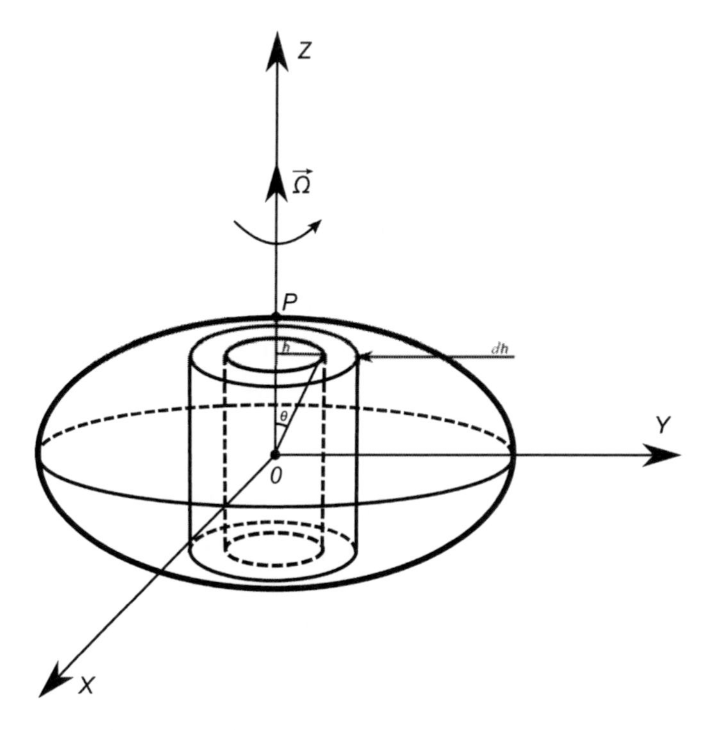

Figure 10.2. Scheme for calculating a share of particles having the same specific angular momentum value in uniform rotating spheroidal body.

On the other hand, according to (10.1.21a) a probability dp_h can be calculated by means of integrating a probability volume density function $\Phi(h, z)$ with respect to all coordinates of cylindrical system z and ε (with the exception of the coordinate h) [52]:

$$dp_h = \int_{-\infty}^{\infty}\int_{0}^{2\pi} \Phi(h, z) h dh dz d\varepsilon = (\alpha / 2\pi)^{3/2}(1-\varepsilon_0^2)\cdot 2\pi h dh \cdot \int_{-\infty}^{\infty} e^{-\alpha(h^2(1-\varepsilon_0^2)+z^2)/2} dz =$$

$$= \frac{\alpha^{3/2}}{\sqrt{2\pi}}(1-\varepsilon_0^2)\cdot \int_{-\infty}^{\infty} e^{-\alpha z^2/2} dz \cdot e^{-\alpha(1-\varepsilon_0^2)h^2/2} h dh = \alpha(1-\varepsilon_0^2) e^{-\alpha(1-\varepsilon_0^2)h^2/2} h dh \,. \quad (10.2.2)$$

By comparing eq. (10.2.1) with eq. (10.2.2) we obtain the share of particles being at distances close to h from the axis of rotation z is

$$dN_h / N = \alpha(1-\varepsilon_0^2) e^{-\alpha(1-\varepsilon_0^2)h^2/2} h dh . \quad (10.2.3)$$

Then we state [52] that the share of particles being at distances close to h from the axis Oz of rotation, i.e. into a volume of an annular cylindrical layer $[h, h+dh]$ at distance h from the axis z, is equal to the share of particles rotating with a constant angular velocity $\vec{\Omega}$ about the axis z and having a specific angular momentum value in an interval $[\lambda, \ \lambda + d\lambda]$:

$$dN_h / N = dN_\lambda / N \,, h \in [h, h+dh], \lambda \in [\lambda, \ \lambda + d\lambda]. \quad (10.2.4)$$

Evidently, the relation (10.2.4) is valid for a rotating spheroidal body in a relative mechanical equilibrium state. Under this condition the rotating particles being at distances h from the axis of rotation z have circular orbits into an uniform rotating spheroidal body and, therefore, the value of z-projection of angular momentum acting on a particle is equal to [61]:

$$L_{0z} = m_0 h^2 \Omega \ . \quad (10.2.5)$$

As follows from eq. (10.2.5) the value of specific angular momentum

$$\lambda = L_{0z} / m_0 = \Omega \, h^2 \quad (10.2.6)$$

is directly proportional to the square distance from the axis of rotation z into an uniform rotating spheroidal body.

Consequently, a share of particles having a specific angular momentum value into an interval $[\lambda, \ \lambda + d\lambda]$ in accordance with eqs. (10.2.3), (10.2.4), (10.2.6) is equal to:

$$dN_\lambda / N = dN_h / N = (\alpha(1-\varepsilon_0^2)/2\Omega) e^{-\alpha(1-\varepsilon_0^2)\Omega h^2/2\Omega} d(\Omega h^2) = (\alpha(1-\varepsilon_0^2)/2\Omega) e^{-\alpha(1-\varepsilon_0^2)\lambda/2\Omega} d\lambda \ , \quad (10.2.7)$$

where λ is a value of specific angular momentum [52]. Analogously, a probability for particles having of values of specific angular momentum belonging to an interval $[\lambda,\ \lambda + d\lambda]$ is equal to:

$$f(\lambda)d\lambda = dp_\lambda = dN_\lambda / N = (\alpha(1-\varepsilon_0^2)/2\Omega)e^{-\alpha(1-\varepsilon_0^2)\lambda/2\Omega}d\lambda, \tag{10.2.8}$$

i.e. a desired probability density function $f(\lambda)$ [52] expressing a mass distribution by a value of specific angular momentum is

$$f(\lambda) = (\alpha(1-\varepsilon_0^2)/2\Omega)e^{-\alpha(1-\varepsilon_0^2)\lambda/2\Omega}. \tag{10.2.9}$$

The probability density function (10.2.9) satisfies the normalization condition $\int_0^\infty f(\lambda)d\lambda = 1$ because $\int f(\lambda)d\lambda = -e^{-\alpha(1-\varepsilon_0^2)\lambda/2\Omega}$. Let us calculate a mean value of specific angular momentum by means of integration by the parts with usage of eq. (10.2.9):

$$\bar{\lambda} = \int_0^\infty \lambda\, f(\lambda)d\lambda = \lambda \int f(\lambda)d\lambda \Big|_0^\infty - \int_0^\infty d\lambda \int f(\lambda)d\lambda = 2\Omega / \alpha(1-\varepsilon_0^2). \tag{10.2.10}$$

According to (10.2.8) the quantity of particles having values of specific angular momentum close to λ is equal to:

$$dN_\lambda = N(\alpha(1-\varepsilon_0^2)/2\Omega)e^{-\alpha(1-\varepsilon_0^2)\lambda/2\Omega}d\lambda. \tag{10.2.11}$$

Hence it is not difficult to find the full angular momentum for uniformly rotating spheroidal body being in a relative mechanical equilibrium [52]:

$$L = \int_0^N m_0 \lambda\, dN_\lambda = m_0 N \int_0^\infty \lambda\, f(\lambda)d\lambda = M\bar{\lambda}\ , \tag{10.2.12}$$

where M is a mass of spheroidal body, i.e. the total mass of star (the Sun) and protoplanetary gas-dust disk. Substuting eq. (10.2.10) in eq. (10.2.12) we obtain that value of the full angular momentum (for an uniformly rotating spheroidal body) is expressed by the formula [52]:

$$L = 2\Omega\, M / \alpha(1-\varepsilon_0^2). \tag{10.2.13}$$

Now let us use the cosmological theory of Schmidt [4] for developing a statistical model of evolution of a rotating and gravitating spheroidal body on the stage of evolution from a protoplanetary flattened gas-dust disk to forming protoplanets. This theory starts from that fact that during the process of origin of a protoplanet each particle (or generally speaking, planetesimal) in a gas-dust protoplanetary cloud (in a swarm of planetesimals) has a chance to get on the protoplanet whose specific angular momentum value is the same as one for the particle/planetesimal (or it differs less than all values for other protoplanets).

We consider a flattened gas-dust protoplanetary cloud as a rotating and gravitating spheroidal body with the specific angular momentum distribution function $f(\lambda)$. According to the Schmidt's hypothesis [3] let us suppose that a domain boundary is determined by the value of specific angular momentum whose magnitude is equidistant from values of specific angular momentums for two neighboring bunches (or planetary embryos).

Let μ_n be a value of specific angular momentum corresponding to the boundary between domains of n -th and $(n+1)$ -th protoplanets (or bunches in a flattened gas-dust protoplanetary cloud) with values of the specific angular momentum λ_n and λ_{n+1} respectively. According to the mentioned above we have

$$\mu_n = (\lambda_n + \lambda_{n+1})/2\,. \tag{10.2.14}$$

As a result of the conglomeration process, the specific angular momentums are averaged, therefore, the specific angular momentum for a forming protoplanet is equal to the ratio [4]:

$$\lambda_n = \left(\int_{\mu_{n-1}}^{\mu_n} \lambda\, f(\lambda) d\lambda\right) / \left(\int_{\mu_{n-1}}^{\mu_n} f(\lambda) d\lambda\right). \tag{10.2.15}$$

Taking into account that the inverse parameter $\bar{\lambda}^{-1}$ for the specific angular momentum of a spheroidal body is very small:

$$\bar{\lambda}^{-1} = \alpha(1-\varepsilon_0^2)/2\Omega \ll 1\;, \tag{10.2.16}$$

if the parameter of gravitational compression $\alpha \ll 1$ [35–49], we can represent the function of specific angular momentum $f(\lambda)$ by Maclaurin's series [52]:

$$f(\lambda) = (\alpha(1-\varepsilon_0^2)/2\Omega)\{1-(\alpha(1-\varepsilon_0^2)/2\Omega)\lambda + [(\alpha(1-\varepsilon_0^2)/2\Omega)\lambda]^2/2 - \ldots\}\,. \tag{10.2.17}$$

Let us limit the series (10.2.17) by the zero-th term, i.e. we suppose that

$$f(\lambda) \approx \alpha(1-\varepsilon_0^2)/2\Omega = const\,. \tag{10.2.18}$$

Taking into account eq. (10.2.18) the formula (10.2.15) becomes

$$\lambda_n = \left(\int_{\mu_{n-1}}^{\mu_n} \lambda \, (\alpha(1-\varepsilon_0^2)/2\Omega) d\lambda \right) / \left(\int_{\mu_{n-1}}^{\mu_n} (\alpha(1-\varepsilon_0^2)/2\Omega) d\lambda \right) = (\mu_n + \mu_{n-1})/2 . \quad (10.2.19)$$

Substituting eq. (10.2.14) in eq. (10.2.19) we obtain the difference equation:

$$\lambda_n = (\lambda_{n+1} + \lambda_{n-1})/2 \quad . \quad (10.2.20)$$

It is clear that eq. (10.2.20) describes the well-known property of an arithmetic progression whose n-th term is calculated by the formula:

$$\lambda_n = a_0 + d \cdot n , \quad (10.2.21)$$

where d is the difference and a_0 is the first (the zero-th) term of arithmetic progression.

Taking into account eq. (10.2.21) and the formula for relation of the specific angular momentum λ_n with the square root of radius R_n for orbit for n-th protoplanet (under condition of circular character of planetary orbit):

$$\lambda_n = \sqrt{\gamma M} \sqrt{R_n} \quad , \quad (10.2.22)$$

we conclude that zero-th approximation of $f(\lambda)$ leads to the well-known Schmidt's law:

$$\sqrt{R_n} = a + b \cdot n , \quad (10.2.23)$$

where a, b are some constants. In 1944 O. Schmidt derived his law for planetary distances. However, for the defined constants a and b this law (10.2.23) does not permit to estimate correctly planetary distances for all planets of the Solar system. In connection with this O. Schmidt [4] proposed to use his law (10.2.23) in the combination: separately for the planets of the Earth group and separately for the planets of the Jupiter group. Moreover, the distribution function of a specific angular momentum of a gas-dust cloud was not derived analytically within framework of the Schmidt's model.

In our view, the cause of the mentioned lack of the Schmidt's law consists in too simplified distribution function as *uniform* one for a specific angular momentum. To modify the evolution model for planet forming let us use the linear approximation of the function $f(\lambda)$ taking into account both zero-th and first terms [52, 53]:

$$f(\lambda) = \frac{\alpha(1-\varepsilon_0^2)}{2\Omega} \cdot \left[1 - \frac{\alpha(1-\varepsilon_0^2)}{2\Omega} \cdot \lambda \right] . \quad (10.2.24)$$

Bearing in mind eq. (10.2.24) we can obtain from the formula (10.2.15) the following:

$$\lambda_n = \left((\mu_n + \mu_{n-1})/2 - (\alpha(1-\varepsilon_0^2)/6\Omega)(\mu_n^2 + \mu_n\mu_{n-1} + \mu_{n-1}^2)\right)/\left(1 - (\alpha(1-\varepsilon_0^2)/4\Omega)(\mu_n + \mu_{n-1})\right) \quad (10.2.25)$$

and then with usage of eq. (10.2.10) and eq. (10.2.14) we have finally the difference equation:

$$\lambda_n = \left(\lambda_{n+1} + 2\lambda_n + \lambda_{n-1} - (\bar{\lambda}^{-1}/3)[(\lambda_{n+1} + 2\lambda_n + \lambda_{n-1})^2 - (\lambda_{n+1} + \lambda_n)(\lambda_n + \lambda_{n-1})]\right)/\left(4 - \bar{\lambda}^{-1}(\lambda_{n+1} + 2\lambda_n + \lambda_{n-1})\right) \quad (10.2.26)$$

To find its solution let us carry out the following substitution:

$$\lambda_n = Z^n, \quad n = 1,\ 2, 3,\ \dots . \quad (10.2.27)$$

As a result of this we obtain from eq. (10.2.26) and eq. (10.2.27) the following characteristic equation [52, 53]:

$$(Z-1)^2 \cdot \left[(\bar{\lambda}^{-1}/3)Z^{n-1}(Z+1)^2 - 1\right] = 0, \quad (10.2.28)$$

which is reduced to two equations:

$$(Z-1)^2 = 0 \quad (10.2.29a)$$

and

$$Z^{n-1}(Z+1)^2 = 3\bar{\lambda}. \quad (10.2.29b)$$

Taking into account that $\bar{\lambda} \to \infty$ at $\alpha \to 0$ in the right part of eq. (10.2.29b), i.e. in the left part $Z^{n-1}(Z+1)^2 \to Z^{n+1}$ at $\alpha \to 0$, the characteristic equation (10.2.29b) can be simplified to the form [52, 53]:

$$Z^{n+1} = 3\bar{\lambda}. \quad (10.2.30)$$

The equation (10.2.29a) has one root with double multiplicity:

$$Z_1^{(1)} = 1, \quad d_1 = 2 \quad (10.2.31a)$$

and the equation (10.2.30) has $n+1$ roots:

$$Z_k^{(2)} = \sqrt[n+1]{3\bar{\lambda}}\, e^{-i2\pi k/(n+1)}, \quad k = 0, \dots, n \quad (\bar{\lambda} >> 1) \quad (10.2.31b)$$

According to the roots (10.2.31a) and (10.2.31b) the general solution of the difference equation (10.2.26) has the form [52, 53]:

$$\lambda_n = (A + B \cdot n)[Z_1^{(1)}]^n + \sum_{k=0}^{n} (A_k + iB_k)[Z_k^{(2)}]^n =$$

$$= A + B \cdot n + (3\overline{\lambda})^{n/(n+1)} \left[A_0 + 2 \sum_{k=1}^{ent[n/2]} A_k \cos(2\pi nk/(n+1) \right], \qquad (10.2.32)$$

where it has been taken into consideration that $A_k = A_{n-k}$, $B_k = -B_{n-k}$ under derivation of eq. (10.2.32). The formula (10.2.32) points to a quantization of the specific angular momentum λ_n as well as a possibility to represent the specific angular momentum as a wave packet: $\lambda_n = \sum_{k=0}^{n} C_k \exp\{i2\pi nk/(n+1)\}$. The similar conclusion has been proposed within framework of the mentioned above quantum mechanical approach [19–34]. The use of such approach is also suggested by the chaotic behavior of the Solar system during its formation and evolution [32, 33], which implies the non-differentiability of trajectories for large time scales [62].

Taking into account eq. (10.2.22) the solution (10.2.32) permits us to obtain a new law for square root of planetary distances in the form [52, 53]:

$$\sqrt{R_n} = a + b \cdot n + (3\overline{\lambda})^{n/(n+1)} \left[a_0 + 2 \sum_{k=1}^{ent[n/2]} a_k \cos(2\pi nk/(n+1) \right], \qquad (10.2.33)$$

where $a, b, a_0, ..., a_{[n/2]}$ are coefficients to be sought for a planetary system (the Solar system) in dependence on $\overline{\lambda}$. Obviously, the proposed law for planetary distances (10.2.33) generalizes the well-known Schmidt's law (10.2.23). Let us note the quantization of specific angular momentum λ_n leads to quantization of planetary orbits in accord with the quantum mechanical approach [19–34, 62].

Let us consider an evolution of our Solar system based on a nebular origin in accord with the theory of Hoyle [6, 7, 11]. Undoubtedly, the nebular origin of the Solar system permits us to describe the proto-Sun together with a flattened gas-dust protoplanetary cloud as a model of a rotating and gravitating spheroidal body. According to the Hoyle's theory [6, 11] we consider an angular momentum value of the presolar nebula (cloud) being equal to $L = =$ 4□1044 (kg m2 /s). This value has been obtained by Hoyle bearing in mind that a primary presolar cloud has a mass density 1 atom per sm3 and angular velocity $\Omega \sim$ 10-15 s-1 as for the Galaxy in the whole [10, 11]. It is well-known that mass of the presolar cloud is approximately $M = 1.988$□1030 (kg). Then it is not difficult to see that the mean value of specific angular momentum for the forming solar system is $\overline{\lambda} = L/M = 2.012$□10 14 (m2/s).

Let us apply the proposed law (10.2.33) to the planetary distances estimation in the forming Solar system [52, 53]. Supposing the case $n = 0$ corresponds to the planet of the Mercury, $n = 1$ conforms to the Venus, $n = 2$ corresponds to the Earth, ..., $n =$ 8 corresponds to the Pluto let us calculate the value $(3\bar{\lambda})^{n/(n+1)}$ in (10.2.33) for different values of n (see Table 10.1).

The proposed law for planetary distances (10.2.33) gives us the calculated formulas to estimate the square root of distances for all planets of the Solar system [52]:

$$\sqrt{R_0} = a + a_0;$$

$$\sqrt{R_1} = a + b + (3\bar{\lambda})^{1/2}(a_0 - 2a_1);$$

$$\sqrt{R_2} = a + 2b + (3\bar{\lambda})^{2/3}(a_0 - a_1);$$

$$\sqrt{R_3} = a + 3b + (3\bar{\lambda})^{3/4} a_0;$$

$$\sqrt{R_4} = a + 4b + (3\bar{\lambda})^{4/5}(a_0 - 2a_1 \cos\frac{3\pi}{5}); \qquad (10.2.34)$$

$$\sqrt{R_5} = a + 5b + (3\bar{\lambda})^{5/6}(a_0 + a_1 - a_2);$$

$$\sqrt{R_6} = a + 6b + (3\bar{\lambda})^{6/7}[a_0 - 2(a_1 \cos\frac{5\pi}{7} + a_2 \cos\frac{3\pi}{7} + a_3 \cos\frac{\pi}{7})];$$

$$\sqrt{R_7} = a + 7b + (3\bar{\lambda})^{7/8}[a_0 + \sqrt{2}(a_1 - a_3) - 2a_4];$$

$$\sqrt{R_8} = a + 8b + (3\bar{\lambda})^{8/9}[a_0 - 2(a_1 \cos\frac{7\pi}{9} + a_2 \cos\frac{5\pi}{9} + \frac{1}{2}a_3 + a_4 \cos\frac{\pi}{9})],$$

where the coefficients are equal to a = 0.622173; b = 0.228316; a_0 = - 0.597348□10^{-12} ; a_1 = 2.773092□10^{-12};

a_2 = 1. 534506□10^{-12}; a_3 = 0.925297□10^{-12}; a_4 = 0.816924□10^{-12} (see Table 10.1).

Taking advantage of coefficient values and substituting their in eqs. (10.2.34) we can find distance square root estimations for planets of the Solar system. A preliminary calculations of $\sqrt{R_n}, n = 0,1,...,8$ in accord with eqs. (10.2.34) and Table 10.1 give us the good agreement with the observable data (see Table 10.2) excepting the case $n = 8$ for the planet of Pluto. To decrease the error of estimation $\sqrt{R_8}$ we can introduce an *additional* coefficient a_5 in the formulas (10.2.34).

In other words, instead of the previous formula $\sqrt{R_8}$ for the Pluto in (10.2.34) we can use the following [52]:

$$\sqrt{R_8} = a + 8b + (3\bar{\lambda})^{8/9}[a_0 - 2(a_1 \cos\frac{7\pi}{9} + a_2 \cos\frac{5\pi}{9} + \frac{1}{2}a_3 + (a_4 + a_5)\cos\frac{\pi}{9})], \quad (10.2.35)$$

where a_5 = 0.7679326□10^{-12}. However, such decision violates the proposed law (10.2.33) because the necessary number of coefficients is equal to $ent[n/2] = 4$ only. In this connection there is no any reason to include the data for the Pluto (10.2.35) in Table 10.2.

Table 10.2 presents the values of square root of planetary distances calculated in accord with the proposed law. In Table 10.3, a comparative analysis for different laws is presented [4, 11, 16]. As it follows from the Table 10.3 the proposed law gives the best results in prediction of planetary distances for the Solar system.

Really, for all planets of the Solar system excepting Earth and Pluto the absolute estimation error and, naturally, relative one are equal to 0% (for the Earth, the absolute estimation error is 11% and relative one is 10%). Thus, the mean error of estimation of planetary distances in the Solar system is equal to 1.4% in accord with the proposed law. For comparison, the Table 10.3 presents also both the well-known Titius–Bode law and its in modification of Wurm [11]:

$$R_n = a + b \cdot 2^n, n = -\infty, 0, 1, 2, ..., 7, \quad (10.2.36)$$

where $a = 0.387$ and $b = 0.293$. However, the formula (10.2.36) indicates additional planets between Mercury and Venus. On this question about "the lost planet" it can be found the answer in the form of the proposition of J.Bailey [63]: "The Moon may be a former planet". According to this proposition the Moon had instable orbit, and then it has been captured by the Earth [11]. However, the modern point of view considers the Moon's origin as a result of a collision of a former planet with the proto-Earth [64–69]. Moreover, the Titius–Bode law predicts existing else a "trans-Martian" planet called Cerera (between the Mars and the Jupiter [11]). However, such planet does not exist in reality although the set of asteroids occurs there.

The empirical formula of Blagg [70]

$$R_n = A \cdot (1.7275)^n [B + f(\alpha + \beta \cdot n)], n = -2, -1, 0, 1, 2, ..., 6, \quad (10.2.37)$$

(where A and B are constants, f is a periodic function, α and β are constant angles) as well as the empirical formula of Richardson [71]

$$R_n = A \cdot (1.728)^n F_n(\theta_n), n = 1, 2, ..., 9, \quad (10.2.38)$$

(where $\theta_n = (4\pi/13)n$ and F_n is a periodic function) also predict existing the so-called "trans-Martian" planet Cerera, i.e. a "gap" between the Mars and the Jupiter because the Cerera is absent in reality. Although the formulas of Blagg and Richardson are pure heuristic and having no a theoretical base, nevertheless, they revealed *the presence of a periodic function* in the planetary distances laws (10.2.37) and (10.2.38). This fact is confirmed by the proposed theoretical law (10.2.33) completely [52, 53].

Table 10.1. The values of coefficients and parameters estimated in accord with the proposed law

Solar system planets	n	$(3\bar{\lambda})^{n/(n+1)}$	$ent[n/2]$	$a, b, a_0, \ldots, a_{[n/2]}$
Mercury	0	1.0	0	a=0.622173;a_0=0.597348·10^{-12}
Venus	1	2.456827·10^{7}	1	a=0.622173;b=0.228316; a_0=-0.597348·10^{-12};a_1=2.773092·10^{-12}
Earth	2	7.142213·10^{9}	1	a=0.622173;b=0.228316; a_0=-0.597348·10^{-12};a_1=2.773092·10^{-12}
Mars	3	1.21776·10^{11}	2	a=0.622173;b=0.228316; a_0=-0.597348·10^{-12}
Jupiter	4	6.677277·10^{11}	2	a=0.622173;b=0.228316; a_0=-0.597348·10^{-12};a_1=2.773092·10^{-12}
Saturn	5	2.076304·10^{12}	3	a=0.622173;b=0.228316; a_0=-0.597348·10^{-12};a_1=2.773092·10^{-12} a_2=1.534506·10^{-12}
Uranus	6	4.668889·10^{12}	3	a=0.622173;b=0.228316; a_0=-0.597348·10^{-12};a_1=2.773092·10^{-12} a_2=1.534506·10^{-12};a_3=0.925297·10^{-12}
Neptune	7	8.573449·10^{12}	4	a=0.622173;b=0.228316; a_0=-0.597348·10^{-12};a_1=2.773092·10^{-12} a_2=1.534506·10^{-12};a_3=0.925297·10^{-12} a_4=0.816924·10^{-12}
Pluto	8	1.375447·10^{13}	4	a=0.622173; b=0.228316; a_0=-0.597348·10^{-12};a_1=2.773092·10^{-12} a_2=1.534506·10^{-12};a_3=0.925297·10^{-12} a_4=0.816924·10^{-12}

Table 10.2. The values of square root of the solar system planetary distances calculated in accord with the proposed law

Solar systemplanets	n	$ent[n/2]$	$\sqrt{R_n}(AU)$	Error, δ (%)
Mercury	0	0	0.622173	0%
Venus	1	1	0.850338	0%
Earth	2	1	1.054732	5%
Mars	3	2	1.234377	0%
Jupiter	4	2	2.280965	0%
Saturn	5	3	3.095158	0%
Uranus	6	3	4.375043	0%
Neptune	7	4	5.495270	0%

Table 10.3. The comparative analysis for different laws of prediction of planetary distances (AU) for the Solar system

Solar system planets	Observable distances	Titius–Bode law	Wurm's modification of Titius– Bode law	Empirical formulas		Schmidt law	Proposed law
				Blagg	Richardson		
Mercury	0.3871	0.4	0.387	0.387	0.3869	0.3844	0.3871
Venus	0.7233	0.7	0.68	0.723	0.7240	0.6724	0.7231
Earth	1.0000	1.0	0.973	1.000	0.9994	1.0404	1.1124
Mars	1.5237	1.6	1.559	1.524	1.5252	1.4884	1.5237
(Cerera)	_	2.8	2.731	2.67	2.8695	_	_
Jupiter	5.2028	5.2	5.075	5.200	5.1935	5.1984	5.2028
Saturn	9.580	10.0	9.763	9.550	9.5053	10.7584	9.580
Uranus	19.141	19.6	19.139	19.23	19.2104	18.3184	19.1410
Neptune	30.198	38.8	37.891	30.13	30.3005	27.8784	30.1980

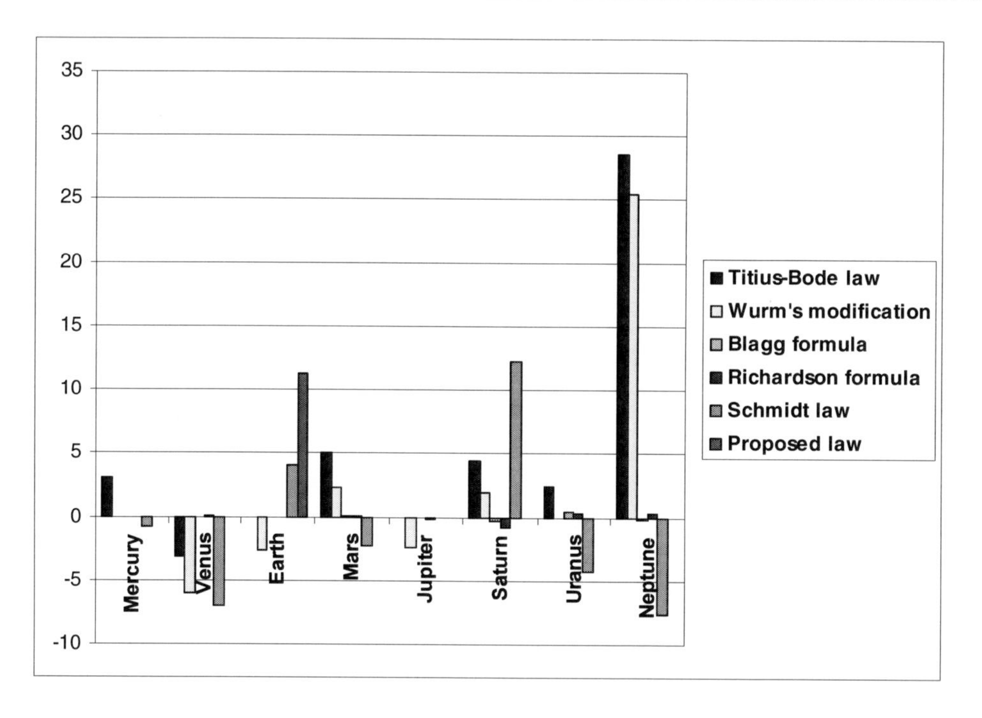

Figure 10.3. The diagram of measure of deviation $\delta = (R_n^{th} - R_n^{obs}) / R_n^{obs}$,% for planets of the Solar system where R_n^{obs} is a value of real (observable) planetary distances in the Solar system, R_n^{th} is an estimation of planetary distances in accord with theoretical laws, n is a number of a planet.

Unlike Blagg's and Richardson's empirical formulas, the Schmidt law (10.2.23) was founded on the basis of scientific hypothesis on a correspondence of a specific angular momentum distribution law of gas-dust protoplanetary cloud to a planetary distances law. The presented (in Table 10.3) law of Schmidt (10.2.23) estimates the distances for exterior planets (beginning from Jupiter) at $a = 2.28$ and $b = 1$ and separately for interior planets (beginning from Mercury) at $a = 0.62$ and $b = 0.20$. In a fact, Schmidt proposed two laws of the Solar system origin [4, 11]. Nevertheless, because of the Schmidt law (10.2.23) is a particular case of the law (10.2.33) then the proposed law generalizes these two laws as unique one. Really, this new law predicts surely the planetary distances for the Solar system with the exception of the Earth and the Pluto.

Let us note this exception is not lack of the proposed theory since the theories of von Weizsäcker [1, 2], ter Haar and Cameron [8] also revealed an exceptional peculiarity for the planets of Earth and Pluto. In particular, ter Haar and Cameron supposed that the best factor in the geometrical progression for planetary distances is equal to 1.89 because this number has been obtained within framework of von Weizsäcker's theory [1, 2]. Namely this geometrical progression gives a divergence with observable planetary data for the Earth and the Pluto only. Thus, the proposed law and von Weizsäcker's and ter Haar–Cameron's theories lead to a similar result.

3 Calculation of Orbits of Planets and Bodies of the Solar System in a Central-Symmetric Gravitational Field of a Rotating Spheroidal Body Based on the Binet's Differential Equation

As it was pointed in previous Section 10.2, the proposed statistical approach to formation of Solar system describes only internal process of self-organization of protoplanets from a gas-dust cloud, and therefore does not include such additional dynamical processes as collisions and huge influences of protoplanets or other big cosmic bodies (asteroids and a swarm of planetesimals or meteorites). Hence, the proposed statistical theory of protoplanet's origin on the basis of evolution of a flattened rotating and gravitating spheroidal body cannot precisely predict a position of planets in the case of giant impacts and, as consequence of it, authentically to estimate their orbits. Really, orbits of moving particles inside a flattened rotating and gravitating spheroidal body initially are circular, however, during evolution of a spheroidal body at formation of protoplanets these orbits can be deformed a little due to collisions with other particles or gravitational influences of forming adjacent planetesimals. As famous astrophysicist V.S. Safronov marked: «The assumption of initial movement of particles on circular orbits looks natural. At small masses of bodies their gravitational variations were weak, and particles moved on the orbits close to circular. In process of growth of a planet, deviations of orbits from circular increased, and all bodies of a zone had an opportunity to be joined in one planet» [15 p.145]. Really, process of evolution of a rotating and gravitating spheroidal body at first leads its flattening, and then this process results in its disintegration on forming protoplanets. Hence, orbits of moving particles at later stages of evolution of a rotating and gravitating spheroidal body are formed mainly under influence of its central gravitational field, i.e. in essence they are Keplerian [61]. In this connection we shall consider more in detail the calculation of orbits of moving bodies and planets in a centrally-symmetrical gravitational field of a rotating and gravitating spheroidal body on the planetary stage of its evolution.

It is well-known [61, 72], that the general method of an orbit finding consists in integration of the differential equations of movement with the subsequent exception of time. Often it is very complex process, therefore a natural question appears, whether it is possible to exclude time before integration so that integration has given directly the required orbit. In particular, it was shown in [72], that it is possible in that case when force does not depend on time. So, we consider movement of the material point subject to action of the central force of a gravitational attraction.

Let f be a specific value of gravitational force, i.e. an acceleration to which the point is subjected. By definition of the central force, directions of this force always pass through a fixed point (or center) which we shall accept for the origin of coordinates. If O is the center of force, then P is some position of a moving point in a plane XY whose rectangular coordinates are x and y, and polar coordinates are r and ε accordingly (Figure 10.4). Then projections of accelerations on axes x and y also are accordingly equal to $-f\cos\varepsilon$ and $-f\sin\varepsilon$, and the differential equations of movement become:

$$\frac{d^2x}{dt^2}=-f\cos\varepsilon=-f\frac{x}{r};\ \frac{d^2y}{dt^2}=-f\sin\varepsilon=-f\frac{y}{r}. \tag{10.3.1}$$

Multiplying the first equation in (10.3.1) on $-y$ and the second equation on x, then adding them we obtain:

$$x\frac{d^2y}{dt^2}-y\frac{d^2x}{dt^2}=0. \tag{10.3.2}$$

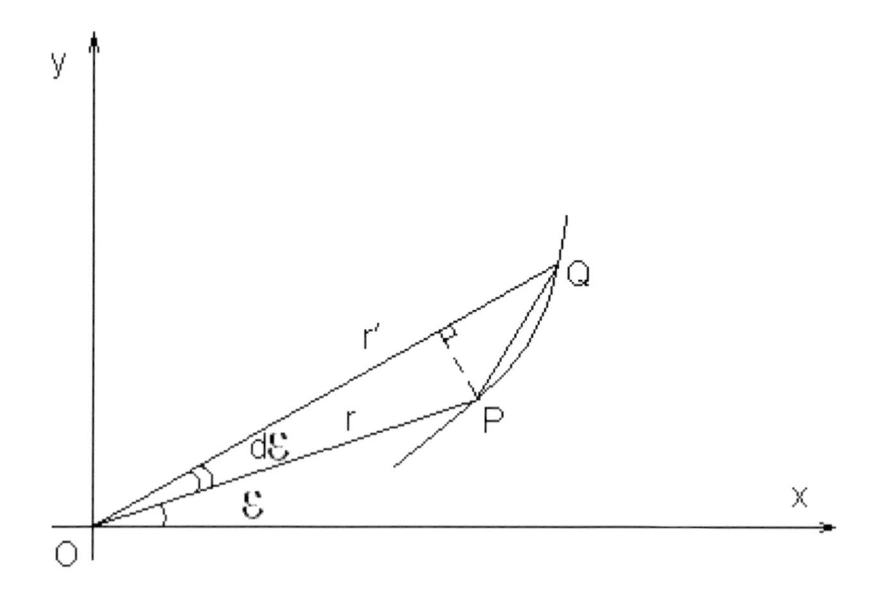

Figure 10.4. Graphic representation of movement of a material point in a field of the central force.

Adding and subtracting the value $\frac{dx}{dt}\cdot\frac{dy}{dt}$ in the left part of equation (10.3.2) we transform it to the kind:

$$x\frac{d^2y}{dt^2}+\frac{dx}{dt}\cdot\frac{dy}{dt}-\frac{dy}{dt}\cdot\frac{dx}{dt}-y\frac{d^2x}{dt^2}=\frac{d}{dt}\left(x\frac{dy}{dt}-y\frac{dx}{dt}\right)=0. \qquad (10.3.3)$$

Integrating the equation (10.3.3) we find that

$$x\frac{dy}{dt}-y\frac{dx}{dt}=C, \qquad (10.3.4)$$

where C is a constant of integration. To find out sense of a constant of integration in (10.3.4) we shall consider movement of a material point in a plane XY for a time interval Δt (Figure 10.4). Let ΔS designate the area of triangle OPQ limited by radius-vector in Figure 10.4 for a time interval Δt :

$$\Delta S=\frac{1}{2}r'r\sin(\Delta\varepsilon)=\frac{r'r}{2}\cdot\frac{\sin(\Delta\varepsilon)}{\Delta\varepsilon}\cdot\Delta\varepsilon. \qquad (10.3.5)$$

Obviously, if an angle $\Delta\varepsilon$ decreases unlimited then the area of triangle OPQ tends to the area of sector, besides the limit r' is r. Passing to the limit at $\Delta t\to 0$ in (10.3.5) we obtain:

$$dS=\lim_{\Delta t\to 0}\left(\frac{r'r}{2}\cdot\frac{\sin(\Delta\varepsilon)}{\Delta\varepsilon}\cdot\Delta\varepsilon\right)=\frac{1}{2}r^2d\varepsilon,$$

whence it directly follows that

$$\frac{dS}{dt}=\frac{1}{2}r^2\cdot\frac{d\varepsilon}{dt}. \qquad (10.3.6)$$

The value (10.3.6) is called an *areal velocity* of a moving point [72]. By substitution

$$r=\sqrt{x^2+y^2}, \qquad \varepsilon=arctg\left(\frac{y}{x}\right),$$

let's write the following expression for the areal velocity in rectangular coordinates:

$$\frac{dS}{dt}=\frac{1}{2}(x^2+y^2)\cdot\frac{d}{dt}[arctg\left(\frac{y}{x}\right)]=\frac{1}{2}(x^2+y^2)\cdot\frac{1}{1+(y/x)^2}\cdot\frac{d}{dt}\left(\frac{y}{x}\right)=\frac{1}{2}x^2\cdot\frac{\dot{y}x-y\dot{x}}{x^2}=\frac{1}{2}\left(x\frac{dy}{dt}-y\frac{dx}{dt}\right) \qquad (10.3.7)$$

So, comparing now the equations (10.3.4) and (10.3.7) we can see that the required constant C is expressed by the areal velocity [72]:

$$C=2\dot{S}=r^2\dot{\varepsilon}, \qquad (10.3.8)$$

where S is the area limited by radius-vector in time t. According to (10.3.8) if the origin of coordinates is chosen by suitable way then movement obeys to the *law of the areas:* $r^2\dot{\varepsilon} = const$. Integrating the equation (10.3.8) we obtain:

$$S = \frac{1}{2} \cdot Ct + c,$$

i.e. the area S changes directly proportional to time (*the Kepler's second law* [61, 72]).

To derive the equation of an orbit of a material point let us return to the equations of movement (10.3.1). As according to a primary assumption the function f does not depend on time then t enters only into derivatives. However, for exception t it is necessary to reduce the order of derivatives preliminary. For convenience we transform the equations (10.3.1) to polar coordinates [72].

As in polar coordinates $x = r\cos\varepsilon$ and $y = r\sin\varepsilon$ then velocity components in rectangular coordinates accordingly are expressed through components in polar coordinates:

$$v_x = \frac{dx}{dt} = \frac{dr}{dt}\cos\varepsilon - r\sin\varepsilon\frac{d\varepsilon}{dt} = v_r\cos\varepsilon - v_\varepsilon\sin\varepsilon; \tag{10.3.9a}$$

$$v_y = \frac{dy}{dt} = \frac{dr}{dt}\sin\varepsilon + r\cos\varepsilon\frac{d\varepsilon}{dt} = v_r\sin\varepsilon + v_\varepsilon\cos\varepsilon, \tag{10.3.9b}$$

where $v_r = dr/dt$ and $v_\varepsilon = r \cdot d\varepsilon/dt$ are the polar components of velocity on radius-vector and along a perpendicular to it. To find acceleration components let us differentiate the formulas (10.3.9a) and (10.3.9b):

$$a_x = \frac{d^2x}{dt^2} = \left[\frac{d^2r}{dt^2} - r\left(\frac{d\varepsilon}{dt}\right)^2\right] \cdot \cos\varepsilon - \left[r\frac{d^2\varepsilon}{dt^2} + 2\frac{dr}{dt} \cdot \frac{d\varepsilon}{dt}\right] \cdot \sin\varepsilon; \tag{10.3.10a}$$

$$a_y = \frac{d^2y}{dt^2} = \left[r\frac{d^2\varepsilon}{dt^2} + 2\frac{dr}{dt} \cdot \frac{d\varepsilon}{dt}\right] \cdot \cos\varepsilon + \left[\frac{d^2r}{dt^2} - r\left(\frac{d\varepsilon}{dt}\right)^2\right] \cdot \sin\varepsilon, \tag{10.3.10b}$$

whence by analogy with (10.3.9a), (10.3.9b) the polar components of acceleration a_r and a_ε (on radius-vector and on perpendicular to it) can be calculated according to formulas [72]:

$$a_r = \frac{d^2r}{dt^2} - r\left(\frac{d\varepsilon}{dt}\right)^2; \tag{10.3.11a}$$

$$a_\varepsilon = r\frac{d^2\varepsilon}{dt^2} + 2\frac{dr}{dt} \cdot \frac{d\varepsilon}{dt} = \frac{1}{r} \cdot \frac{d}{dt}\left(r^2\frac{d\varepsilon}{dt}\right). \tag{10.3.11b}$$

According to Figure 10.4 the polar components of acceleration on radius-vector a_r and on a perpendicular a_ε are equal to $-f$ and 0. Then in accord with formulas (10.3.11a) and (10.3.11b) the differential equations of movement become:

$$\frac{d^2r}{dt^2} - r\left(\frac{d\varepsilon}{dt}\right)^2 = -f; \qquad (10.3.12a)$$

$$\frac{1}{r}\cdot\frac{d}{dt}\left(r^2\frac{d\varepsilon}{dt}\right) = 0. \qquad (10.3.12b)$$

But according to the formula (10.3.8) the integral for the second of these equations is equal:

$$r^2\frac{d\varepsilon}{dt} = C,$$

so, excepting $d\varepsilon / dt$ from the equation (10.3.12a) by means of this integral we obtain:

$$\frac{d^2r}{dt^2} = \frac{C^2}{r^3} - f. \qquad (10.3.13)$$

Supposing $r = 1/q$ we calculate in view of (10.3.8) the following derivatives:

$$\frac{dr}{dt} = -\frac{1}{q^2}\cdot\frac{dq}{dt} = -\frac{1}{q^2}\cdot\frac{dq}{d\varepsilon}\cdot\frac{d\varepsilon}{dt} = -C\cdot\frac{dq}{d\varepsilon}; \qquad (10.3.14a)$$

$$\frac{d^2r}{dt^2} = -C\cdot\frac{d}{dt}\left(\frac{dq}{d\varepsilon}\right) = -C\cdot\frac{d^2q}{d\varepsilon^2}\cdot\frac{d\varepsilon}{dt} = -C^2q^2\cdot\frac{d^2q}{d\varepsilon^2}. \qquad (10.3.14b)$$

Substituting (10.3.14b) in the equation (10.3.13) we obtain the differential equation of the second order relatively q [72]:

$$f = C^2q^2\cdot\left(q + \frac{d^2q}{d\varepsilon^2}\right). \qquad (10.3.15)$$

As the integral (10.3.15) expresses q and consequently r as function from ε, the equation (10.3.8) after integration gives a relation between ε and t. On the other hand, the equation (10.3.15) can be used for a finding of the law of the central force undergoes a material point to move along the given curve. For this purpose it is necessary to write only the equation of a curve in polar coordinates and then to calculate the right part in (10.3.15). This problem is much easier than a direct problem of a finding of an orbit when the law of force is given [72].

In view of that $\vec{f} = -grad\varphi_g$, i.e. for the considered flat movement $f = \left|\vec{f}\right| = d\varphi_g / dr$, and also that $r = 1/q$, we present the equation (10.3.15) in the form of the Binet's formula [62, 72] allowing to define an orbit of a moving material point (planet) in the central gravitational field:

$$\frac{d^2}{d\varepsilon^2}\left(\frac{1}{r}\right) + \frac{1}{r} = \frac{r^2}{C^2} \cdot \varphi_g'(r), \tag{10.3.16}$$

where $\varphi_g(r)$ is function of gravitational potential, $C = r^2 \dot{\varepsilon}$ is an areal (sectorial) constant.

As shown in section 10.1, the gravitational potential of a rotating spheroidal body in a remote zone is described by the formula (10.1.46), i.e.

$$\left.\varphi_g(r,\theta)\right|_{r>>r_*} = -\sqrt{\frac{2\alpha}{\pi}} \cdot \gamma M \cdot \frac{1}{r\sqrt{1-\varepsilon_0{}^2 \sin^2\theta}} \cdot \int_0^{r\sqrt{1-\varepsilon_0{}^2 \sin^2\theta}} e^{-\alpha r'^2/2} dr'. \tag{10.3.17}$$

To take advantage of the Binet' equation (10.3.16) for definition of an orbit of a planet moving in the central gravitational field of a rotating spheroidal body we shall carry out approximation of potential (10.3.17) under condition of $r/r_* \to \infty$ where, as it was designated above, $r_* = 1/\sqrt{\alpha}$.

As a first approximation the gravitational potential (10.3.17) of rotating spheroidal body in a *remote zone* can be estimated by means of the formula:

$$\left.\varphi_g(r,\theta)\right|_{r>>r_*} = -\frac{\sqrt{2}\cdot\sqrt{2}}{\sqrt{\pi}} \cdot \frac{\gamma M}{r\sqrt{1-\varepsilon_0{}^2 \sin^2\theta}} \int_0^{r\sqrt{\alpha\left(1-\varepsilon_0{}^2\sin^2\theta\right)/2}} e^{-\left(r\sqrt{\alpha/2}\right)^2} d\left(r\sqrt{\alpha/2}\right)\Big|_{r/r_*\to\infty} =$$

$$= -\frac{2}{\sqrt{\pi}} \cdot \frac{\gamma M}{r\sqrt{1-\varepsilon_0{}^2 \sin^2\theta}} \int_0^{r\sqrt{\alpha\left(1-\varepsilon_0{}^2\sin^2\theta\right)/2}} e^{-s^2} ds\Big|_{r\sqrt{\alpha}\to\infty} =$$

$$= -\frac{\gamma M}{r\sqrt{1-\varepsilon_0{}^2 \sin^2\theta}} \cdot \frac{2}{\sqrt{\pi}} \lim_{r\sqrt{\alpha}\to\infty} \int_0^{r\sqrt{\alpha}\cdot\sqrt{\left(1-\varepsilon_0{}^2\sin^2\theta\right)/2}} e^{-s^2} ds = -\frac{\gamma M}{r\sqrt{1-\varepsilon_0{}^2 \sin^2\theta}}. \tag{10.3.18}$$

Supposing a condition of smallness $\varepsilon_0{}^2 << 1$ the formula (10.3.18) can be approximated also as follows:

$$\varphi_g(r,\theta)\Big|_{r>>r_*} \approx -\frac{\gamma M}{r}\cdot\left(1+\frac{{\varepsilon_0}^2}{2}\sin^2\theta\right). \tag{10.3.19}$$

Taking into account that the orbit of each forming planet moving in the *central* gravitational field of a rotating spheroidal body (gas-dust cloud) entirely lays in one and the same plane $\theta=\theta_0=const$ near to an equatorial plane of the protoplanetary gas-dust cloud ($\theta=\pi/2$), i.e. $\theta_0 \propto \pi/2$, the formula (10.3.18) becomes:

$$\varphi_g(r,\theta_0)\Big|_{r>>r_*} = -\frac{\gamma M}{r\sqrt{1-{\varepsilon_0}^2\sin^2\theta_0}}. \tag{10.3.20}$$

Substituting (10.3.20) in the Binet's formula (10.3.16) we derive the equation of an orbit of a planet in a remote zone of a spheroidal body:

$$\frac{d^2}{d\varepsilon^2}\left(\frac{1}{r}\right)+\frac{1}{r}=\frac{\gamma M}{C^2\sqrt{1-{\varepsilon_0}^2\sin^2\theta_0}}, \tag{10.3.21}$$

where $C=r^2\varepsilon$ is the areal constant and $\sin^2\theta_0$ is a parameter. Carrying out usual substitution $r=1/q$ in the equation (10.3.21) we obtain:

$$\frac{d^2q}{d\varepsilon^2}=-q+\frac{\gamma M}{C^2\sqrt{1-{\varepsilon_0}^2\sin^2\theta_0}}. \tag{10.3.22}$$

Multiplying both parts (10.3.22) on $2\dfrac{dq}{d\varepsilon}$ in accord with the solution offered in [72]:

$$2\frac{dq}{d\varepsilon}\cdot\frac{d^2q}{d\varepsilon^2}=-2q\frac{dq}{d\varepsilon}+2\frac{\gamma M}{C^2\sqrt{1-{\varepsilon_0}^2\sin^2\theta_0}}\cdot\frac{dq}{d\varepsilon},$$

we transform this equation to the kind:

$$\frac{d}{d\varepsilon}\left[\left(\frac{dq}{d\varepsilon}\right)^2\right]=-\frac{d}{d\varepsilon}\left[q^2\right]+2\frac{\gamma M}{C^2\sqrt{1-{\varepsilon_0}^2\sin^2\theta_0}}\cdot\frac{dq}{d\varepsilon}. \tag{10.3.23}$$

Integrating (10.3.23) we find the first integral of the given equation:

$$\left(\frac{dq}{d\varepsilon}\right)^2 = -q^2 + \frac{2\gamma M}{C^2\sqrt{1-\varepsilon_0{}^2\sin^2\theta_0}} \cdot q + c_1$$

(c_1 is a constant of integration) whence we obtain that

$$\frac{dq}{d\varepsilon} = \pm\sqrt{-q^2 + \frac{2\gamma M}{C^2\sqrt{1-\varepsilon_0{}^2\sin^2\theta_0}} \cdot q + c_1} =$$

$$= \pm\sqrt{-\left(q - \frac{\gamma M}{C^2\sqrt{1-\varepsilon_0{}^2\sin^2\theta_0}}\right)^2 + \frac{(\gamma M / C^2)^2}{1-\varepsilon_0{}^2\sin^2\theta_0} + c_1}$$

and at last,

$$d\varepsilon = \pm\frac{dq}{\sqrt{\frac{(\gamma M / C^2)^2}{1-\varepsilon_0{}^2\sin^2\theta_0} + c_1 - \left(q - \frac{\gamma M}{C^2\sqrt{1-\varepsilon_0{}^2\sin^2\theta_0}}\right)^2}}. \tag{10.3.24}$$

Let be $(\gamma M / C^2)^2 / (1-\varepsilon_0{}^2\sin^2\theta_0) + c_1 = k^2$ and $q - (\gamma M / C^2\sqrt{1-\varepsilon_0{}^2\sin^2\theta_0}) = s$. Choosing the bottom sign in the equation (10.3.24) we write down the given equation in the form:

$$d\varepsilon = -\frac{ds}{\sqrt{k^2 - s^2}} = -\frac{1}{k} \cdot \frac{ds}{\sqrt{1-(s/k)^2}} = -\frac{d(s/k)}{\sqrt{1-(s/k)^2}}. \tag{10.3.25}$$

Then integrating (10.3.25) we obtain:

$$\varepsilon = \arccos\left(\frac{s}{k}\right) + c_2,$$

whence

$$s = k\cos(\varepsilon - c_2), \tag{10.3.26}$$

where c_2 is a constant of integration. Returning to a variable q and then to r, we can write (10.3.26) in the form:

$$\frac{1}{r}-\frac{\gamma M}{C^2\sqrt{1-\varepsilon_0{}^2\sin^2\theta_0}}=k\cdot\cos(\varepsilon-c_2),$$

whence

$$r=\frac{(C^2/\gamma M)\cdot\sqrt{1-\varepsilon_0{}^2\sin^2\theta_0}}{1+(C^2/\gamma M)\cdot\sqrt{1-\varepsilon_0{}^2\sin^2\theta_0}\cdot k\cos(\varepsilon-c_2)}. \quad (10.3.27)$$

Substituting a value of parameter $k=(\gamma M/C^2\sqrt{1-\varepsilon_0{}^2\sin^2\theta_0})\cdot\sqrt{1+c_1\cdot(C^2/\gamma M)^2\cdot(1-\varepsilon_0{}^2\sin^2\theta_0)}$ in the equation (10.3.27) we obtain:

$$r=\frac{(C^2/\gamma M)\cdot\sqrt{1-\varepsilon_0{}^2\sin^2\theta_0}}{1+\sqrt{1+c_1\cdot(C^2/\gamma M)^2\cdot(1-\varepsilon_0{}^2\sin^2\theta_0)}\cdot\cos(\varepsilon-c_2)}. \quad (10.3.28)$$

At a choice of a constant as $c_1=-(\gamma M/C^2)^2$, that directly follows from a designation k^2, the *equation* (10.3.28) *of orbit of a planet in a remote zone* of gravitational field of a rotating spheroidal body goes over to the following:

$$r=\frac{(C^2/\gamma M)\cdot\sqrt{1-\varepsilon_0{}^2\sin^2\theta_0}}{1+|\varepsilon_0\sin\theta_0|\cdot\cos(\varepsilon-c_2)}. \quad (10.3.29)$$

If we assume that $\theta_0=\pi/2-i$ where i an angle of an inclination of an orbital plane then the equation (10.3.29) becomes:

$$r=\frac{(C^2/\gamma M)\cdot\sqrt{1-\varepsilon_0{}^2\cos^2 i}}{1+|\varepsilon_0\cos i|\cdot\cos(\varepsilon-c_2)}. \quad (10.3.30)$$

So, comparing (10.3.29) with the polar equation of conic section with focus in the origin of coordinates $r=\dfrac{p}{1+e\cos\varphi}$ [61, 72] we find that

$$p=(C^2/\gamma M)\cdot\sqrt{1-\varepsilon_0{}^2\sin^2\theta_0},\ e=\varepsilon_0|\sin\theta_0|,\ \varphi=\varepsilon-c_2, \quad (10.3.31)$$

where p is a parameter of an orbit, e is a eccentricity of orbits and $c_2=\varepsilon_*$ is a constant, besides ε_* is an angle between a polar axis and the end of the big axis directed to farther top

[72]. Constants C^2, ε_0 and θ_0 also are defined by initial conditions, and in turn they define p and e by means of (10.3.31). If $e<1$ then the conic section is an ellipse; if $e=1$ then the conic section is a parabola; if $e>1$ then the conic section is a hyperbole; if $e=0$ then the conic section is a circle.

As $\varepsilon_0^2 << 1$ and $|\sin\theta_0| \le 1$ then according to the formula (10.3.31) $e << 1$, i.e. the conic section is an ellipse with small eccentricity. In other words, the formula (10.3.30) expresses the equation of an ellipse in polar coordinates with the origin in focus, i.e. planets of Solar system move on elliptic orbits, in one of which focuses there is the Sun (*the Kepler's first law*) [61, 72].

Let us note that elliptic orbits of planets of Solar system are almost circular, namely $e=0$ at $\theta_0=0$ and $e=\varepsilon_0$ at $\theta_0=\pi/2$, i.e. (10.3.30) is the equation of a circle at $\theta_0=0$ or $i=\pi/2$, and, accordingly, at $\theta_0=\pi/2$ or $i=0$ the relation (10.3.30) is the equation of an ellipse with small eccentricity.

According to the derived formula (10.3.31) and to known formulas of analytical geometry the major and minor semi-axes of an ellipse are accordingly equal:

$$a=\frac{p}{1-e^2}=\frac{(C^2/\gamma M)\cdot\sqrt{1-\varepsilon_0{}^2\sin^2\theta_0}}{1-\varepsilon_0{}^2\sin^2\theta_0}=\frac{C^2}{\gamma M\sqrt{1-\varepsilon_0{}^2\sin^2\theta_0}}, \tag{10.3.32a}$$

$$b=\frac{p}{\sqrt{1-e^2}}=\frac{(C^2/\gamma M)\cdot\sqrt{1-\varepsilon_0{}^2\sin^2\theta_0}}{\sqrt{1-\varepsilon_0{}^2\sin^2\theta_0}}=\frac{C^2}{\gamma M}. \tag{10.3.32b}$$

The least distance called *perihelion* of orbit and the greatest distance called *aphelion* of orbit up to the center of field (which is the focus of an ellipse) are accordingly defined by expressions:

$$r_{\min}=\frac{p}{1+e}=a(1-e)=\frac{C^2\cdot(1-\varepsilon_0|\sin\theta_0|)}{\gamma M\sqrt{1-\varepsilon_0{}^2\sin^2\theta_0}}=\frac{C^2}{\gamma M}\cdot\sqrt{\frac{1-\varepsilon_0|\sin\theta_0|}{1+\varepsilon_0|\sin\theta_0|}}, \tag{10.3.33a}$$

$$r_{\max}=\frac{p}{1-e}=a(1+e)=\frac{C^2\cdot(1+\varepsilon_0|\sin\theta_0|)}{\gamma M\sqrt{1-\varepsilon_0{}^2\sin^2\theta_0}}=\frac{C^2}{\gamma M}\cdot\sqrt{\frac{1+\varepsilon_0|\sin\theta_0|}{1-\varepsilon_0|\sin\theta_0|}}. \tag{10.3.33b}$$

A time of circulation of a planet with mass m on an elliptic orbit, i.e. the period T of its movement, can be convenient defined by means of the law of conservation of the angular momentum in the form of «integral of the areas» [61]:

$$L=L_z=mr^2\dot{\varepsilon}=2m\dot{S}=const, \tag{10.3.34}$$

where $\dot{S}$ is the mentioned above value of areal velocity (10.3.6) of moving body. Integrating this equality on time from 0 up to T we obtain [61]:

$$2mS = LT\,, \tag{10.3.35}$$

where S is the area of an orbit. For an ellipse $S = \pi ab$, so using the formulas (10.3.8), (10.3.31), (10.3.32a,b) we can find that

$$T = \frac{2mS}{L} = \frac{2\pi ab}{L/m} = \frac{2\pi a^2\sqrt{1-e^2}}{2\dot{S}} = \frac{2\pi a^2\sqrt{p/a}}{C} = \frac{2\pi a^{3/2}\sqrt{p}}{C} = \frac{2\pi a^{3/2}\sqrt{(C^2/\gamma M)\cdot\sqrt{1-\varepsilon_0{}^2\sin^2\theta_0}}}{C} =$$

$$= a^{3/2}\cdot\frac{2\pi\cdot\sqrt[4]{1-\varepsilon_0{}^2\sin^2\theta_0}}{\sqrt{\gamma M}} = a^{3/2}\cdot 2\pi\sqrt{\frac{\sqrt{1-\varepsilon_0{}^2\sin^2\theta_0}}{\gamma M}}\,, \tag{10.3.36}$$

The fact that a square of the period should be proportional to a cube of the linear sizes of an orbit expresses *the Kepler's third law* [61, 72], namely, the ratio of cubes of the major semi-axes of orbits to squares of the turning times for all planets of Solar system is equally:

$$\frac{a^3}{T^2} = \frac{\gamma M}{4\pi^2\sqrt{1-\varepsilon_0{}^2\sin^2\theta_0}} = const\,. \tag{10.3.37}$$

Thus, moving bodies (conglomerates of particles, planetesimals, planetary embryos and planets) in a remote zone of a rotating spheroidal body have trajectories in the form of ellipses with the origin in focus. It means that orbits of the Solar system' planets enough distant from the Sun ($r/r_* \to \infty$) are described by ellipses with small eccentricities (this fact occurs for all planets beginning with Venus). Really, the value of *geometrical eccentricity* of orbit $e = \sqrt{a^2 - b^2}/a$ can be defined easy by the condition of a derivation of the equation of an orbit (10.3.29) planets in a remote zone of gravitational field of an uniformly rotating spheroidal body, i.e. according to the formula (10.3.31) $0 \le e << 1$. Besides the geometrical ones, it is considered an *orbital eccentricity* e_o in astrophysics [16]:

$$e_o = (R_a - R_p)/(R_a + R_p)\,, \tag{10.3.38}$$

where $R_a = r_{\max}$ is the aphelion of orbit and $R_p = r_{\min}$ is its perihelion. According to formulas (10.3.33a, b) the orbital eccentricity (10.3.38) is equal to:

$$e_o = \big(a(1+e) - a(1-e)\big)/\big(a(1+e) + a(1-e)\big) = e\,, \tag{10.3.39}$$

i.e. it coincides with geometrical eccentricity for the considered type of orbits of planets moving in a remote zone of gravitational field of an uniformly rotating spheroidal body.

Let's especially note that the derived expression (10.3.37) for the Kepler's third law generalizes the known relation obtained in the theory of Newton [61, 72, 73] in the sense that the constant in the right part includes besides γ and M additional parameters ε_0^2 and θ_0. Thus, the ratio of cubes of the major semi-axes of orbits to squares of the rotation periods for any n-th planet of Solar system is equally, besides a constant coincides with a constant of Newton $\gamma M / 4\pi^2$ up to a very small size $(1/2)\cdot\varepsilon_0^2 \sin^2\theta_0 << 1$:

$$\frac{a_n^3}{T_n^2} = \frac{\gamma M}{4\pi^2} \cdot \frac{1}{\sqrt{1-\varepsilon_0^2 \sin^2\theta_0}} = const\,. \tag{10.3.40}$$

The specific angular momentum of n-th planet in view of formulas (10.3.32a, b), (10.3.35), i.e. the relation $b_n = a_n\sqrt{1-e_n^2}$, is equal:

$$\lambda_n = \frac{L_n}{m_n} = \frac{2S_n}{T_n} = \frac{2\pi a_n b_n}{T_n} = \frac{2\pi a_n^2\sqrt{1-e_n^2}}{T_n}\,. \tag{10.3.41}$$

As follows from the Kepler's third law in the formulation (10.3.40) that

$$T_n = a_n^{3/2} \cdot \frac{2\pi \cdot \sqrt[4]{1-(\varepsilon_0 \sin\theta_0)^2}}{\sqrt{\gamma M}}\,. \tag{10.3.42}$$

Substituting (10.3.42) in (10.3.41) we find the value of specific angular momentum of n-th planet:

$$\lambda_n = \frac{2\pi a_n^2\sqrt{1-e_n^2}}{a_n^{3/2}\cdot 2\pi \cdot \sqrt[4]{1-(\varepsilon_0 \sin\theta_0)^2} / \sqrt{\gamma M}} = \frac{\sqrt{\gamma M a_n (1-e_n^2)}}{\sqrt[4]{1-(\varepsilon_0 \sin\theta_0)^2}}\,. \tag{10.3.43}$$

Let us note that the value of specific angular momentum of n-th planet (10.3.43) obtained on the basis of the statistical theory of spheroidal bodies generalizes the similar formula $\lambda_n = \sqrt{\gamma M a_n (1-e_n^2)}$ [4, 11] derived within framework of the Newton's theory up to very small size $(1/4)\cdot\varepsilon_0^2 \sin^2\theta_0 << 1$:

$$\lambda_n = \sqrt{\gamma M a_n (1-e_n^2)}[1-(\varepsilon_0 \sin\theta_0 / 2)^2]\,. \tag{10.3.44}$$

Taking into account that $e_n^2 = \varepsilon_0^2 \sin^2\theta_0$ in accord with (10.3.31) the formula (10.3.43) becomes:

$$\lambda_n = \frac{\sqrt{\gamma M a_n (1-e_n^{\ 2})}}{\sqrt[4]{1-e_n^{\ 2}}} = \sqrt{\gamma M a_n \sqrt{1-e_n^{\ 2}}}\ . \qquad (10.3.45)$$

According to the formula (10.3.45) the value of specific angular momentum λ_n of n-th planet depends on the eccentricity value e_n of orbit (as function $\sqrt[4]{1-e_n^{\ 2}}$) more weakly than the analogous value μ_n depends on e_n (as $\sqrt{1-e_n^{\ 2}}$) within framework of the theory of Newton first of all due to the account of influence of a flattening parameter ε_0 (geometrical eccentricity of a rotating spheroidal body) and initial value of a polar angle θ_0 (angle of an inclination i of the plane of planetary system [58 p.275] relative to an equatorial plane of a spheroidal body). Thus, within framework of the statistical theory of spheroidal bodies orbits of forming planets are closest to circular when according to the formula (10.2.22) $\lambda_n = \sqrt{\gamma M R_n}$ (R_n is a radius of a circular orbit) that really takes place in Solar system.

As it has been shown by O.Yu. Schmidt [3, 4], formation of planets is possible not only on the basis of gas-dust protoplanetary substance (protosolar nebula) but also by means of capture and join of moving bodies on close orbits (meteorites, asteroids etc.) in a gravitational field of a star. For any body moving on an orbit (with the major semi-axis a and the eccentricity e) in the Solar system O.Yu. Shmidt defines the value of specific angular momentum following the theory of Newton:

$$\lambda = \sqrt{\gamma M a (1-e^2)}\ . \qquad (10.3.46)$$

According to the condensation theory of Schmidt [3, 11] a distribution of orbits and masses of moving bodies (meteorites, planetesimals, planetary embryos) with the angular momentums belonging some interval of values is described by formulas:

$$a = \frac{l}{2} \cdot \frac{1+e}{1-e}, \qquad (10.3.47a)$$

$$dm = \frac{m}{2} \cdot de, \qquad (10.3.47b)$$

where l is a limiting distance on which capture is carried out and which is constant for all system [3] (as a matter of fact l is a parameter of a parabolic orbit which becomes elliptic at capture, i.e. then $l/2 = R_p$ and $a = R_a$ in accord with (10.3.38)). According to formulas (10.3.46), (10.3.47a) we obtain that the specific angular momentum of a moving body is equal:

$$\lambda = \sqrt{\gamma M l/2} \cdot (1+e)\ . \qquad (10.3.48)$$

According to (10.2.14) let μ_n be the value of the specific angular momentum corresponding a *border* between areas of n-th and $(n+1)$-th protoplanets (or planetary embryos) for which specific angular momentums are equal accordingly λ_n and λ_{n+1}. Then for a body moving on the boundary distance for which value e'_n is defined the specific angular momentum can be found by the relation:

$$\mu_n = \sqrt{\gamma Ml/2} \cdot (1 + e'_n). \qquad (10.3.49)$$

Equating the angular momentum of n-th planet to the total angular momentum of moving bodies on close orbits (for example, meteoric substance [3, 11]), we have:

$$m_n \lambda_n = \int_{e'_{n-1}}^{e'_n} \sqrt{\gamma Ml/2} \cdot (1+e)\frac{m}{2} \cdot de = \sqrt{\gamma Ml/2} \cdot \frac{m}{2}[(e'_n - e'_{n-1}) + \frac{1}{2}(e'^2{}_n - e'^2{}_{n-1})] =$$

$$= \sqrt{\gamma Ml/2} \cdot \frac{m}{2}(e'_n - e'_{n-1}) \cdot [1 + \frac{1}{2}(e'_n + e'_{n-1})]. \qquad (10.3.50)$$

Having taken advantage of a designation:

$$m_n = \frac{m}{2} \cdot (e'_n - e'_{n-1}), \qquad (10.3.51)$$

the last equality (10.3.50) in view of (10.3.49) becomes:

$$m_n \lambda_n = m_n \sqrt{\frac{\gamma Ml}{2}} \cdot \frac{(1+e'_n) + (1+e'_{n-1})}{2} = \frac{m_n}{2}(\mu_{n-1} + \mu_n). \qquad (10.3.52)$$

As shown in Section 10.2, starting from formulas (10.2.14), (10.2.22) and (10.3.52), it is easy to obtain

$$\lambda_n = \frac{\lambda_{n+1} + \lambda_{n-1}}{2} \qquad (10.3.53a)$$

and then

$$\sqrt{R_n} = \frac{\sqrt{R_{n+1}} + \sqrt{R_{n-1}}}{2}, \qquad (10.3.53b)$$

i.e. a solution of difference equation (10.3.53b) is the known law of planetary distances of Schmidt (10.2.23).

If instead of (10.3.46) to take advantage of the result (10.3.45) obtained within framework of the statistical theory of spheroidal bodies, i.e. for any body belonging the Solar system and moving on orbit with the major semi-axis a and the eccentricity e the value of specific angular momentum is equal:

$$\lambda = \sqrt{\gamma Ma\sqrt{1-e^2}}\ . \qquad (10.3.54)$$

then analogous substitution (10.3.47a) in (10.3.54) gives us

$$\lambda = \sqrt{\gamma Ml/2} \cdot (1+e)^{3/4} /(1-e)^{1/4}\ . \qquad (10.3.55)$$

Thus, according to (10.3.55) the law of planetary distances for a case of formation of planets by means of capture of bodies in close orbits will have more complex dependence.

4 Calculations of an Orbit Of Planet Mercury and Estimation of Angular Shift of the Mercury' Perihelion Based on the Statistical Theory of Gravitating Spheroidal Bodies

As shown in previous Section 10.3, moving solid bodies in a distant zone of a rotating spheroidal body have elliptic trajectories of the kind (10.3.29). It means that orbits for the enough remote planets from the Sun in Solar system ($r/r_* \to \infty$) are described by ellipses with focus in the origin of coordinates and with small eccentricities. The nearby planet to Sun named Mercury has more complex trajectory. Namely, in case of Mercury the angular displacement of a Newtonian ellipse is observed during its one rotation on an orbit (Figure 10.5), i.e. *a regular (century') shift of the perihelion of Mercury'orbit occurs* [74 p.391].

Figure 10.5. The graphic representation of shift of the perihelion of Mercury' orbit.

After a successful explanation of a deviation of an orbit of Uranus led opening of a new planet of Neptune known astronomer U.J. Leverrier (in occasion of abnormal additional promotion of the perihelion of Mercury on 31" in century) stated in 1859: «Conclusion is clear enough. In vicinity of Mercury, that is more exact between it and the Sun, undoubtedly, there is unknown till now a matter. Whether it itself represents one planet or some small planets or asteroids or, at last, a space dust, the theory cannot give us the answer» [11 p.43]. However, as it has noted by M.M. Nieto [11 p.43], deviations of Mercury' orbit from an orbit calculated theoretically could be caused, firstly, by existence of other planet or, secondly, discrepancy of the law of Newton.

As is known, on a way of specification of the law of Newton the decision of problem of Mercury using the general relativity (GR) theory has been obtained. In this occasion per 1917 A. Einstein in his work «On the special and general theory of relativity» wrote: «Really, astronomers have found that the theory of Newton is insufficient to calculate observable movement of Mercury with accuracy which can be reached at present. After all disturbing influences of other planets on movement Mercury have been accepted into account, it has been found (Leverrier, 1859; Newcomb, 1895) that there is not explained a movement of perihelion of Mercury' orbits which velocity does not differ noticeably from mentioned above +43 angular seconds in century. The error of this empirical result constitutes some seconds only» [75 p.561].

In connection with the above-stated (concerning century angular shift of the Mercury' perihelion of orbit) we shall notice that from a common position of the statistical theory of gravitating spheroidal bodies the points of view as Leverrier (about existence of a unknown matter) and Einstein (about insufficiency of the theory of Newton) practically differ nothing. Really, there exists a plasma and gas-dust substance around of a kernel of a rotating spheroidal body (in this case, the Sun), i.e. the account of that circumstance that forming cosmological bodies have not precise outlines and are represented by means of spheroidal forms demands some specification of the law of Newton in connection with a gravitating spheroidal body [37, 38, 52, 53].

So, with the purpose of Mercury' trajectory finding within the framework of the statistical theory of gravitating spheroidal bodies it is necessary to estimate gravitational potential in nearby removal from the Sun, i.e. *in a remote zone of a gravitational field and in immediate proximity to a kernel of a rotating spheroidal body*. Taking into account that the orbit of planet Mercury entirely lays in one plane $\theta = \theta_0 = const$ let us use, as well as in Section 10.3, the formula (10.3.20):

$$\varphi_g(r)\Big|_{r>>r_*} = -\frac{\gamma M}{r\sqrt{1-\varepsilon_0{}^2 \sin^2\theta_0}}, \qquad (10.4.1)$$

which in view of smallness of parameters $\varepsilon_0{}^2 << 1$ and $|\sin\theta_0| < 1$ becomes the following:

$$\varphi_g(r)\Big|_{r>>r_*} = -\frac{\gamma M}{r}\cdot\left(1+\frac{\varepsilon_0{}^2}{2}\sin^2\theta_0\right), \qquad (10.4.2)$$

where $r_* = 1/\sqrt{\alpha}$, α is a parameter of gravitational compression of a spheroidal body [37], [52].

Let us note that the formula (10.4.2) as well as (10.3.19) can be obtained by Maclaurin-series expansion of function (10.3.17) on degrees of small parameter ${\varepsilon_0}^2$ in linear approximation and under condition of $r / r_* \to \infty$.

To estimate gravitational potential (10.4.2) on closer distance from a kernel of a rotating spheroidal body we shall consider that circumstance that planet Mercury (being the planet of Solar system closest to the Sun) moves around of the Sun on strongly stretched and inclined elliptic orbit (its eccentricity is equal to 0.205 and the inclination of an orbit to a plane of ecliptic is 7°, i.e. approximately 6.3° to the basic plane of the Solar system). Though, as it has been marked in Section 10.3, all formed planets in a remote zone of a rotating spheroidal body have the elliptic and inclined orbits of the kind (10.3.30), however, eccentricities and angles of inclination of orbits for all other planets of the Solar system are rather insignificant (their eccentricities are in an interval 0.0017–0.093 and inclinations of orbits to the basic plane of Solar system belong to an interval 0.3°–2.2°).

The above noted feature of strongly stretched elliptic orbit of Mercury leads to that Mercury is closer to the Sun more than in one and a half time in the perihelion than in the aphelion of its orbit. In view of its greatest proximity on distance to the Sun and essential inclination of its orbit we conclude that *the projection of a point of perihelion of Mercury' orbit can directly get in a nearby vicinity of the Sun, namely, in the visible part of the solar corona* as a kernel of rotating and gravitating orbits of a spheroidal body (see Figure 10.6).

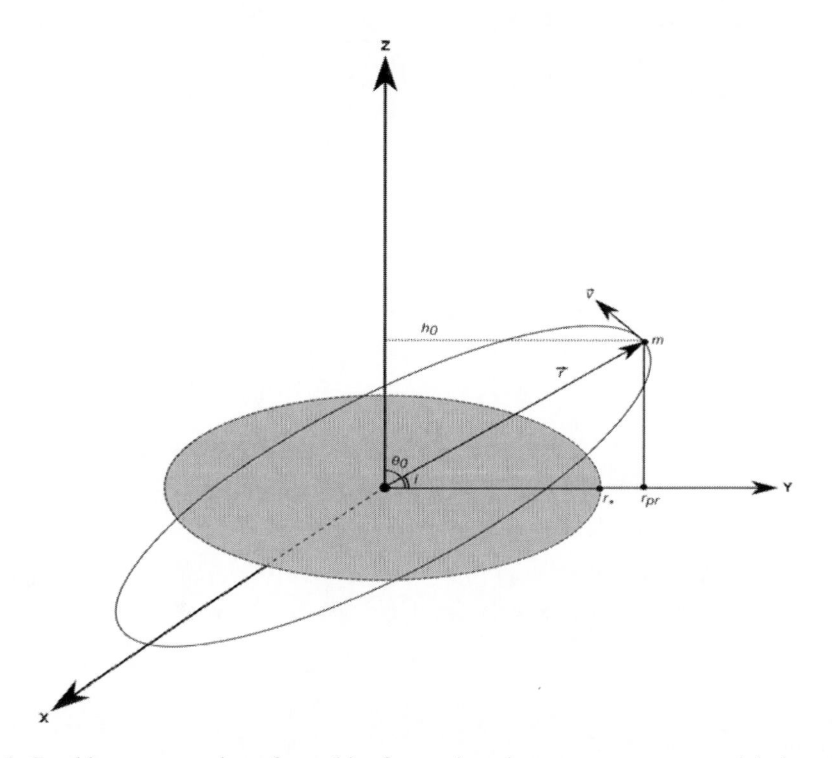

Figure 10.6. Graphic representation of an orbit of a moving planet near to an equatorial plane of a kernel of a gravitating and rotating spheroidal body.

Really, if i is an angle of an inclination of an orbital plane of a planet then the distance h_0 from a moving planet to an axis of rotation of a kernel of a spheroidal body is expressed through distance r from a planet up to the center of a rotating and gravitating spheroidal body by a simple relation:

$$h_0 = r \cos i . \tag{10.4.3}$$

As the orbit of each planet (moving in the central gravitational field of a rotating spheroidal body) entirely lays in one plane characterized by a constant polar angle $\theta_0 = const$ (Figure 10.6) then in view of a relation $\theta_0 = \pi/2 - i$ the formula (10.4.3) takes the form:

$$h_0 = r \sin \theta_0 . \tag{10.4.4}$$

As it has already been marked, for the planet Mercury which is located closest to a visible part of the Solar crown (or a kernel of rotating and gravitating spheroidal body whose equatorial plane is allocated by grey color on Figure 10.6) we estimate with a sufficient accuracy that $h_0 = r_{pr}$, where $r_{pr} = \sqrt{2} \cdot r_*$ [35–40]. Therefore, it follows from (10.4.4) directly that

$$\sin \theta_0 = \frac{r_{pr}}{r} = \frac{\sqrt{2} \cdot r_*}{r} . \tag{10.4.5}$$

So, supposing $\sin \theta_0 = r_{pr} / r = \sqrt{2} \cdot r_* / r$ according to (10.4.5) the formula for estimation of gravitational potential (10.4.2) on closer distance from a kernel of a rotating spheroidal body (in vicinities of Mercury' orbit) can be presented as follows:

$$\varphi_g(r)|_{r >> r_*} = -\frac{\gamma M}{r} \cdot \left(1 + \frac{\varepsilon_0^2 r_*^2}{r^2} \right) = -\frac{\gamma M}{r} \cdot \left(1 + \frac{\delta_0^2}{r^2} \right), \tag{10.4.6}$$

where $\delta_0 = \varepsilon_0 r_* = \varepsilon_0 / \sqrt{\alpha}$ is a value with dimension of length which is small in comparison with distance from a moving planet up to the center of rotating and gravitating spheroidal body (in view of the fact that $r_* << r$ and $\varepsilon_0^2 << 1$).

Substituting (10.4.6) in the Binet formula (10.3.16) we obtain the equation of the *disturbed orbit* of a planet (the Mercury) in a vicinity of a kernel of a rotating spheroidal body:

$$\frac{d^2}{d\varepsilon^2}\left(\frac{1}{r}\right) + \frac{1}{r} = \frac{r^2}{C^2} \cdot \frac{\gamma M}{r^2}\left(1 + \frac{3\delta_0^2}{r^2}\right) = \frac{\gamma M}{C^2} \cdot \left(1 + \frac{3\delta_0^2}{r^2}\right), \tag{10.4.7}$$

where $C = r^2\dot{\varepsilon}$. After traditional substitution $q = 1/r$ the equation (10.4.7) becomes the following:

$$\frac{d^2 q}{d\varepsilon^2} = -q + \frac{\gamma M}{C^2} + \frac{3\gamma M \delta_0{}^2}{C^2} q^2 . \tag{10.4.8}$$

Multiplying both parts (10.4.8) on $2\frac{dq}{d\varepsilon}$ we can transform this equation to the kind:

$$\frac{d}{d\varepsilon}\left[\left(\frac{dq}{d\varepsilon}\right)^2\right] = -\frac{d}{d\varepsilon}\left[q^2\right] + \frac{2\gamma M}{C^2}\cdot\frac{dq}{d\varepsilon} + \frac{2\gamma M \delta_0{}^2}{C^2}\cdot\frac{dq}{d\varepsilon}\left[q^3\right]. \tag{10.4.9}$$

Having integrated (10.4.9) let us calculate the first integral of this equation:

$$\left(\frac{dq}{d\varepsilon}\right)^2 = -q^2 + \frac{2\gamma M}{C^2}\cdot q + \frac{2\gamma M \delta_0{}^2}{C^2}\cdot q^3 + c_1 = c_1 + \frac{2\gamma M}{C^2}\cdot q - q^2 + \frac{2\gamma M \delta_0{}^2}{C^2}\cdot q^3, \tag{10.4.10}$$

whence we find that

$$\frac{dq}{d\varepsilon} = \pm\sqrt{\frac{2\gamma M}{C^2} q \cdot \left(\delta_0{}^2 q^2 + 1\right) - \left(q^2 - c_1\right)}.$$

Separating variables in the given equation we obtain:

$$d\varepsilon = \pm\frac{dq}{\sqrt{\frac{2\gamma M}{C^2} q \cdot \left([\delta_0 q]^2 + 1\right) - \left(q^2 - c_1\right)}}. \tag{10.4.11}$$

Supposing $c_1 = -1/\delta_0{}^2$, $\kappa = \dfrac{C^2}{2\gamma M \delta_0}$ let us rewrite (10.4.11) in the form:

$$d\varepsilon = \pm\frac{dq}{\sqrt{\frac{2\gamma M}{C^2} q \cdot \left([\delta_0 q]^2 + 1\right) - \left(q^2 + \frac{1}{\delta_0{}^2}\right)}} = \pm\frac{dq}{\sqrt{\left([\delta_0 q]^2 + 1\right)\cdot\left(\frac{2\gamma M}{C^2} q - \frac{1}{\delta_0{}^2}\right)}} =$$

$$= \pm\frac{\sqrt{\kappa}\cdot dq}{\sqrt{[\delta_0 q]^2 + 1}\cdot\sqrt{\delta_0 q - \kappa}} = \pm\frac{(2\sqrt{\kappa}/\delta_0)\cdot d(\sqrt{\delta_0 q - \kappa})}{\sqrt{[\delta_0 q]^2 + 1}}. \tag{10.4.12}$$

Introducing a designation $s = \sqrt{\delta_0 q - \kappa}$ and then integrating (10.4.12) we obtain:

$$\varepsilon = \pm \frac{2\sqrt{\kappa}}{\delta_0} \int \frac{ds}{\sqrt{[s^2+\kappa]^2+1}} + c_2 = \pm \frac{2\sqrt{\kappa}}{\delta_0} \cdot \int \frac{ds}{\sqrt{s^4+2\kappa s^2+(\kappa^2+1)}} + c_2, \tag{10.4.13}$$

where где c_2 is a constant of integration.

The relation (10.4.13) is led to elliptic integral of 1-st sort which expresses ε as function from s [59 p.274]. Taking inverse functions and values we can express r as Jacobi elliptic function from ε. As a result, the equation of the disturbed orbit of a planet (near to a kernel of a rotating spheroidal body) describes spirals for which a circle passing through the origin of coordinates and a circle with this origin in the center are limiting cases [72], i.e. *the disturbed trajectories of finite movements of a planet are not closed* [61 p.46].

To obtain the equation of disturbed orbit of the planet Mercury in an explicit form we shall attempt to make some simplification of the initial equation (10.4.10). First of all, let us notice that the similar (10.4.10) equation was derived by A. Einstein within framework of his GR theory. Really, if we introduce designations $2\gamma M\delta_0^2/C^2 = A$, $2\gamma M/C^2 = A/B^2$, $c_1 = 2D/B^2$ then this equation (10.4.10) becomes exactly the equation (11) in Einstein's work [76] (the similar equation (18) was also deduced by K. Schwarzschild in his work [75 p.206], [77]. As A. Einstein indicated in his work, this equation differs from the corresponding equation of the Newton' theory only last term Aq^3 in the right part what allowed him with a sufficient accuracy to replace the decision of this equation in the form of elliptic integral by *pseudo-elliptic* integral (which could be calculated by means of elementary functions) [59 p.105].

In this connection, within framework of the proposed statistical theory of gravitating spheroidal bodies we shall also introduce convenient notations $2\gamma M\delta_0^2/C^2 = \sigma$, $2\gamma M/C^2 = \sigma/\delta_0^2$, and then we shall express a polynomial $K(q) = \sigma q^3 - q^2 + (\sigma/\delta_0^2)q + c_1$ in the right part of the equation (10.4.10) through a corresponding polynomial $N(q) = -q^2 + (\sigma/\delta_0^2)q + c_1$ in the right part of the equation of the Newton' theory on the basis of algorithm of division of two polynomials with a residue:

$$K(q) = L(q)N(q) + R(q), \tag{10.4.14}$$

where $R(q) \equiv K(q) \bmod N(q)$ is a polynomial residue, besides $\deg R(q) < \deg N(q)$ [78]. Direct application of algorithm of division of polynomials with the residue (procedure of division by "corner") gives:

$$L(q) = -\sigma q + 1 - \sigma^2/\delta_0^2, \tag{10.4.15a}$$

$$R(q) = (\sigma c_1 - \sigma^3/\delta_0^4)q + c_1 \cdot \sigma^2/\delta_0^2. \tag{10.4.15b}$$

So, according to formulas (10.4.14) and (10.4.1a, b) the polynomial in the right part of the equation (10.4.10) can be expressed as follows:

$$K(q) = (1 - \sigma q - \sigma^2 / \delta_0^2) \cdot N(q) + (\sigma c_1 - \sigma^3 / \delta_0^4) q + c_1 \cdot \sigma^2 / \delta_0^2 . \tag{10.4.16}$$

Using the results of Section 10.3 we accept a constant of integration c_1 to be proportional of the value $-(\gamma M / C^2)^2 = -(\sigma / 2\delta_0^2)^2 = -\sigma^2 / 4\delta_0^4$. In view of that $\sigma << 1$ we are content by terms not higher than the second order of smallness relative to σ in this formula, i.e. $O(\sigma^2)$:

$$K(q) = (1 - \sigma q - \sigma^2 / \delta_0^2) \cdot N(q) = (1 - \sigma q - \sigma^2 / \delta_0^2) \cdot (-q^2 + (\sigma / \delta_0^2) q + c_1) . \tag{10.4.17}$$

At last, substituting the formula (10.4.17) in the equation (10.4.10) we obtain:

$$\frac{dq}{d\varepsilon} = \pm \sqrt{K(q)} = \pm \sqrt{(1 - \sigma q - \sigma^2 / \delta_0^2) \cdot (-q^2 + (\sigma / \delta_0^2) q + c_1)} , \tag{10.4.18}$$

whence after separation of variables and integration of the equation (10.4.18) we find that

$$d\varepsilon = \pm \frac{dq}{\sqrt{(1 - \sigma[q + \sigma / \delta_0^2]) \cdot (-q^2 + (\sigma / \delta_0^2) q + c_1)}} . \tag{10.4.19}$$

Decomposing the function $(1 - \sigma[q + \sigma^2 / \delta_0^2])^{-1/2}$ in Maclaurin' series we shall write down integral (10.4.19) in the form:

$$\varepsilon = \pm \int \frac{\{1 + (\sigma / 2) \cdot [q + \sigma / \delta_0^2]\} dq}{\sqrt{-q^2 + (\sigma / \delta_0^2) q + c_1}} + c_2 =$$

$$= \pm \left(1 + \frac{\sigma^2}{2\delta_0^2} \right) \cdot \int \frac{dq}{\sqrt{-q^2 + (\sigma / \delta_0^2) q + c_1}} \pm \frac{\sigma}{2} \cdot \int \frac{q dq}{\sqrt{-q^2 + (\sigma / \delta_0^2) q + c_1}} + c_2 , \tag{10.4.20}$$

where c_2 is a constant of integration. By analogy with transformation under a radical in a denominator of the right part of the equation (10.3.24), we shall write down radicals in denominators of subintegral expressions in the right part of the equation (10.4.20) similarly:

$$\varepsilon = \pm \left(1 + \frac{\sigma^2}{2\delta_0^2} \right) \cdot \int \frac{dq}{\sqrt{\left((\sigma / 2\delta_0^2)^2 + c_1 \right) - \left(q - \sigma / 2\delta_0^2 \right)^2}} \pm \frac{\sigma}{2} \cdot \int \frac{q dq}{\sqrt{\left((\sigma / 2\delta_0^2)^2 + c_1 \right) - \left(q - \sigma / 2\delta_0^2 \right)^2}} + c_2 \tag{10.4.21}$$

Introducing previous (as in Section 10.3) notations $k^2 = (\sigma / 2\delta_0^2)^2 + c_1$ and $s = q - \sigma / 2\delta_0^2$ we can note that integrals in the right part of the equation (10.4.21) have meaning only if $k^2 > s^2$, i.e. $\left|\frac{s}{k}\right| < 1$. Taking into account it and choosing the bottom sign in the equation (10.4.21) we obtain:

$$\varepsilon = -\left(1 + \frac{\sigma^2}{2\delta_0^2}\right) \cdot \int \frac{ds}{\sqrt{k^2 - s^2}} - \frac{\sigma}{2} \cdot \int \frac{(s + \sigma / 2\delta_0^2) ds}{\sqrt{k^2 - s^2}} + c_2 =$$

$$= \left(1 + \frac{\sigma^2}{2\delta_0^2}\right) \cdot \arccos\left(\frac{s}{k}\right) + \frac{\sigma^2}{4\delta_0^2} \cdot \arccos\left(\frac{s}{k}\right) - \frac{\sigma}{2} \cdot \left(-\sqrt{k^2 - s^2}\right) + c_2 =$$

$$= \left(1 + \frac{3\sigma^2}{4\delta_0^2}\right) \cdot \arccos\left(\frac{s}{k}\right) + \frac{\sigma}{2} \cdot \sqrt{k^2 - s^2} + c_2 . \tag{10.4.22}$$

Using the mentioned above condition $|s / k| < 1$ let us present the function $\sqrt{k^2 - s^2}$ entering into the right part of the equation (10.4.22) in the form:

$$\sqrt{k^2 - s^2} = k\sqrt{1 - (s / k)^2} = k \sin\left[\arccos\left(\frac{s}{k}\right)\right],$$

and then we shall expand it in the Maclaurin' series:

$$\sqrt{k^2 - s^2} = k \sin\left[\arccos\left(\frac{s}{k}\right)\right] = k \cdot \arccos\left(\frac{s}{k}\right) - k \cdot \frac{[\arccos(s / k)]^3}{3!} + \ldots . \tag{10.4.23}$$

Being restricted by first term in the Maclaurin' series (10.4.23) and substituting it in (7.4.22) this equation can be simplified essentially:

$$\varepsilon = \left(1 + \frac{\sigma \cdot k}{2} + \frac{3\sigma^2}{4\delta_0^2}\right) \cdot \arccos\left(\frac{s}{k}\right) + c_2 , \tag{10.4.24}$$

whence

$$s = k \cdot \cos\left(\frac{\varepsilon - c_2}{1 + k\sigma / 2 + 3\sigma^2 / 4\delta_0^2}\right). \tag{10.4.25}$$

Returning to a variable q and then to r we can write down (10.4.25) in the form:

$$\frac{1}{r}-\frac{\sigma}{2\delta_0^{\,2}}=k\cdot\cos\left(\frac{\varepsilon-c_2}{1+k\sigma/2+3\sigma^2/4\delta_0^{\,2}}\right),$$

whence

$$r=\frac{2\delta_0^{\,2}/\sigma}{1+(2k\delta_0^{\,2}/\sigma)\cdot\cos\left(\dfrac{\varepsilon-c_2}{1+k\sigma/2+3\sigma^2/4\delta_0^{\,2}}\right)}. \tag{10.4.26}$$

Substituting in (10.4.26) the value of parameter $k=\sqrt{(\sigma/2\delta_0^{\,2})^2+c_1}$ we obtain:

$$r=\frac{2\delta_0^{\,2}/\sigma}{1+(2\delta_0^{\,2}/\sigma)\cdot\sqrt{(\sigma/2\delta_0^{\,2})^2+c_1}\cdot\cos\left(\dfrac{\varepsilon-c_2}{1+(\sigma/2)\cdot\sqrt{(\sigma/2\delta_0^{\,2})^2+c_1}+3\sigma^2/4\delta_0^{\,2}}\right)}=$$

$$=\frac{2\delta_0^{\,2}/\sigma}{1+\sqrt{1+c_1\cdot(2\delta_0^{\,2}/\sigma)^2}\cdot\cos\left(\dfrac{\varepsilon-c_2}{1+\left[3+\sqrt{1+c_1\cdot(2\delta_0^{\,2}/\sigma)^2}\right]\cdot\sigma^2/4\delta_0^{\,2}}\right)} \tag{10.4.27}$$

In view of smallness of the parameter σ, i.e. $\sigma<<1$, the equation (10.4.27) becomes:

$$r=\frac{2\delta_0^{\,2}/\sigma}{1+\sqrt{1+c_1\cdot(2\delta_0^{\,2}/\sigma)^2}\cdot\cos\left(\left[1-(3+\sqrt{1+c_1\cdot(2\delta_0^{\,2}/\sigma)^2})\cdot\sigma^2/4\delta_0^{\,2}\right]\cdot(\varepsilon-c_2)\right)} \tag{10.4.28}$$

Taking into account the accepted above notation $\sigma=2\gamma M\delta_0^{\,2}/C^2$ we can determine the values $2\delta_0^{\,2}/\sigma=C^2/\gamma M$ and $\sigma^2/4\delta_0^{\,2}=(\gamma M\delta_0/C^2)^2$, so the equation (10.4.28) for disturbed orbit of the Mercury becomes the following:

$$r=\frac{C^2/\gamma M}{1+\sqrt{1+c_1\cdot(C^2/\gamma M)^2}\cdot\cos\left(\left[1-(3+\sqrt{1+c_1\cdot(C^2/\gamma M)^2})\cdot(\gamma M\delta_0/C^2)^2\right]\cdot(\varepsilon-c_2)\right)} \tag{10.4.29}$$

As a result, comparing (10.4.29) with the polar equation of conic section with focus in the origin of coordinates $r = \frac{p}{1 + e\cos\varphi}$ [61, 72] we find that

$$p = C^2 / \gamma M \quad ; \quad e = \sqrt{1 + c_1 \cdot (C^2 / \gamma M)^2} \quad ;$$
$$\varphi = \left[1 - (3 + \sqrt{1 + c_1 \cdot (C^2 / \gamma M)^2}) \cdot (\gamma M \delta_0 / C^2)^2\right] \cdot (\varepsilon - c_2), \quad (10.4.30)$$

где p is a parameter of an orbit, e is a eccentricity of orbits and $c_2 = \varepsilon_*$ is a constant, besides ε_* is an angle between a polar axis and the end of the big axis directed to farther top [72]. As it has been mentioned above in Section 10.3, the constants C^2, c_1 and c_2 are defined by initial conditions, and in turn, they define p and e by means of (10.4.30).

Taking into account smallness of the parameter $\sigma = 2\gamma M \delta_0{}^2 / C^2$, obviously, the equation (10.4.28) of disturbed orbit of a planet cannot differ so considerably from the equation (10.3.28) of orbits for a planet in a remote zone of a gravitational field of a rotating spheroidal body. In connection with that for not disturbed orbit of a planet in a remote zone of a gravitational field the constant of integration c_1 (in Section 10.3) is chosen to be equal $c_1 = -(\gamma M / C^2)^2$, there is a reason to suppose that $c_1 \to -(\gamma M / C^2)^2$, besides $|c_1| < (\gamma M / C^2)^2$ and $c_1 < 0$. It means that according to (10.4.30) the eccentricity e of disturbed orbit should be $e = \sqrt{1 + c_1 \cdot (C^2 / \gamma M)^2} < 1$, i.e. the formula (10.4.29) expresses the equation of the "disturbed" ellipse in polar coordinates with the origin of coordinates in focus:

$$r = \frac{C^2 / \gamma M}{1 + e \cdot \cos\left([1 - (3 + e) \cdot \delta_0{}^2 \cdot (\gamma M / C^2)^2] \cdot (\varepsilon - \varepsilon_*)\right)}, \quad (10.4.31)$$

i.e. the planet Mercury is moving on a *precessing elliptic orbit* in view of the fact that there is a modulating multiplier of a phase (an azimuth angle $\varepsilon - \varepsilon_*$) in the equation (10.4.31):

$$\eta = 1 - (3 + e)\delta_0{}^2 \cdot \left(\frac{\gamma M}{C^2}\right)^2, \quad (10.4.32)$$

besides the multiplier η is close to unity as $(\gamma M \delta_0 / C^2)^2 = \sigma^2 / (2\delta_0)^2 << 1$. Let us note that according to the derived formula (10.4.31) and the known formula of analytical geometry the major semi-axis a of an ellipse is equal to:

$$a = \frac{C^2 / \gamma M}{1 - e^2},$$

so

$$C^2 = \gamma Ma(1-e^2). \tag{10.4.33}$$

Substituting (10.4.33) in the formula (10.4.32) we obtain:

$$\eta = 1-(3+e)\cdot\left(\frac{\delta_0}{a(1-e^2)}\right)^2. \tag{10.4.34}$$

So, according to formulas (10.4.31) and (10.4.34) at a full turn of a planet on the disturbed orbit the increment of a phase is given by:

$$\Delta\varphi = 2\pi\eta = 2\pi - \frac{2\pi(3+e)\cdot\delta_0{}^2}{a^2(1-e^2)^2}. \tag{10.4.35}$$

The second summand represents the required angular moving of a Newtonian ellipse during one turn of a planet on the disturbed orbit, i.e. displacement of perihelion of orbit for the period is equal to the following angle:

$$\delta\varepsilon = \frac{2\pi(3+e)\cdot\delta_0{}^2}{a^2(1-e^2)^2}. \tag{10.4.36}$$

Taking into account that the value $\delta_0 = \varepsilon_0 r_*$ according to (10.4.6), the formula (10.4.36) becomes:

$$\delta\varepsilon = \frac{2\pi(3+e)\cdot\varepsilon_0{}^2}{\alpha\cdot a^2(1-e^2)^2}, \tag{10.4.37}$$

where through a and by means of e the major semi-axis and the eccentricity of orbit of the Mercury are designated, α is a parameter of gravitational compression and ε_0 is a geometrical eccentricity of kernel of a rotating and gravitating spheroidal body (the Sun) [52]. Taking into account the Kepler's third law [61, 72] and expressing from it a square of the major semi-axis, the formula (10.4.37) takes the form:

$$\delta\varepsilon = \frac{8\pi^3 a(3+e)\cdot\varepsilon_0{}^2}{\alpha\cdot\gamma MT^2(1-e^2)^2}. \tag{10.4.38}$$

where T is the period of a turn of the Mercury around of the Sun.

Before we begin to consider the calculation of Mercury perihelion's advance according to the obtained formula (10.4.37) let us note that the similar formula of an estimation of the angular shift of the Mercury' perihelion by period has been offered by L. Nottale [62]:

$$\delta\varepsilon = \frac{6\pi \in^2}{a^2(1-e^2)^2}, \qquad (10.4.39)$$

where $\in^2 = \delta A / 2M$, besides a value δA is treated as the difference of the polar and equatorial inertial moments of the oblate object, namely, the Solar planetary system.

Moreover, A. Einstein in the mentioned above work [76] obtained that at the whole turn the Mercury' perihelion moves on an angle:

$$\delta\varepsilon = 3\pi \frac{A}{a(1-e^2)} = \frac{6\pi\gamma M}{c^2 a(1-e^2)}, \qquad (10.4.40)$$

where c is the speed of light; in this connection he noted: "Calculation gives for planet Mercury a turn of perihelion on 43" in century whereas astronomers specify 45" ± 5"as an inexplicable difference between observations and the theory of Newton. It means the full consent with observations". In other his work «On the special and general theory of relativity» in occasion of displacement of the Mercury' perihelion A. Einstein wrote:

"If calculate a gravitational field up to higher order values and with corresponding accuracy to calculate movement on an orbit of a material point with infinitesimal mass then the following deviation from laws of movement of planets of Kepler–Newton is obtained: the elliptic orbit of a planet undergoes the slow rotation in a direction of movement of the planet which is equal to

$$\delta\varepsilon = 24\pi^3 \frac{a^2}{T^2 c^2 (1-e^2)}$$

during one full turn of a planet. In this formula a means the major semi-axis, c is the speed of light in usual units, e is the eccentricity of orbit, T is the period of a turn of a planet in seconds.

For planet Mercury, the rotation of an orbit equal 43" in century is obtained that precisely corresponds to the value established by astronomers (Leverrier). Astronomers actually have found that some part of the general movement of perihelion of this planet is not explained by disturbing action of other planets and is equal to the pointed out value" [75 p.195-196].

So, following by the formula (10.4.37) let us calculate of displacement of perihelion of orbit of Mercury based on the statistical theory of gravitating spheroidal bodies. First of all, according to (10.4.37) it is necessary to estimate α (which is a parameter of gravitational compression of a spheroidal body, i.e. of the Sun) on the basis of an estimation of the linear size of its kernel, i.e. *the thickness of a visible part of the solar corona* (see Figure 10.7).

As is known, the solar corona represents the external layers of an atmosphere of the Sun, and it extends, at least, up to borders of the Solar system in the form of "a solar wind", i.e. our Earth as well as other planets of the Solar system are inside of the solar corona. In this connection *the Sun together with the solar corona can be described on the basis of a model of rotating and gravitating spheroidal body.*

The spectrum of the solar corona consists of three various components called L-, K- and F-components (see Figure 10.8). In particular, K-component represents a continuous spectrum of the corona, and on its background one can see an emission L-component within of an apparent interval 9'–10' from a visible edge of the Sun. Beginning with an apparent height nearby 3' and above it is seen a Fraunhoffer' spectrum constituting F-component of the solar corona.

Figure 10.7. The solar corona embodied during the solar eclipse in 1999.

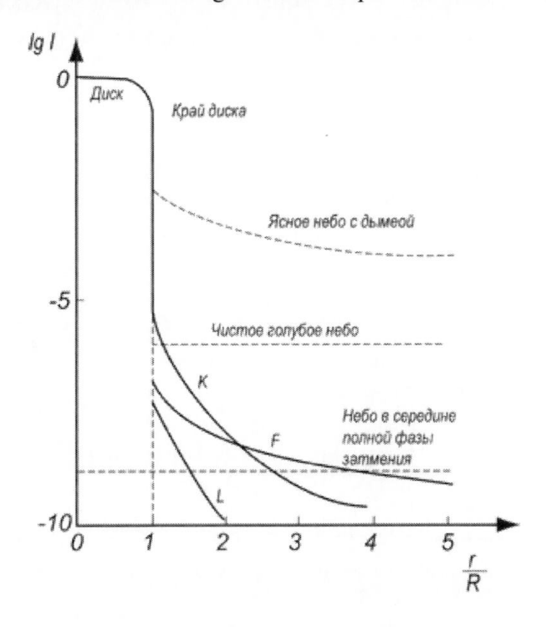

Figure 10.8. Graphic dependence of relative brightness of components of spectrum of the solar corona on distance up to edge of a disk.

The F-spectrum of the corona is formed owing to sunlight scattering on *particles of an interplanetary dust*. In immediate proximity to the Sun the dust cannot exist, therefore F-spectrum starts to dominate in the spectrum of the corona on some distance (at apparent height 20') from the Sun.

As is seen from Figure 10.8, a recession of plots of dependences of relative brightness of components of spectrum of the Solar corona occurs on distance of 3–3.5 radii from the center, i.e. on 2–2.5 radii from the edge of the solar disk. Really, known astronomer from NASA S. Odenwald in his notice «How thick is the solar corona?» wrote: “The corona actually extends throughout the entire solar system as a “wind” of particles, however, the densist parts of the corona is usually seen not more than about 1–2 solar radii from the surface or about 690,000 to 1.5 million kilometers at the equator. Near the poles, it seems to be a bit flatter...” [79].

Thus, accepting thickness of a *visible part of the solar corona* equal $\Delta = 2R$ (here R is radius of the solar disk) we find that $r_* = R + \Delta = 3R$. In other words, the parameter of gravitational compression $\alpha = 1/r_*^2$ of a spheroidal body in case of the Sun with its corona (for which the equatorial radius of disk $R = 6.955 \cdot 10^8$ m) can be estimated by the value:

$$\alpha = \frac{1}{(3R)^2} \approx 2.29701177718 \cdot 10^{-19} \text{(m-2)}. \qquad (10.4.41)$$

If a and b are equatorial and polar radii of the disk then a relative flattening $e_c = (a-b)/a$ of a spheroid (a flattened ellipsoid) can be expressed through its geometrical eccentricity $\varepsilon_0 = \sqrt{a^2 - b^2}/a$ as $e_c = 1 - \sqrt{1 - \varepsilon_0^2}$ whence in view of the value of flatness for the Sun $e_c = 9 \cdot 10^{-6}$ we find that the square of geometrical eccentricity of kernel of a rotating spheroidal body, i.e. the solar disk together with a visible part of the solar corona, is equal:

$$\varepsilon_0^2 = 1 - (1 - e_c)^2 \approx 1.7999919 \cdot 10^{-5}. \qquad (10.4.42)$$

At last, taking into account that the major semi-axis of orbit of planet Mercury is:

$$a = 5.7909068 \cdot 10^{10} \text{(m)}, \qquad (10.4.43)$$

and its orbital eccentricity is equal to:

$$e = 0.20530294, \qquad (10.4.44)$$

let us substitute all the mentioned values (10.4.41) - (10.4.44) in the formula (10.4.37) and calculate the angular displacement of perihelion of Mercury’ orbit for one turn:

$$\delta\varepsilon = \frac{6.28318530718 \cdot 3.20530294 \cdot 1.7999919 \cdot 10^{-5}}{2.29701177718 \cdot 10^{-19} \cdot 3.35346015663 \cdot 10^{21} \cdot 0.91747796891} \approx 0.10580169298''$$

. (10.4.45)

Taking into account that sidereal period of a turn of the Mercury is equal 87.969 terrestrial days then for 1 terrestrial year the Mercury performs 4.15214450545 turns around of the Sun, so angular displacement of perihelion of its orbit for 1 terrestrial year is to $4.15214450545 \cdot \delta\varepsilon = 0.43930391816''$. Thus, according to the statistical theory of gravitating spheroidal bodies *the turn of perihelion of Mercury' orbit is equal to 43.93 angular seconds in century* that well is consistent with conclusions of GR of Einstein and astronomical observations (see Table 10.4) because an accuracy of modern astronomical measurements of movement of perihelion of the Mercury is 0.4500" per century from Clemence's analysis [80, 81].

Let us note that the calculated values of parameter of gravitational compression (10.4.41) and square of geometrical eccentricity (7.4.42) of a rotating spheroidal body allow to estimate a characteristic length δ_0 :

$$\delta_0 = \varepsilon_0 r_* = \sqrt{\varepsilon_0^2 / \alpha} = 8.852249876 \cdot 10^6 \text{ (m)}, \tag{10.4.46}$$

which appears a lot of smaller than average distance from the Sun up to the Mercury $r = 5.79 \cdot 10^{10}$ (m), as it has been supposed initially at the derivation of the initial formula (10.4.6). Thus, the value $r_* = 1/\sqrt{\alpha} = 2.0865 \cdot 10^9$ (m) is almost in 28 times less than average distance r from the Sun up to the Mercury, that has allowed to use competently in the given derivation an estimation (10.4.1) of gravitational potentials in a remote zone of a rotating spheroidal body.

In this connection the obtained formula (10.4.37) is suitable to calculation of angular displacements of perihelia of the subsequent (for the Mercury) next planets though it is quite obvious that because of even greater deviation of characteristic length δ_0 and average distance r from the Sun up to any of these planets, angular displacements of perihelia of their orbits will be absolutely negligible. As A. Einstein marked: "For the Earth and Mars astronomers point to the turn 11" and 9" in century accordingly whereas our formula gives only 4" and 1". However, owing to small eccentricity of orbits of these planets observational data, apparently, are insufficiently exact" [76].

Actually, modern data [75] in Table 10.4 about turns of perihelia of orbits of the neighboring (with the Mercury) planets testify in favour of as GR of Einstein and the proposed statistical theory of gravitating spheroidal bodies.

Table 10.4. Data on angular shifts of perihelia of orbits of planets of the Solar system calculated according to proposed theories and modern astronomical observations

Planets of the Solar system	Angle of displacement of perihelion of orbit of a planet (in angular seconds,")		
	General relativity theory of Einstein	Astronomical observations	Statistical theory of gravitating spheroidal bodies
Mercury	43.03	43.11 ± 0.45	43.93
Venus	8.3	3.4 ± 4.8	4.24
Earth	3.8	5.0 ± 1.2	1.36
Mars	1.4	1.1 ± 0.3	0.34

CONCLUSION

There are theories for exploring the Solar system formation and planetary orbits estimation [4, 8, 10, 11, 15, 16, 25]: electromagnetic theories (Birkeland (1912), Alfvén (1942), e.a.), gravitational theories (Schmidt (1944), Gurevich and Lebedinsky (1950), Woolfson (1964), Safronov (1969), Dole (1970), Vityazev (1978), e.a.), nebular theories (von Weizsäcker (1943, 1947), Berlage (1948), Kuiper (1949, 1951), Hoyle (1960, 1963), ter Haar (1963, 1972), Nakano (1970), Cameron (1963, 1988), e.a.), quantum mechanical theories (Nelson (1966, 1985), Nottale (1993, 1996), El Naschie (1995, 2004), Ord (1996), De Oliveira Neto (1996, 2004), Agnese and Festa (1997), Sidharth (2001), e.a.). In spite of a great amount of work aimed to exploring formation of the Solar system, the mentioned theories are not able to explain all phenomena. In this connection in 1995 *the statistical theory* for a cosmological body forming (so-called spheroidal body model) has been proposed [35–55]. The present work develops this statistical theory relative to the Solar system formation.

In Section 10.1 of this chapter, the proposed theory starts from the conception for forming a rotating spheroidal body as a protoplanetary system from a protoplanetary nebula (or proto-Sun inside presolar nebula) using the derived distribution function and the density mass function for a rotating spheroidal body. As shown in Section 10.1, the derived function of mass density characterizes a flatness process: from initial spherical forms (for a non-rotational spheroidal body case) through flattened ellipsoidal forms (for a rotating spheroidal body) to disks when the squared geometrical eccentricity ε_0^2 varies from 0 till 1. The obtained formulas (10.1.22a-c), (10.1.23) can describe a possible common scenario of formation both a star and a protoplanetary gas-dust disk around it (in particular, the Sun and the solar protoplanetary gas-dust disk). Here the estimation of the gravitational potential (10.1.46) in the remote zone of a uniformly rotating spheroidal body is obtained.

In Section 10.2 of this chapter, the next stage of evolution (from a protoplanetary flattened gas-dust disk to originating protoplanets) is considered. To this end a distribution function (10.2.9) of a specific angular momentum for a rotating uniformly spheroidal body (as a gas-dust flattened protoplanetary cloud) is derived. As the specific angular momentums (for particles or planetesimals) are averaged during a conglomeration process (under a planetary embryo formation) the specific angular momentum for a protoplanet of the Solar system is found in Section 10.2. As a result, a new law (10.2.33) for planetary distances (which generalizes Schmidt's law) is derived theoretically here. Moreover, unlike the well-known planetary distances laws the proposed law is established by a physical dependence of planetary distances from the value of the specific angular momentum for the Solar system.

Section 10.2 also considers an application of the proposed law for planetary distances to the Solar system. As shown in this section (see Tables 10.2 and 10.3), this new law has given a very good estimation of real planetary distances in the Solar system (0% for the relative error of estimation and 1.4% for the absolute error, besides, its maximal value is equal to 11% for the Earth, but for the Pluto the proposed law has given too high error according to the derived rule $ent[n/2]$ determining the *maximal* number of necessary coefficients a_k in the law (10.2.33)).

Thus, this Section has shown:

1. The proposed law of the planetary distances based on a spheroidal body model is in agreement with the Solar system's observable planetary distances.
2. The analysis of Tables 10.2 and 10.3 points to two possible scenarios: a capture of the Moon by the Earth (it is known Bailey's proposition [11, 63]) or an origin of the Moon from rocky debris after a collision of a former planet with the proto-Earth (it is the leading modern Hartmann–Davis's hypothesis and its development within framework of the giant impact theory [64–69]).
3. There is no any gap between Mars's and Jupiter's orbits, i.e. the planet of Cerera did not exist (see Table 10.3).
4. The ninth planet Pluto is not a result of our Solar system forming, and probably, it has been attracted by the Solar system (this proposition was stated by von Weizsäcker, Schmidt, ter Haar and Cameron) [52, 53].

According to the *first* conclusion this paper shows that the proposed law for predicting the distance between the Sun and a planet of the Solar system is relatively accurate for most planets (see Figure 1) but it is not very good for the Earth and especially for the Pluto [52, 53]. The *fourth* conclusion presented also at sessions ST0/PS0 "Plasma processes at Earth and other Solar system bodies" and PS15 "Models of Solar system forming" of the General Assembly of the European Geosciences Union in Vienna, Austria, 02–07 April, 2006 [45, 46] has been confirmed by the decision of the 26th General Assembly of the International Astronomical Union in Prague, Czech Republic, 14–17 August, 2006 (see http://www.astronomy2006.com/).

Let us note the proposed simple statistical approach to investigation of our Solar system forming describes only a natural self-evolution inner process of development of protoplanets from a dust-gas cloud. Naturally, this approach however does not include any dynamics like collisions and giant impacts of protoplanets with large cosmic bodies. Henceforth, the presented statistical theory will only be able to predict surely the protoplanet's positions according to the proposed $ent[n/2]$ rule (see eq. (10.2.33)), i.e. the findings in this chapter are useful to predict if *today's* position of a considered planet coincides with its protoplanet's location or not [52, 53].

Though orbits of moving particles into the flattened rotating spheroidal body are circular ones initially, however, they could be distorted by collisions with planetesimals and gravitational interactions with neighboring originating protoplanets during evolutionary process of protoplanetary formation. In this connection in Section 10.3 of this chapter, the calculation of orbits of planets and bodies of the Solar system in a central-symmetric gravitational field of a rotating spheroidal body based on the Binet's differential equation are carried out. In other words, Section 10.3 considers the calculation of orbits of moving bodies and planets in a centrally-symmetrical gravitational field of a rotating and gravitating spheroidal body on the *planetary stage* of its evolution. In particular, using the derived estimation of the gravitational potential in the remote zone (10.1.46) the equation (10.3.29) of Keplerian orbit of a planet in gravitational field of a rotating spheroidal body is found. Here is shown that formation of planets is possible not only on the basis of gas-dust protoplanetary substance (protosolar nebula) but also by means of capture and join of moving bodies on close orbits (meteorites, asteroids etc.) in a gravitational field of a star.

In Section 10.4, calculations of an orbit of planet Mercury and estimation of angular shift of the Mercury' perihelion based on the statistical theory of gravitating spheroidal bodies are carried out. Here it is noted that in view of greatest proximity on distance to the Sun and essential inclination of orbit of Mercury the projection of a point of perihelion of its orbit can directly get in a nearby vicinity of the Sun, namely, in the visible part of the solar corona as a kernel of rotating and gravitating orbits of a spheroidal body. In this connection, as is well known, the angular displacement of a Newtonian ellipse is observed during one rotation of Mercury on an orbit, i.e. a century'shift of the perihelion of Mercury'orbit occurs. Using the estimation (10.4.6) of gravitational potential on closer distance from a kernel of a rotating spheroidal body (in vicinities of Mercury' orbit) the equation of *precessing elliptic* orbit (10.4.31) of planet Mercury is derived. Taking into account this equation the formula (10.4.36) for calculating displacement of perihelion of Mercury'orbit for the period is proposed in Section 10.4. As a result, according to the proposed statistical theory of gravitating spheroidal bodies the turn of perihelion of Mercury' orbit is equal to 43.93" in century that well is consistent with conclusions of the general relativity theory of Einstein (whose analogous estimation is equal to 43.03") and astronomical observation data (43.11 ± 0.45").

REFERENCES

[1] von Weizsäcker CF. Uber die Enstehung des Planetensystems. *Z. fur Astrophys.* 1943; 22: 319-355.

[2] von Weizsäcker CF. Zur Kosmogonie. *Z. fur Astrophys.* 1947; 24: 181-206.

[3] Schmidt OYu. Meteorite theory of origin of Earth and planets. *Dokl. Akad. Nauk SSSR* 1944; 45(6): 245-249 (in Russian).

[4] Schmidt OYu. The origin of Earth and planets. *Acad. of Sci. USSR Press*, Moscow, 1962 (in Russian).

[5] [Kuiper GP. On the origin of the solar system. In: Hynek JA, editor. *Astrophysics*, ch.8. New York; 1951. p. 357.

[6] Hoyle F. On the origin of the solar nebula. Quart. J. R. Astron. Soc. 1960; 1: 28–55.

[7] Hoyle F. Formation of the planets. In: Jastrow R, Cameron AGW, editors. *Origin of the Solar system*. New York: Academic Press; 1963. p. 63.

[8] ter Haar D and Cameron AGW. Historical review of theories of the origin of the solar system. In: Jastrow R, Cameron AGW, editors. *Origin of the Solar system*. New York: Academic Press; 1963.p. 1-37.

[9] ter Haar D. Some remarks on solar nebula types theories of the origin of the solar system. In: *Origin of the Solar System. Paris*: CNRS; 1972. p. 71-79

[10] Cameron AGW. Origin of the Solar System. Annual Review of Astronomy and *Astrophysics* 1988; 26: 441- 472.

[11] Nieto MM. *The Tutius–Bode law of planetary distances: its history and theory*.Oxford, New York: Pergamon; 1972.

[12] Alfvén H, Arrhenius G. *Evolution of the solar system*. Washington: NASA; 1976.

[13] Gurevich LE, Lebedinsky AI. *Gravitational condensation of dust cloud*. Dokl. Akad. Nauk SSSR 1950; 74 (4): 673-675 (in Russian).

[14] Gurevich LE, Lebedinsky AI. *On the planet formation.–I. Gravitational condensation; – II. Planetary distance law and rotation of planets; – III. Structure of initial cloud and separation of planets by outer and inner ones*. Izvestiya Akad. Nauk SSSR, ser. Fiz. 1950; 14(6): 765-775; 776-789; 790-799 (in Russian).

[15] Safronov VS. *Evolution of protoplanetary cloud and formation of Earth and planets*. Moscow: Nauka; 1969 (reprinted by NASA Tech. Transl. F-677, Washington, D.C.; 1972).

[16] Vityazev AV, Pechernikova GV, *Safronov VS. The terrestrial planets: origin and early evolution*. Moscow: Nauka; 1990 (in Russian).

[17] Dole SH. Computer simulation of the formation of planetary systems. *Icarus* 1970; 13: 494-508.

[18] Reid IN, Hawley SL, editors. *New light on dark stars: red dwarfs, low-mass stars, brown dwarfs*. 2nd ed. Springer; 2005. p.495.

[19] De Oliveira Neto M. Ciência e Cultura (*J Brazil Assoc Advance Sci*) 1996; 48:166.

[20] Nottale L. *Fractal space–time and microphysics: towards a theory of scale relativity*. Singapore: World Scientific; 1993. p. 311.

[21] Nottale L, El Naschie MS, Athel S, Ord G. *Fractal space–time and Cantorian geometry in quantum mechanics*. Chaos, Solitons & Fractals 1996; 7(6) [a special issue].

[22] Agnese AG, Festa R. Clues to discretization on the cosmic scale. *Phys Lett A* 1997; 227:165-171.

[23] Agnese AG, Festa R. *Discretizing μ -Andromedae planetary system,* v2; 1999. Available from: <astro- ph/9910534>.

[24] Arp H. *Seeing red; redshifts, cosmology and academic science*. Montreal: Apeiron; 1998. p. 219.

[25] De Oliveira Neto M, Maia LA, Carneiro S. An alternative theoretical approach to describe planetary systems through a Schrödinger-type diffusion equation. *Chaos Solitons & Fractals* 2004; 21:21-28.

[26] De Oliveira Neto M. Using the dimensionless Newton gravity constant $\overline{\alpha}_G$ to estimate planetary orbits. *Chaos, Solitons & Fractals*, 2005; 24:19-27.

[27] Nelson E. Derivation of the Schrödinger equation from Newtonian mechanics. *Phys Rev* 1966; 4:1079-85.

[28] Nelson E. *Quantum fluctuations*. Princeton (NJ): Princeton University Press; 1985.

[29] El Naschie MS, Rossler E, Prigogine I, editors. *Quantum mechanics, diffusion and chaotic fractals*. Oxford: Pergamon Press; 1995.

[30] Ord G. Classical particles and the Dirac equation with an electromagnetic force. *Chaos, Solitons & Fractals* 1997; 8:727-741.

[31] Sidharth EG. *The chaotic universe*. New York: Nova Science; 2001.

[32] Laskar J. A numerical experiment on the chaotic behaviour of the Solar System. *Nature* 1989; 338: 237-238.

[33] Sussman GJ, Wisdom J. Chaotic evolution of the Solar System. *Science* 1992; 257: 56-62.

[34] El Naschie MS. Quantum gravity from descriptive set theory. *Chaos, Solitons & Fractals* 2004; 19: 1339-1344.

[35] Krot AM. *The statistical model of gravitational interaction of particles*. Uspekhi Sovremennoï Radioelektroniki (special issue "Cosmic Radiophysics", Moscow) 1996; 8: 66-81 (in Russian).

[36] Krot AM. Use of the statistical model of gravity for analysis of nonhomogeneity in earth surface. *Proc. SPIE's 13th Annual Intern. Symposium "AeroSense"*, Orlando, Florida, USA, April 5-9, 1999; 3710: 1248-1259.

[37] Krot AM. Statistical description of gravitational field: a new approach. *Proc. SPIE's 14th Annual Intern.Symposium "AeroSense"*, Orlando, Florida, USA, April 24-28, 2000; 4038: 1318-1329.

[38] Krot AM. Gravidynamical equations for a weakly gravitating spheroidal body. *Proc. SPIE's 15th Annual Intern. Symposium* "AeroSense", Orlando, Florida, USA, April 16-20, 2001; 4394: 1271-1282.

[39] Krot AM. Development of gravidynamical equations for a weakly gravitating body in the vicinity of absolute zero temperature. *Proc. 53rd Intern. Astronautical Congress (IAC) – The 2nd World Space Congress-2002*, Houston, Texas, USA, October 10-19, 2002; Preprint IAC-02-J.P.01: 1-11.

[40] Krot AM. A quantum mechanical approach to description of gravitating body. *Proc. 34th Scientific Assembly of the Committee on Space Research (COSPAR) – The 2nd World Space Congress-2002*, Houston, Texas, USA, October 10-19, 2002.

[41] Krot AM. The kinetic equations for rotating and gravitating spheroidal body. Proc. EGS-AGU-EUG Joint Assembly, Nice, France, April 6-11, 2003; *Geophysical Research Abstracts*, vol.5: EAE03-A-05568.

[42] Krot AM. The equations of movement of rotating and gravitating spheroidal body. *Proc. 54th Intern. Astronautical Congress* (IAC), Bremen, Germany, September 29-October 3, 2003; Preprint IAC-03-J.1.08.

[43] Krot AM. The statistical model of rotating and gravitating spheroidal body with the point of view of general relativity. *Proc. 35th COSPAR Scientific Assembly*, Paris, France, July 18-25, 2004; Abstract-Nr. COSPAR 04-A-00162.

[44] Krot A. A statistical model of beginning rotation of gravitating body. Proc. 2nd European Geoscinces Union (EGU) General Assembly, Vienna, Austria, April 24-29, 2005; *Geophysical Research Abstracts*, vol.7: EGU05-A-04550, SRef-ID: 1607-7962/gra/.

[45] Krot A. The statistical approach to exploring formation of Solar system. Proc. European Geoscinces Union (EGU) General Assembly, Vienna, Austria, April 02-07, 2006; *Geophysical Research Abstracts*, vol. 8: EGU06-A-00216, SRef-ID: 1607-7962/gra/.

[46] Krot AM. Exploring gravitational interaction of particles based on quantum mechanical principles: an oscillator approach. Proc. European Geoscinces Union (EGU) General Assembly, Vienna, Austria, April 2-7, 2006; *Geophysical Research Abstracts,* vol.8: 00220, SRef-ID: 1607-7962/gra/ EGU06-A-00220.

[47] Krot AM. The statistical model of original and evolution planets of Solar system and planetary satellities. Proc. European Planetary Science Congress, Berlin, Germany, September 18-22, 2006; *Planetary Research Abstracts*, ESPC2006-A-00014.

[48] Krot A. On the principal difficulties and ways to their solution in the theory of gravitational condensation of infinitely distributed dust substance. Proc. XXIV IUGG General Assembly, Perugia, Italy, July 2-13, 2007; GS002 Symposium "Gravity Field", Abstract GS002-3598: 143-144.

[49] Krot AM. The main planetary dynamics problem solving based on spheroidal bodies theory. Proc. IAU Symposium No.249 ''Exoplanets: Detections, Formation and Dynamics'', Suzhou, China, October 22-26, 2007; Abstracts of contributions: 33-34. Available from: *http://iaus249.nju.edu.cn/* Abstracts.htm

[50] Krot AM. Investigation of spherical waves propagation in a slow-flowing gravitational compressed spheroidal body. Proc. European Geoscinces Union (EGU) General Assembly, Vienna, Austria, April 13-18, 2008; *Geophysical Research Abstracts,* vol.10: EGU2008-A-00423.

[51] Krot AM. A gravitational potential finding for rotating cosmological body in the context of proto-planetary dynamics problem solving. Proc. 3rd European Planetary Science Congress, Muenster, Germany, September 22-26, 2008; *Planetary Research Abstracts,* vol.3: EPSC2008-A-00077.

[52] Krot AM. A statistical approach to investigate the formation of the solar system. *Chaos, Solitons and Fractals* 2009; 41(3): 1481-1500.

[53] Krot AM. On the principal difficulties and ways to their solution in the theory of gravitational condensation of infinitely distributed dust substance. Proc. of 2007 IAG General Assembly in the book "*Observing our Changing Earth*", vol.133 (Ed. by M.G. Sideris). Berlin, Heidelberg: Springer; 2009, pp. 283-292.

[54] Krot A.M. Self-organization processes in a slow-flowing gravitational compressible cosmological body. Selected papers of the CHAOS-2008 International Conference in the book "*Topics on Chaotic Systems*" (Ed. by C.H. Skiadas, I. Dimotikalis, C. Skiadas). Singapore, New Jersey, London, HongKong: World Scientific; 2009, pp. 190-198.

[55] Krot, A. M. A quantum mechanical approach to description of initial gravitational interactions based on the statistical theory of spheroidal bodies. *Nonlinear Sci. Lett. A.* 2010; 1(4): 329-369.

[56] Landau LD, Lifschitz EM. *Statistical physics, part 1.* Reading (MA): Addison–Wesley Publishing Co.; 1955.

[57] Sretenskii L. *The theory of Newtonian potential.* Moscow-Leningrad: Gostekhizdat; 1946 (in Russian).

[58] Zharkov VN, Trubitsyn VP, Samsonenko LV. *Physics of the Earth and planets: the figures and interior structures*. Moscow: Nauka; 1971 (in Russian).

[59] Gradshtein IS, Ryzhik IM. *The tables of integrals, sums, series and products.* Moscow: Nauka; 1971 (in Russian).

[60] Kenyon SJ. Planet formation in the outer Solar System. *Publ. of the Astronomical Society of the Pacific* 2002; 114: 265–283.
[61] Landau LD, Lifschitz EM. *Mechanics, vol.I.* Moscow: Nauka; 1973 (in Russian).
[62] Nottale L, Schumacher G, Gay J. Scale relativity and quantization of the solar system. *Astronomy & Astrophysics* 1997; 322:1018-1025.
[63] Bailey JM. The Moon may be a former planet. *Nature* 1969; 223: 251-253.
[64] Hartmann WK, Davis DR. Satellite-sized planetesimals and lunar origin. *Icarus* 1975; 24(4): 504-515.
[65] Hartmann WK, Phillips RJ, Taylor GJ, editors. Origin of the Moon. Houston: *Lunar and Planetary Institute*; 1986.
[66] Benz W, Slattery WL, Cameron AGW. The origin of the Moon and the single-impact hypothesis.1 and 2. Icarus 1986; 66(3): 515-535 and *Icarus* 1987; 71(1): 30-45.
[67] Cameron AGW. The origin of the Moon and the single impact hypothesis. 5. *Icarus* 1997; 126(1): 126-137.
[68] Canup RM, Asphaug E. Origin of the Moon in a giant impact near the end of the Earth's formation. *Nature* 2001; 412 (6848): 708-712.
[69] Canup RM. Simulations of a late lunar-forming impact. *Icarus* 2004; 168(2): 433- 456.
[70] Blagg MA. On a suggested substitute for Bode's law. *Mon. Not. Roy. Astron. Soc.* 1913; 73:414-422.
[71] Richardson DE. Distances of planets from the Sun and of satellites from their primaires in the satellite systems of Jupiter, Saturn, and Uranus. Pop. *Astron.* 1945; 53: 14-26.
[72] Moulton FR. *An introduction to celestial mechanics.* New York, London: Macmillan Co.; 1914.
[73] Newton I. *Principia mathematica philosophia naturalis. 1st ed.*, London: Streater; 1686 (reprinted by Berkeley (CA): University of California Press; 1934).
[74] Landau LD, Lifschitz EM. *Classical theory of fields.* Reading (MA): Addison–Wesley Publishing Co.; 1951.
[75] *Albert Einstein and gravitation theory.* Moscow: Mir; 1973 (in Russian).
[76] Einstein A. *Erklärung der Perihelbewegung der Merkur aus der allgemeinen Relativitätstheorie.* Sitzungsber. preuss. Akad. Wiss., 1915; 47(2): 831-839.
[77] Schwarzschild K. *On gravitational field of point mass in Einstein' theory.* Sitzungsber. d. Berl. Akad., 1916; 189.
[78] Krot AM. *Discrete models of dynamical systems based on polynomial algebra.* Minsk: Nauka i tekhnika; 1990 (in Russian).
[79] Odenwald S. How thick is the solar corona? *http://www.astronomycafe.net/qadir/q1612.html 1997.*
[80] Clemence GM. *Rev. Mod. Phys.* 1947; 19: 361.
[81] Weinberg S. *Gravitation and cosmology: Principles and applications of the general theory of relativity.* New York: John Wiley and Sons; 1972.

INDEX

A

B

C

D

E

F

G

H

I

J

K

L

M

N

O

P

Q

R

S

T

U

V

W

Z